Plant Disease Management

Principles and Practices

Authors

Hriday S. Chaube, Ph.D.
Associate Professor
Department of Plant Pathology
College of Agriculture
G. B. Pant University of Agriculture and Technology
Pantnagar, India

Uma S. Singh Ph.D.
Division of Plant Pathology
International Rice Research Institute
Manila, Philippines

CRC Press
Boca Raton   Ann Arbor   Boston

Library of Congress Cataloging-in-Publication Data

Plant disease management: principles and practices / authors Chaube, H.S. and Singh, U.S.

p. cm.
Includes bibliographical references.
ISBN 0-8493-5758-6 ✓
1. Plants — Disease management. II. Principles
III. Practices.
I. Chaube, H. S. II. Singh, U. S.
SB813.N89   1990   90-23273
623'.13—dc21   CIP

Direct all inquiries to CRC Press, Inc., 2000 Corporate Blvd., N. W., Boca Raton, Florida, 33431.

© 1991 by CRC Press, Inc.

International Standard Book Number 0-8493-5758-6

Library of Congress Card Number 90-23273
Printed in the United States

PREFACE

This book attempts to provide concise, critical, synthetic, and up-to-date coverage of different aspects of plant disease management. The first eleven chapters are devoted to principles and related aspects and the remaining seven to management practices based on them. This book also attempts to capture some of the images of such rapidly expanding fields as host-parasite recognition and biotechnology (Chapters 3, 4, and 16) even at the risk of making the subject a bit conceptual. A special section is included in Chapter 16 on Genetically Engineered Cross-Protection.

The book is intended to serve as a text for advanced undergraduate and graduate students of plant pathology and related disciplines and as a reference source for teachers, researchers, students, and technologists. Considering the cost factor, size is a great limitation for such type of publications and the subject of the book "Plant Disease Management" is obviously too broad to be covered comprehensively in a brief text. In selecting materials to be included in certain chapters, emphasis has been placed on recent developments in rapidly expanding areas of research, a hazardous approach, as exciting new ideas and concepts often have very short half-lives. Nevertheless, being thought provocative they stimulate the readers to be more critical, apply their own mind, and be a part in development of conceptual plant pathology.

Although we fully agree with the words of Cervantes that "he that publishes a book runs a great hazard, since nothing can be more impossible than to compose one that may secure the approbation of every reader", we sincerely believe that criticism by readers is essential for a text book to march closer to perfection. Students and teachers, especially the latter, are requested to comment and give suggestions for further improvements in the book. We shall welcome them and in the light of that, rectify our mistakes, omissions, and shortcomings in future editions.

H. S. Chaube
U. S. Singh

THE AUTHORS

Hriday S. Chaube, Ph.D., is Associate Professor of Plant Pathology, in the Department of Plant Pathology, College of Agriculture, G. B. Pant University of Agriculture and Technology, Pantnagar, India.

Dr. Chaube received his B. Sc. (Ag) degree from the University of Gorakhpur in 1964 and his M. Sc. (Ag) Plant Pathology degree from Banaras Hindu University in 1966. He completed his Ph. D. (Plant Pathology) as staff candidate from G. B. Pant University of Agriculture and Technology, Pantnagar in 1978. Dr. Chaube has been on the staff of G. B. Pant University of Agriculture and Technology, Pantnagar since 1966. He was appointed Senior Research Assistant, Assistant Professor/Subject Matter Specialist in 1970 and Associate Professor/Senior Research Officer in 1980.

Dr. Chaube is a member of Indian Phytopathological Society and Society of Mycology and Plant Pathology. He has taught 10 under graduate and post-graduate courses in Plant Pathology and has published above 60 research papers, 15 research abstracts, over 40 popular articles, 6 reviews, 2 bulletins and co-edited 2 books. Dr. Chaube has been engaged in phytopathological research since 1964. His major areas of research are ecology and management of soil-borne pathogens.

Uma S. Singh Ph.D., is currently in the Division of Plant Pathology at the International Rice Research Institute in Manila, Philippines.

Dr. Singh obtained his B. Sc. (Biology) degree from Delhi University in 1975. He received his M. Sc. (Ag) and Ph. D. in Plant Pathology from G. B. Pant University of Agriculture and Technology, Pantnagar in 1977 and 1983, respectively. He joined G. B. Pant University of Agriculture and Technology, Pantnagar in 1983 as a Research Associate and was promoted to Assistant Professor in December 1983.

Dr. Singh is a member of the Indian Phytopathological Society, Society of Mycology and Plant Pathology and Biological Chemist of India. He is associated with the teaching of five post-graduate courses of department of Plant Pathology and Molecular Biology and biotechnology and has published above 30 research papers, 6 research abstracts, 10 reviews/conceptual articles and co-edited 2 books.

His area of research includes fungicides particularly their uptake and translocation in plants; biochemistry and molecular biology of plant-parasite interaction, and plant tissue culture particularly its use in the study of host-parasite interaction. His research works on uptake, translocation, and distribution of fungicides in plants have been highly acclaimed through three prestigious awards. These are "Pesticide India Award" of the Society of Mycology and Plant Pathology, twice for the years 1982 and 1987 and "Prof. M. J. Narasimhan Award" of Indian Phytopathological Society for the year of 1985. Two M. Sc. students who worked under his guidance were also awarded with "Pesticide India Award" for the years 1986 and 1988.

DEDICATED
to the following great theoreticians, researchers, and teachers of plant pathology

J. G. Horsfall
E. Gaumann
Y. L. Nene
R.S. Singh
&
J. E. Van der Plank
and
to our beloved parents
late Mrs. & Mr. A. N. Chaube
&
Mrs. & Mr. Ram Sahay Singh

ACKNOWLEDGMENT

We sincerely thank the Vice-Chancellor and University Librarian, G. B. Pant Unversity of Agriculture and Technology, Pantnagar for permitting us to write this book and making the most up-to-date literature available to us. We are equally thankful to all those who granted permission for reproduction of dates, illustrations, figures, etc. We also wish to express our gratitude to all the pioneers in the field of phytopathology whose work has been freely used in the preparation of this book. Thanks are also due to our colleagues and students especially to Dr. Abdul Bari, Dr. B. K. Pandey, and Mr. P. K. Mukherjee, who have helped us in various ways. Mr. S. C. Prajapati for photography and Mr. R. K. Arora for art work and line drawings and Mr. P.P. Saxena and Mr. R. C. Saxena for typing also deserve our sincere thanks.

TABLE OF CONTENTS

Chapter 7
Survival of Plant Pathogens

Chapter 8
Dissemination of Plant Pathogens

Chapter 9
Pathometry-Assessment of Disease Incidence and Loss

Chapter 13
Physical Methods 161

Chapter 14
Biological Control 175

Chapter 16
Host Resistance and Immunization

Chapter 1

THE PLANT PATHOGENS

I. INTRODUCTION

Organisms suffer from disease or disorders due to some abnormality in the functioning of their system. These abnormalities may be due to factors which have no biological activity of their own (abiotic factors) or those entities which show some biological activity (mesobiotic agents) and those that are established cellular organisms. A pathogen can be broadly defined as any agent or factor that induces *pathos* or disease in an organism, but the term is generally used to denote biotic and mesobiotic causes.

Which factor, entity or agent should be called abiotic and which one biotic or living? By institution and experience we have known that a thing that does not grow, reproduce, move, or show response to external stimuli is nonliving and those things that show these properties are living. However, when viruses appear in the picture, the whole concept of living vs. nonliving becomes somewhat confused.

II. CELLULAR ORGANISMS AND VIRUSES

Although viruses are acellular, they always have one type of nucleic acid and in those that have RNA, the latter performs the same functions as the DNA in cellular organisms. Consequently, there is reproduction of different type, regulation of the activities whatever they may be, and existence of the phenomena of variability. They do have proteins but only of structural type. They synthesize the enzyme for replication with the help of host ribosomes; some have the replicative enzyme in the capsid, and some use the host enzymes. One major nonliving character of viruses is the ability of the nucleic acid to retain activity even after chemical purification and crystallization. These entities are, therefore, considered not true organisms but at the threshold of life. The basic differences between viruses and cellular organisms are summarized in the Table 1.

III. DIVISION OF ORGANISMS

When the living organisms were divided into plant and animal kingdoms, fungi were included among the plants. The complex and heterogenous behavior of many lower organisms created limitations in this system. To solve this problem, E. Haeckel, a German biologist proposed a third kingdom Protista *(Protisto* = very first) about a century ago (1866) to include organisms which lacked tissues. Protists themselves were later divided into lower and higher protista. The lower protista which included bacteria and blue green algae, was named the fourth kingdom, Monera. Thus, the living world came to have four kingdoms, i.e., Plantae, Animalia, Protista, and Monera (synonym = Mychota).[1,2] Whittaker[3] proposed a five-kingdom system and placed fungi in a separate kingdom, coordinate with higher plants and animals. The two primitive kingdoms are the Monera (prokaryotic-organisms bacteria and blue-green algae) and Protista (unicellular-eukaryotic organisms).

The development of electron microscope and associated preparative techniques for biological materials made it possible to study the ultra-structures in the cell such as intranuclear-intranucleolar structures, cell organelles other than nuclei, membrane of the cell and intracellular organelles, etc. These observations led to the recognition of two kinds of cells in the living systems. In the more complex *eukaryotic cell*, which is the unit of structure in plants, metazoan animals, protozoa, fungi, and most algae, the nucleus is bound by a nuclear membrane and divides by mitosis.

TABLE 1
Differences Between Viruses and Cellular Organisms

Trait	Viruses	Organisms
Basic unit of structure and function	Nucleocapsid consisting or nucleic acid core and protein coat of shell	Cell
Nucleic acid components	Either RNA or DNA	Both RNA and DNA
Enzyme content	One or a few enzymes	Many enzymes
Reproductive process	Nucleic acid core separates from capsid, replicates forming own replica and proteins separately which assemble to form a new capsid; nucleo-capsides never yield progeny capsids directly	Progenitor cells yield cells directly by different processes of reproduction; parents retain identity; cellular identity never lost
Dependence on another organism	Always obligately dependent on a commandeered cellular organism	Usually independent, microbes sometimes facultatively dependent, rarely obligately dependent
Location in other organism	Always within the cell; intracellular, intramembranous, etc.	Only parasitic forms have different types of inter- and intracellular relationship

The less complex *prokaryotic cells* are the unit of structure in two microbial groups, the bacteria including mycoplasmas and rickettsiae, and the blue-green algae or blue-green bacteria. The prokaryotic nucleus (nucleoid) is not membrane bound and its division is nonmitotic. This unambiguous bipartite division of organisms, exclusively based on cellular properties still retains algae, fungi, and protozoa in the protists as eukaryotes distinct from plants and animals on the basis of little or no differentiation of cells and tissues as given below.

The primary sub-division of cellular organisms

Eukaryotes   1. Multicellular, extensive differentiation of cells and tissues
- A. Plants
  - a. seed plants
  - b. ferns
  - c. mosses
  - d. liverworts
- B. Animals
  - a. invertebrates
  - b. vertebrates

2. Unicellular, coenocytic or multicellular with little or no differentiation of cells and tissues
- C. Protists
  - a. algae
  - b. protozoa
  - c. fungi

Prokaryotes   Bacteria including photo bacteria or blue-green algae, mollecutes including mycoplasmas and spiroplasmas

IV. CLASSIFICATION OF PLANT PATHOGENS

The groups of pathogens or causes of plant diseases are given in the chart:

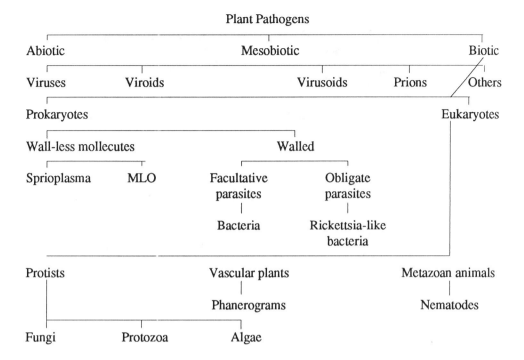

A. ABIOTIC CAUSES (INANIMATE CAUSES)

The inanimate causes include (1) adverse climatic conditions such as very high or very low temperatures, unfavorable intensity of light, excess of humidity or rains; (2) chemical injuries caused by (a) faulty application of pesticides and plant nutrients, and (b) atmospheric impurities or pollutants such as phytotoxic components of smog, ozone, sulfurdioxide, ethylene, etc.; and (3) adverse soil conditions including low, high, or unbalanced soil moisture; poor soil structure affecting root growth, aeration, water holding capacity, and proper oxygen supply; deficiency, excess or imbalance of nutrients, injurious salts, and soil reaction.

B. MESOBIOTIC CAUSES

Mesobiotic (*meso* = middle) agents occupy an unique position between abiotics and biotics as they possess characteristics of both the systems. They include viruses, viroids, virusoids, prions, and others.

1. Viruses
a. *General Characteristics*
Viruses are entities that are too small to be seen with the light microscope, multiply

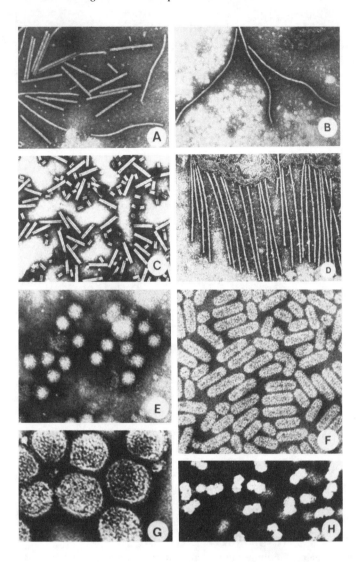

FIGURE 1.   Electron micrographs of negatively stained particles of viruses of some important morphological groups. (A) Mixtures of particles of tobacco mosaic virus (tobamovirus) and potato virus X (potexviruses), (B) *Wisteria* vein mosaic virus (potyviruses), (C) tobacco rattle virus (tobraviruses), (D) red clover vein mosaic virus (carlaviruses), (E) carnation mottle virus (tombusviruses) and carnation etched ring virus (large particle, Caulimoviruses), (F) alfalfa mosaic virus (alfalfa mosaic virus group), (G) tomato spotted wilt virus (tomato spotted wilt virus group), (H) bean golden mosaic virus (geminiviruses) (From Boss, L., *Introduction to Plant Virology*, Longman, London, 1983. With permission.)

only in living cells, and have the ability to cause disease.[4] All viruses are parasitic in cells and cause a multitude of diseases to all forms of living organisms. More than 600 viruses are known to cause diseases of plants.[5] Viruses are mostly either rod shaped or polyhedral, or variants of these two basic structures (Figure 1). Viruses behave as chemical molecules.[6] At their simplest, viruses consist of nucleic acid and proteins, with the protein coat wrapped around the nucleic acid. There is always only RNA or only DNA in each virus and, in most plant viruses, only one kind of protein. Viruses multiply by inducing host cells to form more viruses. They cause diseases not by consuming cells or killing them with toxins, but by upsetting metabolism of the cells which in turn, leads to development by the cell of abnormal substances and conditions injurious to the functions and the life of the cell or the organism.

Another virus like pathogen associated with plant diseases and popularly known as *satellite viruses*, are viruses associated with certain typical viruses but depend on the latter for multiplication and plant infection and reduce the ability of the typical viruses to multiply and cause disease, i.e., satellite viruses act like parasite of the associated typical viruses.

b. *Nomenclature and Classification*

Viruses are broadly classified on the basis of shape of their capsid and the type of nucleic acid they contain. One of the first systems of classification of viruses into groups that proved useful for elongated viruses was proposed. It was based on particle morphology, size, and on serological affinities. This system was further improved and now the International Committee on Taxonomy of Viruses (ICTV), established in 1966, recognizes 26 groups which include nearly 200 precisely described plant viruses. The grouping is mainly based on particle morphology and size and such criteria as whether the nucleocapsid is naked or enveloped, the number of virion types and number of genome segments, and type and strandedness of the nucleic acid. The recognized groups, as described by Gibbs and Harrison,[7] Bos,[8] and Agrios[9] are given in the following chart:

```
Plant Viruses

    –Elongated, helical (all SS RNA)
            –Rigid                      –Tobra Viruses (Tobacco rattle virus, Pea early
                                          browning virus)
                                        –Tobamo Viruses (Tobacco mosaic, Cucumber
                                          green mottle mosaic, Cucumber virus 4, Sunhemp mosaic,
                                          Tomato mosaic)
                                        –Hordei Viruses (Barley stripe mosaic virus)
            –Flexuous                   –Potex Viruses (Potato virus X, Papaya mosaic, Cassava
                                          common mosaic, Clover yellow mosaic, White clover
    –Bacilliform                          mosaic)
    (SS RNA)
    –Without envelope
    (Alfalfa mosaic)
            With envelopes              Poty Viruses (Potato virus Y, Bean common mosaic,
                                          Bean yellow mosaic, Lettuce mosiac
            (Rhabde Viruses)            Pea seed borne mosaic, Soybean mosaic,
            (Lettuce necrotic           –Sugarcane mosaic, Turnip mosaic)
            yellow mosaic)
            (Potato yellow              –Carla viruses (Carnation latent virus, Potato virus
            dwarf virus)                 S. Potato virus M, Pea streak, Red clover vein mosaic)
    –Isometric Viruses                  –Clostero viruses (Citrus tristeza, Beet yettlow stunt,
                                          Carrot yellow leaf, Clover yellows, Wheat yellow leaf)
    –SS RNA (Monopartite)
    –Maize chlorotic dwarf
    –Tymo Viruses (Turnip yellow mosaic, Okra mosaic, Peanut yellow mottle,
      Cacao yellow mosaic)
    –Tombus Viruses (Tomato bushy stunt virus, Egg plant mottled crinkle virus)
    –Sobemo Viruses (Southern bean mosaic virus, turnip rosette virus)
    –Luteo Viruses (Barley yellow dwarf virus, Legume yelow, Carrot red leaf,
      Potato leaf roll, Pea and bean leaf roll, Turnip yellows, Soybean dwarf)
    –SS RNA (multi partite)
    –Como Viruses (Cowpea mosaic virus, Bean rungose mosaic, Cowpea severe
      mosaic, Radish mosaic, Squash mosaic)
    –Nepo Viruses (Tobacco ring spot virus, Arabis mosaic, Grape vine fan leaf,
      Mulberry ring spot, Tomato ring spot, Cherry leaf roll)
    –Pea enation mosaic
    –Diantho Viruses (Carnation ring spot virus, Sweet clover necrotic mosaic, Red
      clover necrotic mosaic)
    –Cucumo Viruses (Cucumber mosaic virus)
    –Bromo Viruses (Brome mosaic, Broad bean mottle, Cowpea chlorotic mottle)
    –Ilar Viruses (Tobacco streak virus, Apple mosaic, Citrus leaf rugose, Citrus
      variegation)
    –Tomato spotted wilt
    –ds RNA
            Reo Viruses
            Phtto reoviruses (Wound tumor virus, Rice dwarf virus)
            Fiji Viruses (Sugarcane Fiji disease virus, Maize rough dwarf virus,
              Oat sterile dwarf virus)
    SS DNA
            Gemini Viruses (Maize streak virus, Bean golden mosaic, Tomato
            golden yellow, Mungbean yellow mosaic)
    ds DNA
            Caulimo Viruses (Cauliflower mosaic virus, Dahalia mosaic, Carnation
            etched ring, Strawberry vein banding virus.
```

TABLE 2
Plant Diseases, Their First Occurrences, Viroid Etiology and
Abbreviation of Corresponding Viroids[12,15]

Disease	Year of first observation	Viroid etiology	Abbreviated names
Potato spindle tuber	1917	1971	PSTV
Citrus exocortis	1943	1972	CEV
Chrysanthemum stunt	1945	1973	CSV
Chrysanthemum chlorotic mottle	1967	1975	CCMV
Cucumber pale fruit	1963	1974	CPFV
Coconut cadang-cadang	1927	1975	CCCV
Hop stunt	1952	1977	HSV
Columnea viroid	1978	1978	COV
Avocado sunblotch	1931	1979	ASB
Tomato apical stunt	1981	1981	TASV
Tomato planta macho	1974	1982	TPMV
Burdock stunt	1983	1983	BSV
Carnation stunt	1983	1983	CarSV
Coconut tinangaja	—	1988	CTiV
Australian grapevine	—	1988	AGV

TABLE 3
List of Virusoids[13]

Virusoid	Abbreviation	No. of nucleotides
Lucern transient streak virus RNA2	vLSTV	324
Solanum nodiflorum mottle virus RNA2	vSNMV	378
Subterranean clover mottle virus RNA2	v(388)SCMoV	388
Subterranean clover mottle virus RNA2'	v(332)SCMoV	332
Valvet tobacco mottle virus RNA2	vVTMoV	366, 367

2. Viroids

Since the viroid nature of the spindle tuber disease of potatotes was first reported,[10,11] 14 additional viroid-infected plant diseases have been described (Table 2). Viroids are molecular parasite of higher plants composed of naked, single stranded, low molecular weight, circular RNA (M_r 0.8 to 1.3 $\times$ 10^5) which utilizes only host component for its replication.[12,13] There is no evidence that any viroid can code for one or more polypeptides *in vivo*. They exist in solution as rod-like structures arranged in a series of short base-paired and nonbase-paired regions.[12,13]

3. Virusoids

In addition to viroids, there is a second group of low molecular weight ssRNAs associated with plant diseases; these are the plant virus satellite RNAs.[13] Although some of these RNAs also do not show mRNA activity, they differ from viroids in two significant aspects: (1) they are dependent on a helper virus for their replication; and (2) they are encapsidated by either viral-coded or satellite RNA-coded coat protein. Amongst the satellite RNAs, the virusoids show most similarity to viroids in being covalently closed circular RNA molecules in the same size range, varying from 324 to 388 nucleotides.[13] Virusoids are single stranded, circular, low molecular weight, viroid-like RNA (M_r 1.1 to 1.3 $\times$ 10^5). These satellite RNAs are dependent on a helper virus for replication and are encapsidated in the circular form by that virus. They exist in solution as rod-like structures with secondary structures similar to viroids. The list of virusoids is given in Table 3.

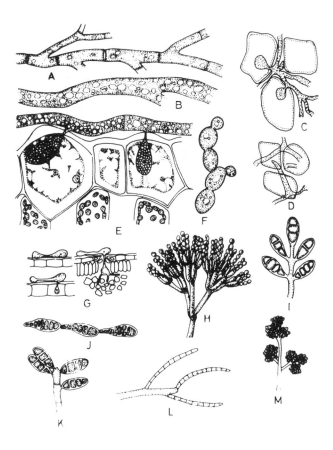

FIGURE 2. Fungal structures. (A) Aseptate hyphae, (B) aseptate hyphae, (C) intercellular hyphae and haustoria, (D) intracellular hyphae, (E) ectophytic hyphae and haustoria, (F) budding (yeasts), (G) development of appressoria, (H—M) sporophores and spores.

4. Others

The *satellite RNAs*, are small, linear RNAs found in virions of certain multicomponent viruses. Satellite RNAs may be related to the RNA of the virus or they may be related to those of the host. Satellite RNAs generally attentuate the effects of viral infection and may, possibly, represent a protective response of the host to viral infection.[9]

In 1982, Prusiner[16,17] described another infectious agent called "*prions*." These are composed of a small proteinaceous infectious particle and has been shown to cause the "scrapie disease of sheep and goat." Prions are believed to cause several chronic degenerative diseases of animals and humans. Although prions have so far been found only in animals, it is not unreasonable to expect that they will soon be shown to affect plants also.[9]

C. BIOTIC AGENTS
1. Eukaryotes
a. Fungi
i. General Characteristics

Fungi are microorganisms having chlorophyll-less, nucleated, unicellular, or multicellular filamentous thallus and reproduce by asexual and sexual spores. The body of a fungus is usually composed of distinctive elongated cells called "hyphae" (Figure 2) which aggregate together to form a "mycelium". The structure of fungal cell is unusual. In all but a few primitive species the protoplasm is surrounded by a cell wall. However, this usually consists mainly of "chitin" rather than cellulose as in true plants. In Mastigomycotina the hyphae typically lack cross walls (septa), resulting in a completely "coenocytic" mycelium (Figure 2). In Ascomycotina and Basidiomycotina cross-walls are produced, but even here there may be many nuclei per cell and the septa are usually perforated by pores so that the cytoplasm is still continuous. Among parasitic fungi, the hyphae may be "ectophytic" or "endophytic". The ectophytic mycelium lives on the surface of the host while endophytic hyphae are found within the host tissues. The endophytic mycelium may be inter or intra-cellular (Figure 2). Parts of mycelium may be organized into specialized absorptive structures (haustoria) or into dispersive or reproductive structures. Another specialized structure called "appressorium" (L. *appremere* = to press against) is formed from germ tubes of several fungi (Figure 2). This is simple or lobed mucilaginous swelling which helps fungus structures to attach to the host surface.

Fungi are heterotrophic organisms. They have holophytic or absorptive type of nutrition. They produce extracellular enzymes to degrade insoluble substrate into smaller fragments and finally into soluble units which are then absorbed by hyphae. Fungi may be biotrophs, hemibiotrophs, or perthotrophs.

ii. Asexual Reproduction

In asexual reproduction the hyphae cut-off minute spores. The structure and origin of these spores varies greatly and each type has been given a different name. *Chlamydospores* are thick walled, usually round, spores formed by the direct transformation of certain cells of a hypha, or spores. The term *conidium* (pl. conidia) is applied to spores which are pinched off as distinct bodies from the ends of special spore-bearing hyphae, the *conidiophores* (Figures 2 and 3). Such spores may be single or in clusters or in chains. Occasionally there is very little or no difference between the shape of spores and structure of the hypha producing it. Such spores have been called *oidia*. In many fungi, the asexual spores are produced in spore-mother sac or in different types of fruiting bodies or are arranged in groups in different types of stroma. In the lower groups of fungi which have aseptate hyphae, spherical, tubular or ovoid sacs, the *sproangium* (pl. sporangia) (Figure 3) are formed by swelling of the hyphal tip or any part of the hypha. In aquatic forms the sporangia liberate naked protoplasmic bodies, the *zoospores*, which are provided by delicate vibratile or undulating filaments or thread-like processes, the *flagella*, by means of which they are able to swim about in the surrounding film of water (Figure 3). Hickman and Ho[18] have described the characteristics of zoospores in different groups of lower fungi. In the higher forms of the phycomycetes the sporangia (Figure 3) liberate walled nonmotile or aplanospores (sporangiospores). There are occassions when the sporangium, instead of liberating spores, itself acts as an independent reproductive unit and functions as a conidium. Moist conditions favor multiplication by zoospores while in dry conditions mostly the sporangia show direct germination. The asexual spores may be formed in special structures (spore-fruits) variously known as pycnidia, sporodochia, acervuli, etc. (Figure 3).

iii. Sexual Reproduction

Sexual reproduction, i.e., nuclear fusion and meiosis, usually in that sequence, is wide-spread in fungi. Fertilization can be brought about in a variety of ways. In aquatic forms, the fusion of free-living gametes or of a free-living gamete with a gametangium is common. In other fungi there may be fusion of gametangia (Figures 3 and 4) which either differ clearly in form and function (as antheridium and oogonium) or in size only, or not at all. In the vast majority of fungi, however, tip fusion of undifferentiated hyphae enables potentially conjugative nuclei to migrate through the hyphae, frequently reciprocally, until fusion ul-

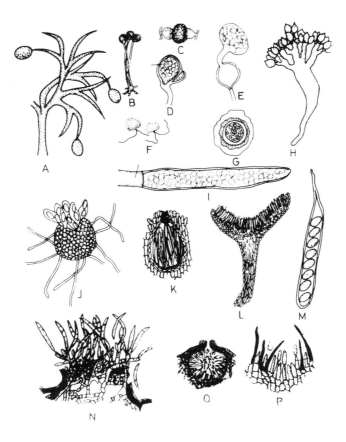

FIGURE 3. Fungal structures, (A) Conidiophores and conidia, (B) sporangiophore and sporangia, (C) Zygospores, (D) antheridium-oogonium, (E) germinating sporangium and formation of vesicle (F) zoospores, (G) oospores, (H, I) Sproangiophores and sporangia, (J) Cleistothecium, (K) perithecium, (L) apothecium (M) ascus and ascospores, (N) sporodochium, (O) pycnidium, (P) acervulus.

timately occurs in a cell of a specialized and often preformed structure. Meiosis is, thereafter, confined to a specialized cell type and is frequently followed by the production of spores from it or, in apomictic forms, its equivalent without meiosis. Two such cell types, are the ascus and the basidium. Both are normally terminal, or subterminal, somewhat ovoid cells. The four nuclei derived by meiosis in an ascus usually undergo a further somatic division and the eight products become incorporated in endogenously formed ascospores which are subsequently discharged or released. A similar sequence of fusion followed by meiosis occurs in a basidium but here the basidiospores generally arise immediately after meiosis at the tips of four tapering processes, the sterigmata (Figure 4).

During the development of asci, the surrounding monokaryotic hyphae organize a thick protective coat around the developing asci. The wall is formed by prosenchymatous or pseudoparenchymatous fungal tissues. The protective coat and the asci are jointly called *ascocarp*. Ascocarps are the sexual fruit bodies that protect the asci and also regulate release of the ascospores. Ascocarps are of three types: *cleistothecium*, which is completely closed and usually globular, *perithecium*, which is ostiolate and globular or more commonly flask-shaped, and the *apothecium*, which is cup or saucer-shaped bearing exposed hymenium (Figure 3).

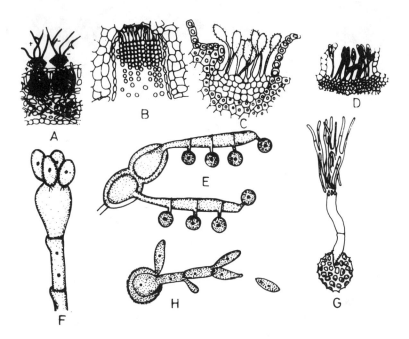

FIGURE 4.   Fungal structures. (A) Pycnium and receptive hyphae, (B) aecial cup, (C) uredial stage (D) telial stage, (E) germinating teleutospore of rust fungus, (F) basidium and basidiospores, (G-H) germinating teleutospores of smut fungi.

iv. Classification of Plant Parasitic Fungi

The fungi that cause disease on plants are a diverse group, and because of their large numbers and diversity, only a broad, sketchy classification of some of the most important genera is presented here.[19]

Kingdom—Protista (Eukaryotic); Subkingdom: Mycota (Fungi)
Division of Mycota: Myxomycota and Eumycota
Classes of Myxomycota: out of 4 classes plant pathogens only in class Plasmodiophoro-
                       mycetes, order Plasmodiophorales
Genus: *Plasmodiophora, Polymyxa, Spongospora*
Subdivisions of Eumycota: Mastigomycotina, Zygomycotina, Ascomycotina, Basidiomy-
                       cotina, and Deuteromycotina. Classes, orders, and important
                       genera of Mastigomycotina
Class: Chytridiomycetes; order: Chytridiales, Genus: *Olpidium, Physoderma, Synchytrium, Urophlyctis*
Class: Oomycetes; order: Peronosporales;
Genus: *Pythium, Phytophthora, Albugo, Plasmopara, Peronospora, Bremia, Sclerospora,
       Pseudoperonospora Sclerophthora*
Subdivision: Ascomycotina
Class: Hemiascomycetes, order: Taphrinales;
Genus: *Taphrina, Protomyces*
Class: Pyrenomycetes; order: Erysiphales
Genus: *Erysiphe, Microsphaera, Podosphaera, Sphaerotheca, Uncinula, Phyllactinia*
Order: Sphaeriales; Genus: *Ceratocystis, Diaporthe, Endothia, Glomerella, Rosellinia, Valsa,
       Xylaria*
Order: Hypocreales; Genus: *Claviceps, Gibberella, Nectria*

Class: Loculoascomycetes; Order: Myriangiales; Genus: *Elsinoe*. Order: Dothidiales; Genus: *Guignardia, Mycosphaerella*. Order: Pleosporales; Genus: *Ophiobolus, Venturia, Botryosphaeria, Cochliobolus*

Class: Discomycetes; Order: Helotiales; Genus: *Monilinia, Sclerotinia*. Order: Pezizales; Genus: *Pseudopeziza*

Subdivision: Basidiomycotina

Class: Teliomycetes; Order: Uredinales; Genus: *Cronartium, Gymnosporangium, Melampsora, Phragmidium, Puccinia, Uromyces*

Order: Ustilaginales; Genus: *Sphacelotheca, Tilletia, Neovossia, Urocystis, Ustilago*

Class: Gasteromycetes; Order: Exobasidiales; Genus: *Corticium, Exobasidium*. Order: Polyporales; Genus: *Fomes, Pellicularia, Polyporus, Poria, Stereum, Thanatephorus, Typhula*

Order: Agaricales; Genus: *Armillaria, Lenzites, Marasmius, Pemophora, Pholiota, Pleurotus, Schizophyllum*

Subdivision: Deuteromycotina

Class: Hyphomycetes; Genus: *Alternaria, Aspergillus, Botrytis, Cephalosporium, Cercospora, Cladosporium, Fusasrium, Helminthosporium, Phymatotrichum, Pyricularia, Rhizoctonia, Sclerotium, Thielaviopsis, Verticillium*

Class: Coelomycetes; Genus: *Ascochyta, Cytospora, Diplodia, Phoma, Phomopsis, Phyllosticata, Septoria, Colletotrichum, Coryneum, Cylindrosporium, Marssonina, Melanconium, Sphaceloma*

b. The Protozoa

The protozoa are mostly unicellular, microscopic animals, generally motile, and have typical nuclei. They may live alone or in colonies, may be free-living, symbiotic, or parasitic. Some protozoa subsist on other organisms such as bacteria, algae, yeasts, and on other protozoa, some saprophytically on dissolved substances in the surroundings, and some by photosynthesis as in plants.

Flagellates were found to be associated with plants in 1909 when *Phytomonas davidi* was found parasitizing the latex-bearing cells—the laticifers—of the laticiferous plant *Euphorbia*.[9] Since then several other species of *Phytomonas* have been reported from plants belonging to the families Asclepiadaceae (e.g., *P. elmassiani* on milk weed), Moraceae (e.g., *P. bancrofti* on a ficus species), Rubiaceae (e.g., *P. leptovasorum* on coffee), etc. All plant flagellates belong to the order Kinetoplastida, subfamily Trypanosomatidae. The plant-infecting Phytomonads are apparently transmitted by insects but so far insect vectors are known only for *P. elmassiani*.

c. The Algae

Algae are unicellular or multicellular, filamentous, thalloid, and chlorophyllus organisms which multiply by cell division, detachment of a portion from other cells, asexual, and sexual processes.

The parasitic algae may be endophytes or epiphytes. The endophyte algae include the species of *Chlorochytrium, Rhodochytrium,* and *Phyllosiphon*. These induce different types of symptoms on hosts for example, yellow raised spots and galls on leaves, stunting of plants, downward curling of leaves with bright red galls, large chlorotic lesions on leaves, etc.

The species of genus *Cephaleuros* are epiphytes. It has a wide host range, infecting members of 54 families. The alga produces red rust pustules on stem, leaves, and fruits and causes disease on mango, citrus, orange, guava, cacao, coffee, tea, oil-palm, avocado, pepper, and papaya.

d. Phanerogamic Plant Parasites

A number of flowering plants are parasitic on economic plants and cause considerable damage. These parasitic plants produce flowers and seeds similar to those produced by the plants they parasitize. They belong to several widely separated botanical families and vary greatly in their dependence on their host plants. Some of these parasites attack roots (Root parasites) of the host while others parasitize the stem (Stem parasites). Some are devoid of chlorophyll and depend entirely upon their hosts for food supply (Holoparasites) while others have chlorophyll and obtain only the mineral constituents of food from the host (Semi parasites). The common parasitic flowering plants may be grouped as follows.

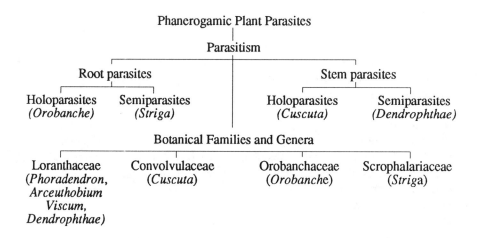

e. Nematodes
i. General characteristics

Nematodes (the name is derived from the greek word for thread) are elongate, tubular organisms, somewhat spindle shaped, which move like snakes. With some exceptions, adult plant parasitic nematodes are elongated worms ranging in length from about 0.30 mm to over 5.00 mm. The anterior end of a typical plant parasitic nematode tapers to a rounded or truncated lip region, the body proper is more or less cylindrical, and the posterior end tapers to a terminus which may be pointed to hemispherical. Proportions of the elongated body vary greatly, some species being more than 50 times longer than wide, others being only about 10 times longer than wide. The adult males are always slender worms. The females of some species become swollen at maturity and have pear-shaped or spheroid bodies (Figure 5). The mouth of the nematode is at the "anterior" end, and the terminus is at the "posterior" end. The excretory pore, vulva, and anus are on the "ventral" side; and the opposite side is called "dorsal". In plant-parasitic nematodes of the order Tylenchida, the

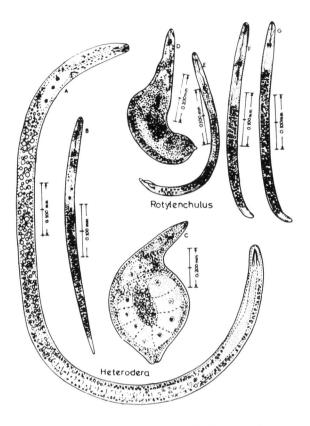

FIGURE 5. Plant parasitic nematodes. (A) *Heterodera* (cyst. ne-matode) male, (B) larva, (C) female, (D) *Rotylenchus* (reniform nematode) female, (E) male, (F) immature female, (G) larva.

mouth contains a stylet or mouth spear, a hardened, hollow, cuticular structure (Figure 5) similar to a hypodermic needle.

Reproduction of plant parastic nematodes is of three general types, varying with species. In bisexual species the female is fertilized by the male. In hermaphroditic species both eggs and sperms are reproduced by the female. In parthenogenetic species the eggs develop without fertilization. The life history is usually very simple, with five distinct stages, the first four of which end in a molt. After the final molt the nematode differentiate into adult males and females.

In terms of habitat, plant parasitic nematodes are either *ectoparasites* or *endoparasites*. Both these can be either migratory or sedentary.

ii. Classification

Taxonomically, plant parasitic nematodes are classified in two large groups. About 1000 of the approximately 1100 described species, belong to the Tylenchida group. The remainder belong to the order Dorylaimida. Differentiation between Tylenchida and Dorylaimida is based essentially on the structure of stylet and of the esophagus. A broad classification with pertinent examples is given.[9,20]

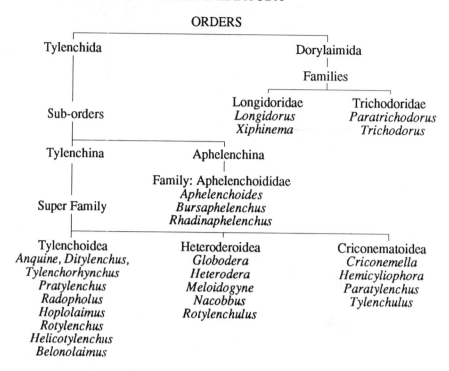

PHYLUM:NEMATODA

ORDERS

Tylenchida   Dorylaimida

Families

Longidoridae   Trichodoridae
Longidorus   *Paratrichodorus*
Xiphinema   *Trichodorus*

Sub-orders

Tylenchina   Aphelenchina

Family: Aphelenchoididae
Aphelenchoides
Bursaphelenchus
Rhadinaphelenchus

Super Family

Tylenchoidea   Heteroderoidea   Criconematoidea
Anquine, Ditylenchus,   *Globodera*   *Criconemella*
Tylenchorhynchus   *Heterodera*   *Hemicyliophora*
Pratylenchus   *Meloidogyne*   *Paratylenchus*
Radopholus   *Nacobbus*   *Tylenchulus*
Hoplolaimus   *Rotylenchulus*
Rotylenchus
Helicotylenchus
Belonolaimus

2. Prokaryotes

a. Bacteria

i. General Characteristics

Bacteria are simple microorganisms usually consisting of single prokaryotic cells, i.e., cells containing a single circular chromosome but no nuclear membranes or internal organelles comparable to mitochrondria or chloroplasts. Almost all plant pathogenic bacteria are rod shaped, (Figure 6) the only exception being the species of *Streptomyces*, which are filamentous. The rod shaped bacteria are more or less short or cylindrical and in young cultures they range from 0.6 to 3.5 μm in length and from 0.5 to 1.0 μm in diameter. The bacterial cells are surrounded by a wall made up of mucopeptide, which is peculiar to bacteria and is not found elsewhere. Its amount varies in the cell walls of the two main divisions of the eubacteria (true-bacteria): Gram positive and Gram negative. In Gram-positive bacteria it is

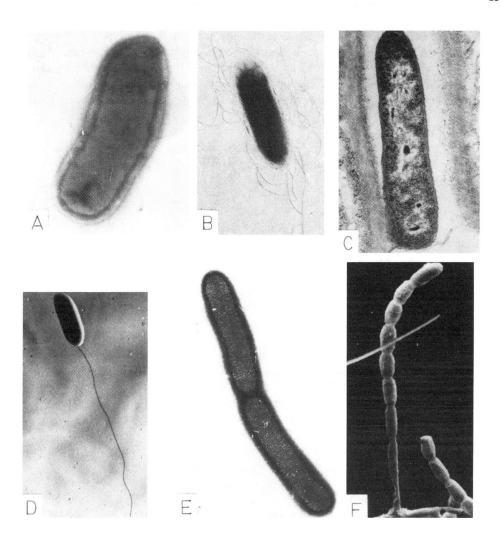

FIGURE 6. Electron micrographs of important genera of plant pathogenic bacteria. (A) *Agrobacterium*, (B) *Erwinia*, (C) *Pseudomonas* in intercellular spaces of tobacco mesophyll cells (Agrios, G. N., *Plant Pathology*, Academic Press, 1988. With permission.). (D) *Xanthomonas campestris* pv. *citri* (courtsey, Dr. M. Goto), (E) *Xylella fastidiosa* (Wells, J. M. *et. al., Int. J. Syst. Bact.,*, 3, 1987. With permission.), F) *Streptomyces ipomoea* (Clark, C. A. and Matthews, S. W., *Phytopathology*, 77, 1987. With permission.)

the major cell-wall component. However, in Gram-negative bacteria it is present in small quantities, the major portion being formed by lipoprotein and lipopolysaccharide.

Plant pathogenic bacteria are non-motile or motile by means of one or more flagella (Figure 6). Bacterial flagella are made up of flagellin molecules. On the basis of the number and position of flagella the bacteria can be grouped in five categories. They are *atrichous*—no flagellum present; *monotrichous*—a single flagellum at one end of the cell; *amphitrichous*—one flagellum at each end of the cell; *lophotrichous*—two or more flagella at one or both ends of the cell, and *peritrichous*—large number of flagella all over the cell.

ii. Reproduction

Bacteria multiply with astonishing rapidity and their significance as pathogens stems particularly from the fact that they can produce tremendous numbers of cell in a short period of time. Rod-shaped plant pathogenic bacteria reproduce by the asexual process known as "binary fission". This is a very quick process and is completed in 20 to 30 min. At this rate one bacterial cell could produce 1 million bacteria in 6 to 10 h. Fission involves formation of a septum and chromosome division. Both the events occur simultaneously. The cells are divided by a simple division into two halves. Each half becomes an independent individual.

iii. Classification

The kingdom-Prokaryoteae comprised two divisions: Photobacteria and Scotobacteria. The following is the condensed form of the grouping which lists only the parts having plant pathogenic bacteria.[21]

Kingdom: Prokaryoteae

Division I: Photobacteria—sensitive to light, no plant pathogens
Division II: Scotobacteria—indifferent to light
Class I: The Bacteria — Parts 2 to 6, 9 to 14, and 16 have no plant pathogens
Part 7:   Gram-negative, aerobic rods and cocci. Five families with 14 genera and 6 genera of uncertain affiliation. Plant pathogens in the families—Family: Pseudomonadaceae (order Psudomonadales); Genera: *Pseudomonas, Xanthomonas, Xylella*[22]
Xylella: rod shaped, under some cultural conditions filamentous, nonmotile, aflagellate, nonpigmented, nutritionally fastidious, earlier known as RLB or Xylem-imited fastidious bacteria
Family: Rhizobiaceae; Genus: *Agrobacterium*
Part 8:   Gram-negative, facultatively anaerobic rods. Two families with 17 genera and 9 genera of uncertain affiliation.
Family: Enterobacteriaceae (order-Eubacteriales); Genus: *Erwinia*
Part 15:   Endospore forming rods and cocci; Genera: *Clostridium* and *Bacillus*
Part 17:   Gram-positive, irregular rods and filamentous bacteria. Five genera described in arbitrary sequence, one family with two genera, and one order comprising of 8 families and 31 genera. Plant pathogens in:
A.   Irregular rods. Genera: *Corynebacterium* and *Clavibacter*. *Clavibacter*—contains most important phytopathogenic bacteria formerly classified as *Corynebacterium*. A few phytopathogenic former *Corynebacterium* species are still listed as *Corynebacterium* but they, too, are expected to be transferred to other genera.
B.   Filamentous bacteria. Order: Actinomycetales; i. Family: Nocardiaceae; Genus: *Nocardia*.ii Family: Streptomycetaceae; Genus: *Streptomyces*
Class III: Mollecutes; pleomorphic scotobacteria devoid of cell wall; collectively known as mycoplasmas
Part 19:   One order and 3 families; Order: Mycoplasmatales; Family I: Mycoplasmataceae; Genus: *Mycoplasma* with 51 named species; none plant pathogenic; organisms resembling *Mycoplasma* in plants not yet cultured, known as MLO. Genus: *Ureoplasma* (no plant pathogen); Family II: Spiroplasmataceae; Genus: *Spiroplasma* with single species *S. citri*

This system of classification has been further revised. Kingdom Prokaryotae has been classified into four divisions (Gracilicutes, Firmicutes, Tenericutes, and Mendosicutes) based on their phylogenic evolution. Phytopathogenic bacteria are placed in division Gracilicutes and Mycoplasmas in Tenericutes.[23] Sections have been created based on specific characteristics of a group of related organisms. There are 11 sections. Phytopathogenic bacteria have been described under sections 4 and 5. Rickettsias and chlamydias have been placed in section 9 while mycoplasmas in section 10. There is no change, however, in family and genera. They have been retained as such.

In the following Table 4 salient features of plant pathogenic prokaryotes are summarized.

b. The Fastidious Prokaryotes
i. The Ratoon Stunting Disease (RSD) Bacterium

In 1973, a small bacterium was found associated with diseased sugar cane plants. This coryneform bacterium was seen in extracts from diseased but not from healthy plants, and was found not to pass through a 0.22 μm filter. The RSD bacterium measures 0.25 to 0.5 × 1 to 4 μm with occasional lengths of 10 μm or more reported. In 1980, the RSD-associated bacterium was isolated in axenic culture, and Koch's postulates were proved to establish its causal role to RSD.[24] In culture, the bacterium, which measures 0.25 to 0.35 × 1 to 4 μm, divides by septum formation. The bacterium is coryneform, having straight or slightly curved cells with occasional swellings at the tip or in the middle. Cells frequently contain mesosomes. In 1981, another group working independently also reported the axenic culture of the bacterium as well as its ability to colonize sorghum-sudangrass following inoculation.[25]

The RSD bacterium is Gram-positive, nonspore-forming aerobic, oxidase-negative, catalase-positive, and nonacid fast. The cell walls of the bacterium contain 2,4-diaminobutyric acid, glutamic acid, glycine and alanine as the major aminoacids, and fucose and rhamnose as the major sugars.[26] The only bacterium with which the RSD bacterium is serologically related is *Clavibacter xyli* sp. *cynodontis,* a morphologically similar bacterium isolated from bermuda grass.[24]

Taxonomic studies have designated the RSD bacterium as *C. xyli* sp. *xyli.*[25] This new genus groups all of the coryneform plant pathogenic bacteria that contain 2,4 diaminobutyric acid in their cell walls, and the type species if *Clavibacter michiganense* sp. *michiganense,* formerly classified in the genus *Corynebacterium*. The guanine-plus-cystosine contents of *C. xyli* ssp. *xyli* is 66 mol%, and the whole cell, fatty acid extracts contain 17 to 24% 15:0 anteiso, 5 to 14% 16:0 iso, and 62 to 72% 17:0 anteiso acids.[27]

ii. The Xylem-Limited Gram-Negative Bacteria

All Gram-negative xylem-limited bacteria which have been studied so far are similar in morphology suggesting close phylogenetic relationship. Pierce's disease, alfalfa dwarf, and almond leaf scorch are presumably caused by same organism.[28] They are serologically related. The agent of Pierce's disease and almond leaf scorch have been isolated on same culture medium and pathogenicity proved.[29] Transmission of these bacteria is almost entirely by xylem-feeding insects or by vegetative propagation. In general, the xylem-limited, Gram-negative bacteria have elongated cells of 0.2 to 0.5 × 1 to 4 μm. The Pierce's disease and almond leaf scorch bacteria measure 0.25 to 0.5 × 1 to 4 μm while the bacterium associated with phony disease of peach is slightly smaller, being 0.25 to 0.4 × 1 to 3 μm. The cells usually have a well-defined cell wall and plasma membrane, both triple layers in structure.

iii. The Phloem-Limited Bacteria

Plant phloem is habitat of three types of pathogenic prokaryote organisms, viz., the MLO, *Sprioplasma* and Rickettsia-like bacteria (RLB), and the flagillated protozoa of the genus *Phytomonas* which are eukaryotes though fastidious in nature. So far these phloem limited bacteria have not been cultured *in vitro* and Koch's postulates for their association with respective diseases have not been proved. Therefore, not much is known about their nature, taxonomy, and serological relationships. Clover club leaf, white clover disease, and

TABLE 4
Salient Features of Plant Pathogenic Genera of Bacteria

Genus or Trivial name	Gram stain	Flagellation	Morphology and size	Colony and pigmentation	G-C mol%	Habitat and disease symptom	Reactions
Agrobacterium	Gram-negative	Peritrichous	Rods 0.8 × 1.5—3 μm	Smooth, nonpigmented	60-63	Rhizosphere and soil inhabitants, causes hypertrophy	When growing on carbohydrate containing media abundant polysaccharide slime produced
Clavibacter (Corynebacterium)	Gram-positive	None or Polar	Straight or slightly curved rods, pleomorphic V-form 0.5—0.9 × 1.5—4 μm	Irregularly stained pigments or granules and clubshaped swellings	53-55	Soil-borne, produce canker, wilt, rots, and fasciation	
Erwinia	Gram-negative	Peritrichous	Straight rods, 0.5—1.0 × 1.0—3.0 μm	None or yellow blue pink pigmentation	50-57	Facultative anaerobes, Necrosis, wilt diseases and soft rots	The "amylovora" group do not produce pectic enzymes while "carotovora" group has strong pectolytic activity
Pseudomonas	Gram-negative	One or many polar flagella	Straight to curved rods 0.5—1.0 × 1.5—4.90 μm	None or green blue	57.7-67	Inhabitants of soil, fresh water and marine environment, cause leaf spots, galls, wilt, blight, canker, etc.	Soluble pigments produced, no acid from lactose, some (e.g. *P. syringae*) produce yellow green diffusable fluorescent pigments on a medium of low iron content
Streptomyces	Gram-positive	None	Slender, branched hyphae, 0.5—1 μm in diameter, spores formed	On nutrient media colonies are small (1-10 mm in diameter) smooth surface, with a weft of aerial mycelium that may appear granular, powdery or velvety, wide variety of pigments formed	69-73	Soil inhabitants, cause scab disease	Produce one or more antibiotics active against bacteria, fungi, etc.

Xanthomonas	Gram-negative	Polar	Straight rods, 0.4—1.0 × 1.2—3.0 μm	Growth on agar media usually yellow	63-69	All species are plant pathogenic and found only in association with plants or plant materials, cause leaf spot blight, canker, etc.	Produce acid from lactose
Xylella	Gram-negative	Nonmotile, aflagellate	Straight rods, 0.3 × 1-4.0 μm	Colonies small, with smooth or finely undulated margins, nonpigmented	—	Strictly aerobic, nutritionally fastidious, habitat is xylem of plant tissue	—

clover rugose leaf curl are supposed to be caused by the same or very similar RLB. They show very similar symptoms and are sensitive to same antibiotics.[30] However, all phloem-limited RLB have several common structural features which may indicate a common phylogeny and, eventually, recognition as a single taxon.[28]

Two evidences have been cited as proof of pathogenicity of phloem inhabiting RLB and their distinction from MLO. These are (1) the high correlation between symptom expression and the presence of RBL in plants, and (2) the remission of symptoms accompanied by disappearance of the RLB after treatment with penicillin in contrast to penicillin insensitivity of MLO. The symptoms produced by phloem limited RLB are characteristics of yellows type which include stunting, yellowing of young leaves, virescence of floral parts, premature flowering and fruit drop, witches' broom, and often, premature death of the entire plant. Transmission is by leaf hopper, dodder, and grafting.

The bacterial cells are found primarily in mature sieve elements, irregularly distributed among the vascular bundles. They are mostly rigid rods (not pleomorphic) and nonmotile. These features distinguish them from *Mycoplasma* and *Spiroplasma,* which are also found in the phloem. The cells measure 0.2 to 0.5 $\times$ 1.0 to 2.0 (0.3 $\times$ 1.3) μm and are bound by a double membrane, or a cell wall and cytoplasmic membrane. Both membranes are triple layered and are separated by an electron lucent zone.

Hopkins[30] has listed following diseases caused by nontissue restricted, Gram-negative fastidious bacteria: infectious necrosis of grapevines, apple proliferation, carrot proliferation, chlorosis, and aspermy of wheat and yellows of grapevine. Not much is known about these agents.

iv. Mycoplasma-Like Organisms (MLO)

The organisms observed in plants and insect vectors, with the exception of sprioplasmas, resemble the mycoplasmas of the genera *Mycoplasma* or *Acholeplasma* in all morphological aspects. In the absence of *in vitro* cultivation, extensive studies of these organisms have been carried out only by thin section electron microscopy of diseased tissues or from their insect (leaf hopper) hosts. The plant MLO are pleomorphic, small rounded (60 to 100 nm = 0.06 to 0.1 μm in diameter), large globular, (Figure 7) and branched filaments, 1 to 2 to several micrometers long. Average diameter is 0.3 to 0.8 μm. The cells are bounded by a single trilaminar unit membrane and contain ribosomes and chromatin. No membrane-bound cell organelles are present. Small rounded and large globular bodies predominate in late season or in the advanced pathological stage while branched filamentous forms are seen in early season or in early stages of disease.

The reproduction of MLO is not clear. Variable methods of reproduction have been proposed for *Mycoplasma.* These include binary fission, budding, release of inclusion bodies, and formation of intracellular "elementary bodies".

The disease caused by nonhelical, noncultivated MLO include aster yellows, elm yellows, or phloem necrosis, coconut lethal yellowing, stolbur, paulownia witches' broom, potato witches' broom, potato phyllody, clover phyllody, sesamum phyllody, little leaf of egg plant and legumes, mulberry dwarf, bunchy top of banana, bunchy top of papaya, pear decline (degeneration), X-disease of peach, grassy shoot of sugarcane, sandal spike, purple top roll of potato, etc.

v. Spiropiasmas

After the discovery of MLO in yellows diseased plants by Japanese workers in 1967. The etiology of corn stunt and citrus stubborn diseases was reexamined. Davis et al.[31] examined the sap from corn stunt infected plants and observed motile and helical organism that are today recognized as corn stunt spiroplasma (CSS).

Sprioplasmas are pleomorphic cells that vary in shape from spherical to slightly ovoid,

21

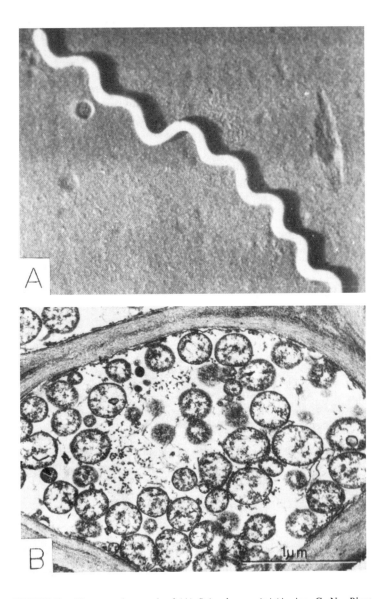

FIGURE 7. Electron micrograph of (A) *Spiroplasma citri* (Agrios, G. N., *Plant Pathology*, Academic Press, 1988. With permission.), (B) MLO of aster yellows in *Nicotiana rustica* (Heitefuss, R. and Williams, P. H., Eds., *Physiological Plant Pathology*, Springer Verlag, Berlin, 1976. With permission.)

100 to 250 nm or larger in diameter, to helical (Figure 7) and branched nonhelical filaments that are about 120 nm in diameter and 2 to 4 μm long during active growth and considerably longer in later stages of their growth. Colonies of spiroplasmas on agar have a diameter of about 0.2 mm; some have a typical "fried egg" appearance, but others are granular. Electron microscopy reveals the ultrastructure of cell membrane (single triple-layered membrane) and internal contents such as ribosomal granules and DNA strands. Spiroplasmas are Gram-positive or Gram-variable.

The reproduction mechanism of sprioplasmas are incompletely known. It is believed that *S. citri* multiplies by classic binary fission through appearance of one or several constrictions in relatively long helices followed by the release of two or more shorter daughter helical cells having one or two turns. Besides helical morphology, motility is the other most distinct property of spiroplasmas except the one nonhelical variant of *S. citri*. Spiroplasmas exibit two types of motion of their body. One consists of flexional movement and the other is a rapid spinning about the long axis of the spiral. Spiroplasmas require sterol of for growth. They are resistant to penicillin but inhibited by erythromycin, tetracycline, neomycin, and amphotericin.

Spiroplasmas are known to cause stubborn disease of citrus, corn stunt, rice yellow dwarf, witches' broom of cactus, yellows of lettuce and aster, and many other unconfirmed hosts. Transmission of plant pathogenic spiroplasmas and MLO from plant to plant is similar. Natural transmission is by leaf hopper vectors and by parasitic dodder.

REFERENCES

1. **Copeland, H. F.**, *The Classification of Lower Organisms,* Pacific Books, Palo Alto, California, 1956, 180.
2. **Berkeley, F. A.**, *Outline Classification of Organisms,* Hoppkins Press, Providence, Massachusetts, 1968, 205.
3. **Whittaker, R. H.**, New concepts of kingdoms of organisms, *Science,* 163, 150, 1969.
4. **Waterson, A. P. and Wilinson, L.**, *An Introduction to the History of Virology,* Cambridge Univ. Press, Cambridge, 1978, 237.
5. **Matthews, R. E. F.**, *Plant Virology,* Academic Press, New York, 1970, 778.
6. **Frankel-conrat, H.**, *The Molecular Basis of Virology,* Reinhold, New York, 1968, 656.
7. **Gibbs, A. H. and Harrison, B. D.**, *Plant Virology—The Principles,* Edward Arnold, London, 1976, 292.
8. **Bos, L.**, *Introduction to Plant Virology,* Pudoc, Wageningen, 1983, 160.
9. **Agrios, G. N.**, *Plant Pathology,* Academic Press, New York, 1988, 803.
10. **Diener, T. O.**, Potato spindle tuber "Virus". IV. A replicating low-molecular weight RNA, *Virology,* 45, 411, 1971.
11. **Singh, R. P. and Clark, M. C.**, Infectious low molecular weight ribonucleic acid, *Biochem. Biophys. Res. Commun.,* 44, 1077, 1971.
12. **Singh, R. P.**, Pathogenesis and host-parasite relationship in viroids, in *Experimental and Conceptual Plant Pathology,* Vol. 2, Singh, R. S., Singh, U. S., Hess, W. M., and Weber, D. J., Eds., Gordon and Beach Science Publishers, New York, 1988, 599.
13. **Keese, P. and Symons, R. H.**, The structure of viroids and virusoids, in *Viroids and Viroid like Pathogens,* Semancik, J. S., Ed., CRC Press, Boca Raton, FL, 1987, 177.
14. **Keese, P., Osorio-Keese, M. E., and Symons, R. H.**, Coconut tinangaja viroid: sequence homology with coconut cadangcadang viroid and other potato spindle tuber viroid related RNAs, *Virology,* 162, 508, 1988.
15. **Pezian, M. A., Koltunow, A. M., and Krake, L. R.**, Isolation of three viroids and a circular RNA from grape vines, *J. Gen. Virol.,* 69, 413, 1988.
16. **Prusiner, S. B.**, Novel proteinaceous infectious particles cause Scrapie, *Science,* 216, 136, 1982.
17. **Prusiner, S. B.**, Prions: novel infectious pathogens, *Advances Virus Res.,* 29, 1, 1984.
18. **Hickman, C. J. and Ho, H. H.**, Behaviour of zoospores in plant pathogenic phycomycetes, *Annu. Rev. Phytopathol.,* 4, 195, 1966.
19. **Ainsworth, G. C.**, Introductions and keys to higher texa, in *The Fungi: An Advanced Treatise* Vol. IV Ainsworth, G. C., Sparrow, F. K., and Sussman, A. S., Eds., Academic Press, New York, 1973.
20. **Reddy, P. P.**, *Plant Nematology,* Agricole Publishing, New Delhi, 1983, 287.
21. **Buchanan, R. E. and Gibbons, N. E.**, *Bergey's Manual of Determinative Bacteriology,* 8th Eds., Williams and Wilkins, Ballimore, Md. 1974, 1268.
22. **Wells, J. M., Raju, B. C., Hung, H. Y., Weisburg, W. G., Mandelco-Paul, L., and Brenner, D. J.**, *Xylella fastidiosa,* gen. nov. sp. nov., Gram-negative, xylem limited, fastidious plant bacteria related to *Xanthomonas* spp., *Int. J. Syst. Bacteriol.,* 37, 136, 1987.
23. **Krieg, N. R., and Holt, J. G.**, *Bergey's Manual of Systematic Bacteriology,* Vol. I., William and Wilkins, Baltimore, MD., 1984, 964.
24. **Davis, M. J., Gillaspie, A. G., Jr., Harris, R. W., and Lawson, R. H.**, Ratoon stunting disease of surgarcane: isolation of the causal bacterium, *Science,* 240, 1365, 1980.

25. **Davis, M. J., Gillaspie, A. G., Jr., Vidaver, A. K., and Harris, R. W.,** Clavibacter: a new genus containing some phytopathogenic coryneform bacteria, including *Clavibacter xyli* sub sp. *xyli* sp. nov., sub sp. nov. and *Clavibacter xyli* sub sp. *cynodontis* sub sp. nov., pathogens that cause ratoon stunting disease of sugarcane and bermudagrass stunting disease, *Int. J. Syst. Bacteriol.,* 34, 107, 1984.
26. **Davis, M. J., Lawson, R. H., Gillaspie, A. G., Jr., and Harris, R. W.,** Properties and relationships of two xylem limited bacteria and a mycoplasma like organism infecting bermuda grass, *Phytopathology,* 1973, 341, 1983.
27. **Gillaspie, A. G., Jr., Sasser, M., and Davis, M. J.,** Fatty acid profiles of bacteria causing ratoon stunting disease (RSD) of sugarcane and bermuda grass stunting disease (B.S.D), Abstr., *Phytopathology,* 74, 880, 1984.
28. **Davis, M. J., Whitcomb, R. F., and Gillaspie, A. G., Jr.,** Fastidious bacteria of plant vascular tissue and invertibrates (including so called rickettsia-like bacteria) in *The Prokaryotes: A Handbook on Habitats, Isolation and Identification of Bacteria* Vol. 2, Starr, M. P., Stolp, H., Truper, H. G., Balows, A., and Schegel, H. G., Eds., Springer-Verlag, Berlin, 1981, 2172.
29. **Davis, M. J., Purcell, A. H., and Thomson, S. V.,** Pierce's disease of grapevine: isolation of the bacterium, *Science,* 199, 75, 1978.
30. **Hopkins, D. L.,** Disease caused by leafhopper-borne rickettsia-like bacteria. *Annu. Rev. Phytopathol.,* 15, 277, 1977.
31. **Davis, R. E., Worley, J. F., Whitcomb, R. F., Ishiyama, T., and Steere, R. L.,** Helical filaments produced by a mycoplasma-like organism associated with corn stunt disease, *Science,* 176, 521, 1972.

Chapter 2

PLANT DISEASES

I. INTRODUCTION

Plants not only sustain the man and animals, they are also the source of food for multitudes of organisms living in the ecosystem. A conflict of interest, therefore, between the contenders of suppliers from plants is inevitable. Thus, while man has been able to subjugate plants and animals for his own use, the competing microorganisms still defy his efforts and claim a major share of resources which man would like to use for himself. Plant diseases have been considered as stubborn barriers to the rapid progress of food production.

II. THE SCIENCE OF PLANT PATHOLOGY

A. CONCEPT

Plant pathology or phytopathology (*phyton* = plant; *pathos* = ailments; *logos* = knowledge) is that branch of agricultural, botanical, or biological science which deals with the cause, etiology, resulting losses, and control of plant diseases.

B. RELATIONSHIP WITH OTHER SCIENCES

The objectives of plant pathology is to identify the cause(s) of disease, the mechanism of disease development, the factors affecting disease development and finally economic and efficient management of diseases. Knowledge of basic biological and physical sciences, as well as comprehension of agricultural, environmental, and social sciences, are the foundation stones upon which the science of plant pathology rests.[1] For instance, to understand and manage disease, plant pathologists must understand the biology, physiology, reproduction, dispersal, survival, and ecology of all the multiple pathogens and parasites of plants. They also must understand the concepts of stress and strains in plants and how environmental factors induce disease when the limits of tolerance are exceeded.[2] The interrelations of different sciences with plant pathology are illustrated in the Figure 1.

III. CONCEPT OF DISEASE IN PLANTS

A. AN OVERVIEW

Disease is one of those terms that is very difficult to define. It is realized that disease (literally *dis-ease*) implies lack of comfort and, therefore, involves deviation from normal functioning. According to Agrios,[3] a plant is healthy when it can carry out its physiological functions to the best of its genetic potential. However, it is difficult to determine genetic potential of a plant because gene expression itself is influenced by environmental factors. Nevertheless, one can grow the plants under different sets of environmental and/or nutritional conditions and can find out the best combination under which its growth and reproduction are optimum. This very state of the plant which cannot be improved further by manipulation of environmental factors, is considered as healthy and any deviation, of course negative, from this state is considered as abnormal or "diseased". It is, thus, seen that any deviation from normalcy results in the diseased state.

From time to time several definitions,[4-10] which have been proposed, are in fact descriptive but not simultaneously exclusive. All these definitions indicate that disease: (1) is related to poor functioning of growth and reproduction in the plant, (2) is malfunctioning physiology of plant, and (3) reduces the plant's ability to survive and maintain its ecological

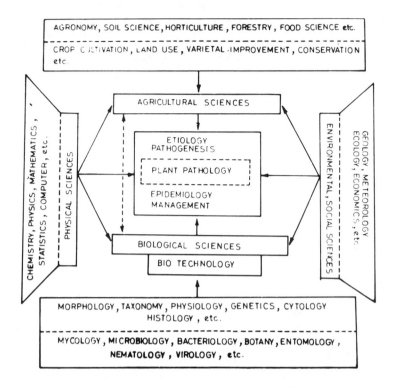

FIGURE 1.   The relation of plant pathology to other sciences.

niche. Horsfall and Dimond[11] arrived at a descriptive explanation which embodies the concept of a disease. According to them disease: (1) is not a pathogen but it is caused by a pathogen, (2) is not the symptom but results in symptoms, (3) is not a condition as the condition results from disease and is not synonymous with it, (4) is not an injury, (5) cannot be catching or infectious, it is actually the pathogen which is catching or infectious, (6) results from continuous irritation, and (7) is a malfunctioning process and this must result in some suffering and hence disease is a pathological process.

Based on the concept of 'Dialectical Materialism,' Das Gupta[12] defined disease as a "malfunctioning process involving more or less continuous interaction between host and pathogen occurring through a large number of quantitative changes between qualitative states of healthy and diseased conditions".

Thus, most of the recent defintions emphasize on (1) continuous interaction/irritation between host and parasite/pathogen and (2) disease is a malfunctioning process. These two points require in depth analysis.

B. CONTINUOUS VS. TRANSIENT INTERACTION

The concept of continuous interaction, was emphasized by Horsfall and Dimond[11] mainly for the purpose of differentiating disease from injury. Disease signifies a dynamic process, whereas injury represents a static state. However, there are situations where injury does not result in instantaneous death of the plant but it is of sufficient magnitude to lead to the malfunctioning of physiological processes of plants beyond their easy tolerance resulting in gradual development of symptoms at cellular and/or morphological level, e.g., injuries caused by air pollutants, herbicides, sun scald, cold, etc. Similarly, damage to the vascular system caused by stem borer is purely an injury. If this injury is of such a magnitude that plant is unable to compensate resulting losses in its water and nutrient transports, it would suffer ultimately. The resulting effect would gradually manifest at morphological level as a wilt or dieback symptom. Therefore, there is little justification in excluding these situations from realm of disease.

Another major question is continuous interaction between whom? Of course, host is one partner but whether other partner is primary incitant or its product? What about crown gall, nutritional deficiencies, insect toxemias, toxicities caused by air pollutants like ozone, SO_2, PAN etc., sun scald or cold injury or symptoms induced by host-specific pathotoxin rather than the pathogen itself? What should be considered as a threshold point to demarcate between transient and continuous interaction? We have already discussed the situations where injury might result in disease development. This leads to the conclusion that there is not much justification in including the concept of continuous interaction or association in definition of plant disease.

C. PHYSIOLOGICAL VS. BIOCHEMICAL PROCESS

What are those processes whose malfunctioning results in a disease? What are those malfunctioning processes which constitute disease? Studies conducted during the last two decades have revealed some of the finer details of host-parasite interaction. It is now accepted that first step in the interaction, particularly in biotrophs, is mutual recognition of host and parasite as a compatible partner. This recognition is essential for establishment of genetic and subsequent physiological synchrony. Positive recognition for basic compatibility leads to a cascade of biochemical changes in the host. Several biochemical reactions are intensified, inhibited, induced, or altered in integrated manner to support growth of the pathogen. The pathogen not only draws nutrient from host to support its growth and reproduction but its interaction with the host results in the synthesis and/or release of hydrolytic enzymes and toxic metabolites detrimental to the host resulting in shift of balance more and more toward parasitism. If alterations are beyond the tolerance limit of the host, the latter suffers at its physiological, cytological, and/or morphological level resulting in different types of symptoms and syndromes depending upon host and parasite involved. Similar to biotic causes, abiotic factors like nutritional deficiencies, toxins, pollutants, herbicides, etc., also bring about a cascade of biochemical alterations in host plant resulting in malfunctioning of its physiological processes which are gradually manifested at morphological level.

Biochemical alterations precede physiological changes which in turn may result in cytological and morphological alterations recognized as disease symptoms. Since symptoms are not synonymous to disease but are results of disease, latter must precede the former. Hence, disease becomes synonymous to altered and induced biochemical changes in host brought about by an invading pathogen.

D. A DEFINITION IS REPRESENTATIVE OF TIME

Lack of an adequate definition of "disease" is not at all an impediment to the progress of plant pathology. A definition does not result in an advancement of our understanding of the phenomenon. A better understanding of a phenomenon enables us to define it more accurately and precisely. Definition of any phenomenon given at a particular time must represent the status of our understanding of the same at that time. Therefore, a definition must be representative of time.

Analysis presented so far leads us to the conclusion that whenever development of symptom is gradual (as opposed to sudden death), the mechanism involved is more or less the same (i.e., alternation in biochemical reactions precedes malfunctioning of the physiological processes followed by expression at the morphological level) irrespective of the nature of the cause (biotic or abiotic) or association (i.e., transient or continous). At the same time there is growing realization that malfunctioning of the physiological processes should be considered as a symptom rather than disease itself.

Keeping in view the above discussion a moderately precise definition of "disease" is proposed as "a sum total of the altered and induced biochemical reactions in a system of a plant or plant part brought about by any biotic, mesobiotic or abiotic factor(s) leading to malfunctioning of its physiological processes, and ultimately manifesting gradually at cytological and/or morphological level. All these alterations should be of such a magnitude that they become a threat to the normal growth and reproduction of the plant''.

IV. CLASSIFICATION OF PLANT DISEASES

A. GENERAL CLASSIFICATION

Various schemes of classifying plant diseases have been proposed. A disease may be *localized* or it may be *systemic*. The diseases are *soil borne, seed borne,* or *air borne*. The symptoms or signs which appear on the affected plant parts, also form a basis for grouping the plant diseases. Thus, we find diseases known as rusts, smuts, root rots, wilts, blights, cankers, mildew, fruit rot, etc. According to the host plants the diseases may be grouped as cereal disease, forage crop diseases, flax diseases, etc.

McNew[13,14] proposed a system based upon seven physiological processes. Depending upon which vital functions are being adversely affected, a plant disease would be classified in one of the following groups: (1) soft rots and seed decays, (2) damping off and seedling blights, (3) root rots, (4) gall diseases and others in which meristematic activity is impaired, (5) vascular wilts, (6) disease affecting photosynthesis (bacterial and fungal spots and blights, downy and powdery mildews, and rusts), and (7) diseases interfering with translocation (viral diseases and diseases caused by MLO).

B. BASED ON OCCURRENCE

1. Endemic Diseases

The word endemic means ''prevalent in, and confined to a particular district or country'' and is applied to disease. When a disease is more or less constantly present from year to year in a moderate to severe form, in a particular country or part of the earth, it is classed as endemic.

2. Epidemic or Epiphytotic Diseases

The term ''epidemic'' is derived from a Greek word meaning ''among the people'' and in true sense applies to those diseases of human being which appear very virulently among a large section of the population. To carry the same sense in the case of plant diseases the term ''epiphytotic'' has been coined. An epiphytotic disease is one which occurs widely but periodically. It may be present constantly in the locality but assumes severe form only on occassions.

3. Sporadic Diseases

Sporadic diseases are those diseases which occur at very irregular intervals and locations and in relatively few instances.

4. Pandemic Diseases

These occur all over the world and result in mass mortality, e.g., late blight of potato.

C. ACCORDING TO MAJOR CAUSAL AGENTS

1. Noninfectious or Nonparasitic or Physiological Diseases

These diseases with no biotic or mesobiotic agents associated, remain noninfectious and cannot be transmitted from one diseased plant to another healthy plant.

2. Infectious Diseases

These are diseases which are incited by biotic and/or mesobiotic agents under a set of suitable environments. Association of a definite pathogen is essential with such diseases.

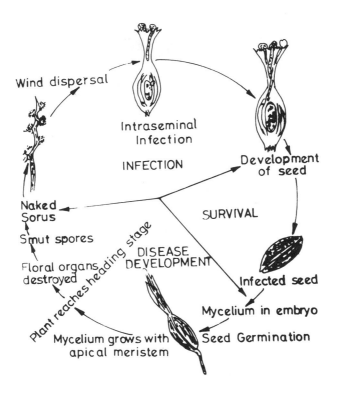

FIGURE 2.   Simple interest disease (single-cycle disease), e.g., loose smut of wheat.

D. BASED ON PRODUCTION OF INOCULUM

1. Simple Interest Disease (Single-Cycle Disease)

When the increase of disease is mathematically analogous to simple interest in money, it is called simple interest disease. There is only one generation of disease in the course of one epidemic. Such diseases develop from a common source of inoculum, i.e., the capital is constant, and often there is one generation of infection in a season (Figure 2).

2. Compound Interest Disease (Multicycle Disease)

When the increase in disease is mathematically analogous to compound interest in money, the disease is called compound interest disease. There are several or many generations of the pathogen in one life of the crop, i.e., the capital is increased by the amount of interest (Figure 3).

E. BASED ON HOST-PATHOGEN DOMINANCE SYSTEM

Kommedahl and Windels[15] have divided diseases, based on the "Host-Pathogen Dominance System" described below.

Pathogen Dominant Diseases—The pathogen is dominant over the host, but the relationship is transitory because the resistance of the host is less initially than it become, eventually. Such pathogens are tissue nonspecific and attack young, immature root tissues or senescent tissues of mature plant roots. They seldom damage a rapidly growing, maturing root, so the period of disease development is short. Some times such pathogens are macerative; some times they are toxicogenic, or sometimes both can occur. The pathogenesis is due primarily to the viruence of the pathogen. Physiological specialization is relatively uncommon. Important pathogens are *Aphanomyces, Macrophomina, Phytophthora, Pythium, Rhizoctonia, Sclerotium*, etc.

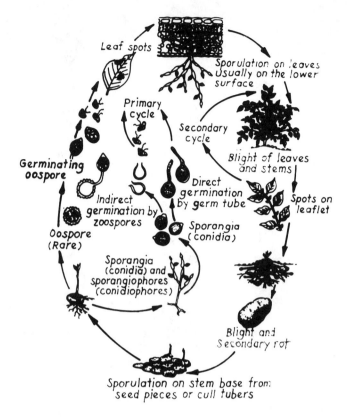

FIGURE 3. Compound interest disease (multi-cycle disease), e.g., late blight of potato.

Host-Dominant Diseases—In host-dominant diseases the host is dominant and the pathogen is successful only when factors favor the pathogen over the host. The resistance of the host is strong enough to keep the pathogen from advancing too rapidly against the host defenses during the vegetative growth phase and the host thereby prolongs the relationship. Damage is most severe in plants in the reproductive and senescent phases. In this group are some pathogens which are tissue nonspecific but most are tissue specific. The pathogen may be macrative, toxicogenic, or both. Important pathogens are the species of *Armillaria, Polyporus, Poria, Helminthosporium, Fusarium,. Verticillium,* etc.

V. IMPORTANCE OF PLANT DISEASES

A. PLANT DISEASES AND HUMAN AFFAIRS

In the history of mankind plant diseases have been connected with a number of important events. The late blight of potato (*Phytophthora infestans*) is a famous example of what a plant disease can do to change the course of history. In 1845 this disease devastated millions of acres in Europe, U.S., and Canada. So sudden and so complete was the catastrophe that in only a few days, fields which had promised abundant harvests, were transformed into blackened waste of vegetation overlying fowl and putrifying masses of rotten tubers. In the early 1870s another fungus disease, the coffee rust (*Hemileia vastatrix*), wiped out the coffee[1] plantations of Sri Lanka. The tea and rubber industry replaced coffee in Sri Lanka and other eastern areas. In the 1930s the entire banana industry in Central and South America was threatened with extinction by Sigatoka (*Mycosphaerella musicola*) and the industry could be saved only by 1940 when Bordeaux mixture was used on a massive scale.[16]

The eating habits of people are very difficult to change and plant diseases appear to responsible for many deeply implanted eating habits. While in northern and southern Europe, wheat bread is more common, in central Europe, people eat rye bread. In the southern part of the U.S. people eat corn bread while in the north wheat bread is more common. Horsfall[17] has suggested that this situation resulted due to the ravages of wheat rust. In France, between 1878 and 1882, the wine industry was threatened due to the introduction of downy mildew (*Plasmopara viticola*) from the U.S. Brown spot of rice (*Helminthosporium oryzae*) caused human suffering of a very high magnitude in Bengal, India during the 1940s. Rice prices rose so high that many people could not buy it. People from rural areas migrated to cities in the hope of finding work and rice. "Finding neither, they slowly died of starvation."[18] In 1970, southern corn leaf blight (*Helminthosporium maydis*), destroyed about 15% of the U.S. corn crop, causing a loss of about $1 billion. Chestnut blight (*Endothia parasitica*) wiped out the American chestnut. The dramatic biological and human events of this epiphytotic and the attempts to stop it have been documented by Hepting.[20] Similarly from the 1930s through the 1970s, Dutch elm disease, caused by *Ceratocystis ulmi* (= *Ophiostoma ulmi*), destroyed elm trees in residential neighborhoods and in forests in the eastern two thirds of the U.S. The disease caused economic hardship to individuals and municipalities and lowered the quality of their environment.[21] Catastrophic plant diseases have been ably summarized by several workers.[22-26]

Some examples of the plant diseases that have caused severe losses in the distant and/or recent past or which may cause severe losses in the near future are listed in Table 1.

B. DISEASE MANAGEMENT AND WORLD POPULATION

The world's human population as of 1981 is greater than 4.0 billion and, at the present rate of 2.14% annual growth it is expected to be 5.4 billion by 1990 and 6.7 billion by the year 2000. According to Agrios[3] the developing countries, in which 57.6% of the population is engaged in agriculture, have the lowest agricultural output, their people are living on a substandard diet, and they have the highest population growth rates (2.64%). It is estimated that even today some 800 million people are undernourished and 2.0 billion suffer from hunger or malnutrition or both.[3]

Suppression of plant diseases and the reduction of yield losses due to diseases are a necessary part of increasing food supply. For example, it is estimated that in the U.S. alone each year, crop worth $9.1 billion are lost to diseases, $7.7 billion to insects, and $6.2 billions to weeds.[3] An estimate of the world crop production and of the preharvest losses[27] to diseases are given in Table 2.

C. PLANT DISEASES AND MODERN AGRICULTURE

The "green revolution" which mainly involved raising of crops to boost production in yield by introducing high-yielding varieties have resulted in many instances, susceptibility of these varieties to new diseases.[28,29] Of the several factors that have resulted in addition to already existing major diseases, certain minor and new diseases have assumed serious proportions as to cause considerable damage to the crop.

High-yielding dwarf and semidwarf wheat varieties produced and distributed by CIM-MYT, Mexico not only increased wheat production in Mexico in the mid 1960s, 6.5 times that in 1945, thus changing Mexico from a wheat importing to a wheat-exporting country, they also behaved very similarly and were just as productive in Africa and Asia. To produce high yields with these varieties, many agronomic practices had to be altered drastically. Monoculture of these wheats in India, West Pakistan, Afghanistan, and Turkey increased from about 23,000 acres in 1966 to 30 million acres in 1971, replacing hundreds of local varieties.[28] Taking the example of Indian agriculture alone, the impact of these varieties can be judged by the fact that from the total production of 11 to 12 million tons of wheat in

TABLE 1
Examples of Plant Diseases of Historical and Economic Significance

Disease	Caused by	Area	Remarks
Fungal Diseases			
Ascochytosis of chickpea	*Ascochyta* rabiei	Bulgaria, Greece, India, Morocco, Pakistan, Spain, USSR	Frequent epidemics, continuous loss
Banana leaf spot or sigatoka disease	*Mycosphaerella musicola*	Previously central and South America, now world wide	1930s to date several epidemics, great annual loss
Botrytis grey mold of several crops	*Botrytis cinerea*	Worldwide	Frequent epidemics, continuous loss
Brown spot of rice	*Helminthosporium oryzae* (= *Cochliobolus miyabeanus*)	Asia (The 1943 Great Bengal Famine)	Continuous loss
Cereal rusts	*Puccinia* spp.	Worldwide	Frequent epidemics, huge annual loss
Cereal smuts	*Ustilago* spp.	Worldwide	Continuous loss on all grains
Chestnut blight	*Endothia parasitica*	U.S.	Destroyed all American chestnut trees (1904—1940)
Coffee berry	*Colletotrichum coffeanum*	Kenya and other African states	Substantial annual loss
Coffee rust	*H. vastatrix*	Southeast Asia, Brazil	Annihilated all coffee in Southeast Asia (1870s—1880s), since 1970, present in Brazil
Downy mildew of grapes	*Plasmopara viticola*	U.S., Europe	European epidemics (1870s—1880s)
Downy mildew of maize, pearl millet and sorghum	*Sclerophthora* spp. *Sclerospora* spp.	Southeast Asia	Frequent local epidemics, great annual loss
Downy mildew of tobacco	*Peronospora tabacina*	U.S., Europe	European epidemics (1950s—1960s), epidemic in North America (1979)
Dutch elm	*Ceratocystis ulmi*	U.S., Europe	Highly destructive to American elm trees from 1930 to date
Ergot of rye, pearl-millet and wheat	*Claviceps* spp.	Worldwide	Great annual loss, poisonous to humans and animals
Karnal bunt	*Neovossia indica*	India	Spreading, severe annual loss to grain yield and export
Late blight of potato	*Phytophthora infestans*	Widespread, especially in cool humid climate	Substantial annual loss; epidemics-Irish famine (1845—46)
Mango malformation	Several causal agents reported	India	Spreading, severe annual loss
Monilia pod rot	*Monilia* sp.	South America	Very destructive, spreading elsewhere
Powdery mildew of grapes	*Uncinula necator*	Worldwide	European epidemics (1840s—1850s), very destructive if uncontrolled

TABLE 1 (continued)
Examples of Plant Diseases of Historical and Economic Significance

Disease	Caused by	Area	Remarks
Red rot of sugarcane	*Glomerella tucumanensis* (*Colletotrichum falcatum*)	Worldwide	Endemic, occassionally epidemic
Rust of ground nut	*Puccinia arachidis*	Caribbean and Central America	Of great economic importance in most areas of the world
Rust of sugarcane	*Puccinia* spp.	U.S.	Highly destructive and spreading
Southern corn leaf blight	*Helminthosproium maydis*	U.S.	Epidemic 1970, $1 billion lost
Soybean rust	*Phakopsora pachyrhizi*	Southeast Asia, USSR	Spreading

Bacterial Diseases

Disease	Caused by	Area	Remarks
Bacterial leaf blight of paddy	*X. campestris* pv. *oryzae*	Southeast Asia	Very destructive, spreading fast since 1960s
Bacterial wilt of banana	*Pseudomonas solanacearum*	America	Spreading elsewhere
Cassava bacterial blight	*X. campestris* pv. *manthotis*	Asia, Africa, Latin, America	Destructive, spreading elsewhere
Citrus canker	*X. campestris* pv. *citri*	Worldwide	Killed millions of trees in Florida, 1910s and again in the 1980s
Fire blight of apple and pear	*Erwinia amylovora*	North America, Europe	Very destructive, kills numerous trees annually
Ratoon stunting of sugarcane	*Clavibacter xyli* ssp. *xyli*	In more than 50 countries	Spreading, annual loss up to 20%

Viral and Viroid Diseases

Disease	Caused by	Area	Remarks
African cassava mosaic	Virus	Africa	Threatening Asia and the Americas; destructive in Africa
Barley yellow dwarf	Virus	Widespread	Causes substantial loss on small grains
Bunchy top of banana	Virus	Asia, Australia, Egypt, Pacific island	Highly destructive
Citrus quick decline (Tristeza)	Virus	Africa, Americas	Millions of trees being killed
Coconut cadang-cadang	Viroid	Southeast Asia	Killed millions of trees (Approx. 15 million) in Phillipines to date
Hoja blanca of rice	Virus	Americas	Destructive, great annual loss
Maize streak	Virus	Africa	Spreading on sugarcane, corn, wheat, etc.
Plum pox or Sharka	Virus	Europe	Destructive to plums, peaches, apricot, spreading
Potato leaf roll	Virus	Widespread	Huge annual loss
Rice Tungro	Virus	Southeast Asia	Substantial annual loss
Sterility mosaic of pigeon pea	Virus	India	Spreading
Sugarbeet yellows	Virus	Worldwide	Huge annual loss
Sugarcane mosaic	Virus	Widespread	Great losses on sugarcane and maize

TABLE 1 (continued)
Examples of Plant Diseases of Historical and Economic Significance

Disease	Caused by	Area	Remarks
Swollen shoot of cocoa	Virus	Africa	Huge annual loss
Yellow mosaic of pulses	Virus	Southeast Asia	Huge annual yield loss

Mycoplasmal Diseases

Disease	Caused by	Area	Remarks
Lethal yellowing of co-conut plants	MPO	Cetral America	Spreading to U.S., highly destructive
Peach yellows	MPO	Eastern U.S., U.S.S.R.	10 million peach trees killed so far
Pear decline	MPO	Canada, Europe, Pacific Coast states	Destructive, millions of trees killed

Nematode Diseases

Disease	Caused by	Area	Remarks
Burrowing nematode	*Radopholus* spp.	Widespread	Destroyed numerous trees in Florida, U.S., severe on banana also
Root knot	*Meloidogyne* spp.	Worldwide	Continuous losses on vegetables and other crops
Sugarbeet cyst nematode and other crops	*Heterodera* spp.	Widespread	Annual loss

TABLE 2
Estimated 1982 World Crop Production and Preharvest Losses
(In millions of Tons)[27]

Crop	Potential production	Actual production	Estimated losses to Diseases
Cereal	2588	1695	238
Coffee, cocoa, tea	15	8	3
Fibre crop	58	40	6
Fruits	394	302	50
Legumes	67	45	8
Natural rubber	5	4	0.6
Oil crops	346	240	34
Potatoes	376	255	82
Root crops (other than potatoes)	976	556	163
Sugarbeets	423	319	44
Sugarcane	1802	811	346
Tobacco	9	6	1
Vegetables	509	368	51

India in 1965, it has today exceeded 35 million tons, with this intensive and extensive cultivation of new varieties, the disease position has also changed from time to time. A semidwarf wheat variety "Kalyansona" released in 1967, was resistant to most of the then prevailing races of the black stem rust, completely resistant to yellow rust, loose smut, hill bunt, and tolerant to foliar blights. It was however, susceptible to some races of brown rust and under field conditions the disease intensity was only traces to light.[30] But in 1971 and 1972 and again in 1973, its resistance broke down to yellow rust because of the appearance of new strains namely, 14-A, 20-A, and 38-A.[31] At the same time, it also became much more susceptible to brown rust in the field. Cultivation of the same gene type in the hills possibly gave a chance to few virulences (to which this variety was susceptible) to develop at a much faster rate.[32]

Another example, showing direct influence of changing the varietal position on disease situation is provided by the fluctuations in frequency of occurrence of loose smut of wheat (*Ustilago nuda* var. *tritici*). With the increase in the area of Kalyansona between 1968 and 1975, the loose smut of wheat gradually dwindled down and its frequency was less than 0.5%.[33] With its replacement by Sonalika in 1974, however, the disease once again reappeared in severe form in 1976. In the crop season of 1980—1981, the disease intensity in northern India was more than 4%.[34]

Among the minor diseases of wheat, Karnal bunt needs special mention. During the last 10 to 15 years, there has been a progressive increase in its incidence. This is due to intensive cultivation of some varieties, changes in crop management, and also frequent exchange of seed material.[29]

A similar "green revolution" with respect to improvement of rice varieties has been carried out by IRRI, Phillipines. New nonlodging dwarf rice varieties that respond favorably to high nitrogen fertilization and produce high yields were developed and distributed widely in southeast Asia and elsewhere.[28] Soon, however, many of these varities became susceptible to diseases, such as bacterial blight (*Xanthomonas campestris* pv. *oryzae*), bacterial leaf streak (*X. campestris* pv. *oryzicola*), sheath blight, false smut, etc., that were either unknown or unimportant when old local varieties were planted, but which now, due to high nitrogen fertilization and double cropping of large expanses of genetically homogenous varieties, reached catastrophic proportions.[28,29]

Expansion of irrigation in Venezuela made it possible to produce two rice crops in a year where only one was grown before. This resulted in serious outbreak of the virus disease *hoja blanca* because new conditions favored the multiplication and spread of the insect vector of the virus from one crop to another. Irrigation also increases the population and distribution of many fungal bacterial, and nematode pathogens. Similarly, to sustain intensive cultivation, herbicides, insecticides, fungicides, etc., are increasingly being used. The use of pesticides to control plant diseases and other pests has been increasing steadily at an annual rate of about 14% since the mid-1950s.[28] In 1970 approximately 1 billion lbs. of pesticides, including fungicides were produced alone by U.S. companies for domestic and foreign use. Pesticides increased yields of crops in most cases in which they were applied. However, these chemicals have also been found to influence the susceptibility of the host and alter the microbial equilibrium in the ecosystem.

REFERENCES

1. **Roberts, D. A. and Boothroyd, C. W.,** *Fundamentals of Plant Pathology,* W. H. Freeman and Co., San Francisco, 1972, 402.
2. **Horsfall, J. G. and Cowling, E. B.,** Prologue: how disease is managed in *Plant Diseases—An Advanced Treatise,* Vol. I, Horsfall, J. G. and Cowling, E. B., Eds., Academic Press, New York, 1977, 465.
3. **Agrios, G. N.,** *Plant Pathology,* 3rd Academic Press, San Diego, 1988, 803.
4. American Phytopathological Society, Report of the Committee on technical words, *Phytopathology,* 30, 361, 1940.
5. **Heald, F. D.,** *Introduction to Plant Pathology,* McGraw Hill, New York, 1943.
6. British Mycological Society, Definition of some terms used in plant pathology, *Trans. Br. Mycol. Soc.,* 33, 154, 1950.
7. **Stakman, E. C. and Harrar, J. C.,** *Principles of Plant Pathology,* Ronald Press, New York, 1957, 581.
8. **Walker, J. C.,** *Plant Pathology,* 2nd ed., McGraw-Hill, New York, 1957, 693.

9. **Wheeler, H.,** *Plant Pathogenesis,* Springer-Verlag, Berlin, 1975, 104.

10. **Robinson, R. A.,** *Plant Pathosystem,* Springer-Verlag, Berlin, 1976, 184.

11. **Horsfall, J. G. and Dimond, A. E.,** Prologue: the diseased plant, in *Plant Pathology—An Advanced Treatise,* Vol. I, Horsfall, J. G. and Dimond, A. E., Academic Press, Eds., New York, 1959.

12. **Das Gupta, M. K.,** Concept of disease in plant pathology and its application elsewhere, *Phytopathol. Z.,* 88, 136, 1977.

13. **McNew, G. L.,** Out line of a new approach in teaching plant pathology, *Plant Dis. Rep.,* 34, 106, 1950.

14. **McNew, G. L.,** The nature, origin, and evolution of parasitism, in *Plant Pathology—An Advanced Treatise,* Vol. II, Horsfall, J. G. and Dimond, A. E., Eds., Academic Press, New York, 1960, 16.

15. **Kommedahl, T. and Windels, C. E.,** Fungi: pathogen or host dominance in disease, in *Ecology of Root Pathogens,* Krupa, S. V. and Dommergues, Y. R., Eds., Elsevier, Amsterdam, 1979, 291.

16. **Wardlaw, C. W.,** *Banana Diseases Including Plantains and Abaca,* Longmans, London, 1961, 648.

17. **Horsfall, J. G.,** The fight with the fungi or the rusts and the rots that rob us, the blast and the blights that beset us, *Am. J. Bot.,* 43, 522, 1956.

18. **Padmanabhan, S. Y.,** The great Bengal famine, *Annu. Rev. Phytopathol.,* 11, 11, 1973.

19. **Ullstrup, A. J.,** The impact of the southern corn leaf blight epidemics of 1970-71, *Annu. Rev. Phytopathol.,* 10, 37, 1972.

20. **Hepting, G. H.,** Death of the American chestnut, *Forest History,* 18, 60, 1974.

21. **Sinclair, J. B., Campana, R. J.,** Dutch elm disease: perspectives after 60 years, *Search Agric.,* 8(5), 1, 1978.

22. **Riker, A. J.,** Internationally dangerous tree diseases and Latin America, *J. Forestry,* 62, 229, 1964.

23. **Klinkowsky, M.,** Catastrophic plant diseases, *Annu. Rev. Phytopathol.,* 8, 37, 1970.

24. **Holliday, P.,** Some tropical plant pathogens of limited distribution, *Rev. Plant Pathol.,* 50, 337, 1971.

25. **Meredith, D. S.,** Epidemiological considerations of plant diseases in the tropical environment, *Phytopathology,* 63, 1446, 1973.

26. **Horsfall, J. G. and Cowling, E. B.,** The sociology of plant pathology, in *Plant Diseases—An Advanced Treatise,* Vol. I, Horsfall, J. G. and Cowling, E. B., Eds., Academic Press, New York, 1977, 465.

27. **F.A.O.,** *Production Year Book,* Food and Agriculture Organization, Rome, 1982.

28. **Agrios, G. N.,** *Plant Pathology,* 2nd ed., Academic Press, New York, 1978, 703.

29. **Govindu, H. C.,** Green revolution—its impact on plant diseases with special reference to cereals and millets, *Indian Phytopathol.,* 35, 363, 1982.

30. **Joshi, L. M., Gera, S. D., and Saari, E. E.,** Extensive cultivation of Kalyansona and disease development, *Indian Phytopathol.,* 26, 370, 1973.

31. **Joshi, L. M., Srivastava, K. D., Singh, D. V., and Ramanujam, K.,** Wheat rust epidemics in India since 1970, *Cereal Rusts Bull.,* 8, 17, 1980.

32. **Nagarajan, S. and Joshi, L. M.,** Further investigations on predicting wheat rust appearance in central and Peninsular India, *Phytopathol. Z.,* 98, 84, 1980.

33. **Joshi, L. M., Nagarajan, S., and Srivastava, K. D.,** Epidemiology of brown and yellow rusts of wheat in North India. I. Place and time of appearance and spread, *Phytopathol. Z.,* 90, 116, 1977.

34. **Joshi, L. M., Srivastava, K. D., and Singh, D. V.,** Wheat disease, *Newslett. I.A.R.I.,* New Delhi, 14, 1981.

Chapter 3

DISEASE DEVELOPMENT

I. INTRODUCTION

Plant diseases result from the interaction of a pathogen with its host but the intensity of this interaction is markedly affected by the environmental factors. The role of environments in pathogenesis is as important as susceptibility of the host and pathogenicity of the causal agent. Any consideration of disease in the crop, therefore, involves the "disease-triangle".

The pathogen interacts with the host and vice versa. Both influence each other, the host providing nutrition to pathogen and the latter causing disease in the former. The host also influences the environment through crop canopy, root exudates in soil, withdrawl of nutrients from soil, and other activities mediated through these effects. Environment affects the host through physical, chemical, and biotic factors involved in plant growth and metabolism. Pathogen can also influence environment through such effects as defoliation, addition of dead crop residue to soil, changing the host physiology and, thereby, host root exudation, etc. Environment may affect the pathogen in the same manner as the host but may be against the pathogen and in favor of the host or vice versa.

Since these interactions and their effects are not spontaneous and time of interaction influences the result of interaction the "disease triangle" can be modified as shown in Figure 1.

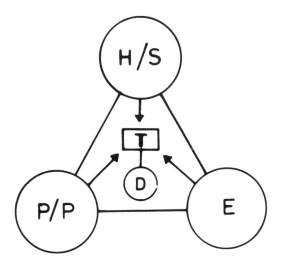

FIGURE 1. A combination of factors required for the disease to appear: H/S = host/suscept, P/P = pathogen/parasite, E = environment, D = disease, T = time.

II. PATHOGENESIS

Development of disease in the plant is not a sudden effect. A chain of events are responsible for causation of any disease. The symptoms and manifestation of injury to the plant is the last link in the chain of events. Before symptoms appear the pathogen independently or with the host has to pass through several stages. These stages, reactions, and interreactions arranged in a sequence lead to disease development and the entire chain of events leading to disease development is known as "pathogenesis" (Figure 2).

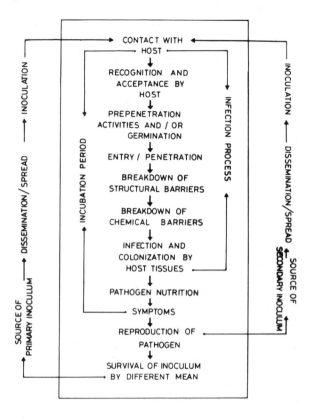

FIGURE 2.   Chain of events leading to disease development.

The infection chain can be "continuous" or "intermittent". Further, these chains may be "homogenous" (survival on single plant species), or "heterogenous" (survival on many plant species).

A.   STAGES IN DISEASE DEVELOPMENT
1.   Inoculum and Inoculation

The infective propagules coming in contact with the host are known as *inoculum* and the process which ensures this contact is called *inoculation*. In fungal pathogens the inoculum may be hyphal fragments, asexually and/or sexually produced spores, specialized vegetative organs, etc.; in bacteria, MLO, RLO, viruses, viroids, and virusoids, inoculum is always whole individuals; in nematodes, inoculum may be adults, larvae, or eggs; in phanerogams it may be plant parts or seeds.

The means of survival are the first link in infection chain or disease cycle. The initial infection that occurs from these sources in the crop is primary infection and the propagules that cause this infection are called primary inoculum. After initiation of disease in the crop, the spores or other structures of the pathogen are sources of secondary inoculum and cause secondary infection, thereby spreading the disease in the field.

2.   Contact

Pathogenesis caused by infectious agent begins as soon as the inoculum comes into contact with targeted/infected organs of the plant. Most of the bacterial and fungal pathogens come into contact with their hosts accidently in the form of wind-borne or water-borne propagules. Some fungi, many bacteria and most of the viruses are brought to their hosts by insect or other vectors. Motile propagules (zoospores) of fungi are attracted to plants by root exudates (chemotaxis) and get accumulated behind the root tip[1] (Figure 3).

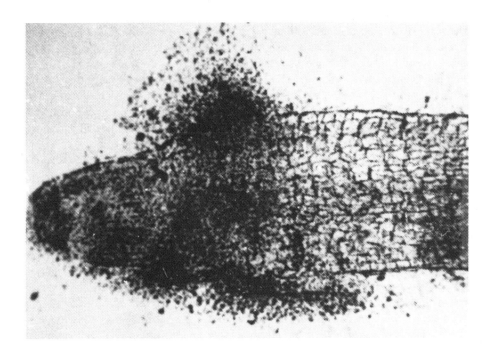

FIGURE 3. Attraction of zoospores of *Phytophthora cinnamomi* to roots. (Milholland, R. D., *Phytopathology*, 65, 789, 1975. With permission.)

Plant parasitic nematodes like zoosporic fungi have been studied for their oriented movement and accumulation at the root zone. According to Klink et al.,[2] the directed movement and accumulation of nematodes to plant roots is possibly in response to localized stimuli. Carbon dioxide, in terms of the volume produced and the distance over which its effects are expressed, is possibly the most important potential attractant produced by growing roots.[3] A number of plant parasitic nematodes such as *Meloidogyne* spp., *Aphelenchoides fragariae, Ditylenchus dipsaci,* and *Pratylenchus penetrans* are reported to react positively in a CO_2 gradient and move to the source of this gas.[4,5] Other gradients that are produced by or associated with growing roots include water, O_2, pH, root exudates such as amino acids, organic acids, and electrical. Among the amino acids, glutamic acid has been demonstrated to attract plant parasitic nematodes under certain conditions.[6]

3. Recognition and Specificity

Majority of plant pathogens are known to have a limited host-range. For example, *Pyricularia oryzae* which causes rice blast can not infect tomato, while *Alternaria solani* causing early blight of tomato cannot infect rice. This type of host-pathogen specificity is termed as "basic compatibility". The limited host-range of most pathogens suggests that the nonhost defense mechanisms are not easy to overcome. Thus, knowledge of specific processes involved in this type of resistance would not only provide information about how compatibility is determined, but may also provide ways for devicing highly efficient control measures. It is speculated that (1) lack of specific nutrients required by the pathogen, (2) inability of the pathogen to break or by pass general defense barriers of the host, and/or (3) nonrecognition of the pathogen and host as compatibile partners required for the establishment of genetic and physiological synchrony between them, may be possible factor(s) responsible for the nonestablishment of the basic compatibility.

4. Germination and Prepenetration Activities

The plant viruses are particulate in nature and wholly passive in transmission and entry into the host. They reach the interior of host cells only with the activity of organs of their vector or by some other mechanical means, not by themselves. Transmission of bacteria and their entry into the host cells is passive in the sense that either they enter through wounds or through natural openings. They have no dormant structures hence no prepenetration activity. Nematodes cannot multiply outside the living host or away from the host as they depend for nutrition solely on their host. It is only in fungi that complicated activities do occur before penetration.

According to the nature of the spore and the environmental conditions, germination of fungal spores occurs in various ways. Sporangia and oospores of fungi produce zoospores in wet conditions and germ tubes in dry conditions, while those of some other appear always to germinate by zoospores (*Plasmopara viticola*) or always by germtubes (e.g., *Peronospora* sp.). Teliospores of rust and smut fungi normally germinate by producing a promycelium on which basidiospores (sporidia) are borne. Some spores germinate easily on release and sometimes even before release, while resting spores have a "dormancy". Several environmental and nutritional factors (temperature, moisture, pH, light, oxygen, CO_2, nutrients, etc.) affect spore germination.

Fungal invasion is chiefly by germ tubes or structures derived from them. There are two well-known situations involving hyphae acting in a concerted way to achieve host penetration. Hyphae of *Rhizoctonia solani* often aggregate to form an "infection cushion" from which multiple penetrations occur by means of appresoria and penetration pegs. Another type of penetration is exemplified by the penetration of intact root periderm by the "rhizomorphs" of *Armillaria mellea*. In the ectotrophic infection habit among the specialized pathogens such as *Ophiobolus graminis,* the fungus progresses epiphytically over the root system as a sparse network of dark hyphae known as runner hyphae from which hyaline branches quickly penetrate and infect the underlying cortex.

5. Penetration

a. Penetration Structures

In zoosporic fungi there are six different modes of penetration.[7] Zoospore penetration typically involves encystment of zoospore on the outer surface of host cell and subsequent development of an expanding vacuole (Figure 4) within the cyst. Zoospores of *Rozella* encyst upon hyphae of *Allomyces,* and germinate to produce a subglobose germ tube from which the parasite protoplast enters the host through a tiny pore in the host wall[8] (Figure 4). Penetration process by *Olpidium* includes growth of a penetration tube from the cyst followed by movement of the entire fungal protoplast[7] into the host cell. Host penetration by *Phytophthora parasitica* is almost similar to that of *Olpidium* up to the point of passage of protoplast into the host cell.[9]

The formation of specialized accessory structures to facilitate passage through the cell wall has been observed in the *Pythium*-type of penetration. Kraft et al.[10] who studied the penetration of *Pythium aphanidermatum* of bent grass, (Figure 4) observed that (1) the cyst germinates to produce a germ tube, and (2) the germ tube differentiates into an appressorium to achieve penetration. Host penetration by members of the Plasmodiophorales represents one of the most highly specialized penetration type.[7] The process involves several unique structural aspects including (1) the development of a tube, or Rohr, within the zoospore cyst, (2) development of a sharply pointed rod, or Stachel within the Rohr, (3) slender tubular extension of the Rohr, the Schlauch, and (4) an adhesorium which develops during inversion of the Rohr by evagination.

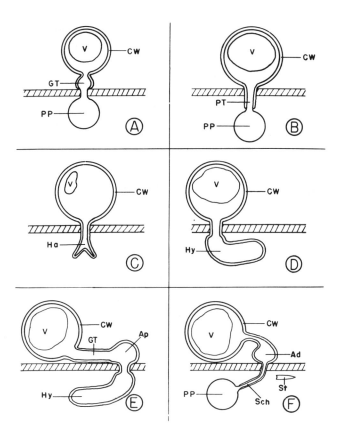

FIGURE 4. A diagrammatic comparison of penetration types exhibited by zoospores. (A) *Rozella allomycis,* (B) *Olpidium brassicae,* (C)
Chytridium sp., (D) *Phytophthora parasitica,* (E) *Pythium aphanidermatum,* (F) Plasmodiophorales, (Ad—adhesorium, Ap—Appressorium, CW—cyst wall, GT—germ tube, Ha—haustorium, Hy—hypha,
PP—parasite protoplast, PT—penetration tube, Sch—schlauch, St—
stachel, V—Vacuole - (Aist, J. R. in *Physiological Plant Pathology,*
Heitefuss, R. and William, P. H., Eds., Springer-Verlag, Berlin, 1976.
With permission.)

At least four types of penetration are exhibited by nonmotile fungus spore.[7] The simplest
has been described for conidia of *Cladosporium* and involves growth of germtube directly
through the cuticle and into the middle lamella. Penetration by rust uredospore requires two
distinct stages. The first stage involves penetration by a primary penetrating hypha from an
appressorium through a stomata and into substomatal cavity. The primary hypha develops
into a substomatal vesicle from which secondary hyphae arise and grow in contact with
neighboring host cells.

There are variations in the behavior of the germ tube at the time of penetration through
the stomata, as in *Peronospora destructor* infecting onion leaves; the germ tube continues
to grow after the formation of the first appressorium. In *Pseudoperonospora cubensis* the
zoospores swim toward stomata and encyst above the line separating the guard and epidermal
cells. Hyphae penetrate the stomatal aperture and swell to form a substomatal vesicle from
which, in turn, other hyphae may grow to form haustoria in the adjacent cells of the leaves.

b. Mechanisms of Penetration

Pathogens may enter plants through wounds, natural openings, or by direct penetration.

In nature viruses, viroids, MLO, RLB, etc., enter plants through wounds made by their vectors. Bacteria enter plants mostly through wounds, less frequently through natural openings, and never directly. Nematodes with the help of stylet (Figure 5) enter plant directly and, sometimes, through natural openings. Fungi enter their hosts either directly or through natural openings and wounds.

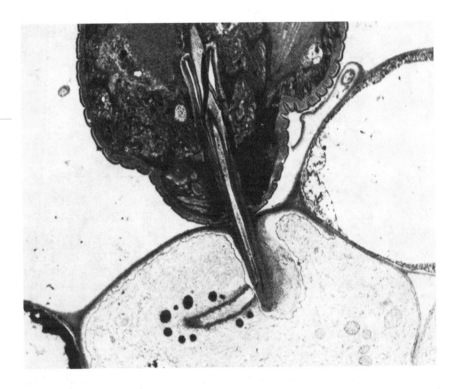

FIGURE 5.   Electron micrograph showing penetration of host cells by plant parasitic nematode *Rotylenchulus reniformis*. (Dropkin, V. H., *Introduction to Plant Nematology*, John Wiley and Sons, New York, 1980, 293. With permission.)

In fungal pathogens direct penetration through cutinized epidermal wall is achieved by mechanical means or enzymatic action or by both. In some fungi germ tube tip swells to form "appressoria" (Figure 4), while in many others even hyphae as such penetrate. Appressoria formed in different species vary morphologically. They may be swollen hyphal tips to well-defined melanized, thick-walled structure as found in *P. oryzae* and *Colletotrichum* spp. With the latter type, the architecture and mechanical forces are certainly critical factors in penetration process, but enzymes may also have an equally important role in weakening the cuticle and in digesting pectin and cellulose. Melanization of the appressorial wall appears to be necessary for the architecture and rigidity needed to support and focus the mechanical forces involved in penetration process as the hyaline appressoria of *P. oryzae* or *Colletotrichum* spp. formed in the presence of melanin biosynthesis inhibitors like tricyclazole, pyroquilone, etc., fail to penetrate the host surface and cause infection.[11]

Enzymatic penetration of host surface has now been conclusively demonstrated at least in certain host-parasite systems like papaya fruit—*Colletotrichum gloesporioides,* pea epicotyl—*Fusarium solani,*[12,13] etc. In these systems cutinases are found to be essential for penetration of intact host surface and subsequently to cause infection. Cutinase inhibitors like anticutinase antibodies, DFP, (diisopropyl flurophosphate) organophosphorus insecticides, carbendazim, etc., provide protection against infection. Mutants incapable of producing cutinase are nonpathogenic. Moreover, they become pathogenic with exogenous application of cutinase. Cutinase gene is induced by cutin. Soliday et al.[14] used CDNA strategy; isolated mRNA for cutinase from induced cultures of *F. solani* f. sp. *lini;* prepared cDNA by reverse transcription, and cloned cutinase gene. By sequencing the gene they were able to predict the primary structure of cutinase, which, as they point out, should help in the effective development of inhibitors for use as antipenetrants. This is the first fungal gene to be cloned which is directly associated with pathogenicity.

Electron microscopic evidences suggest that in general cuticular penetration involves both mechanical as well as enzymatic activity. Cuticular membranes usually appear depressed inward during penetration suggesting involvement of mechanical force. On the other hand, clear holes produced without any sign of torn edges suggest that softening or erosion of the rather brittle cuticle probably occurred prior to penetration. However, at least in those cases where direct penetration occurs and no specialized structure is formed, it would seem that enzymatic action must be almost exclusively the mode of penetration.[15]

6. Infection and Colonization

In viral infection the coat protein is removed during passage and nucleic acid is released in the host cell. Viruses do not absorb nutrients from cell but their nucleic acid replicates to produce more viruses nucleic acid and by using host translational machinery, form the protein coat. Assembly of nucleic acid and coat protein gives rise to virus particles. Intercellular movement of viral nucleic acid/particles takes place through plasmodesmata.

Bacteria and fungi dissolve the cell walls by enzymes after entry into the host and thus absorb nutrients. Nematodes use force as well enzymes to break or dissolve cell walls to reach the cell protoplasm. Enzymes and hormones produced by them cause tissue disintegration and other abnormalities.

7. Chemical Weapons

The main groups of substances secreted by pathogens are enzymes, toxins, growth regulators, polysaccharides, and antibiotics. Of these enzymes and toxins play important role in necrotrophs while growth hormones are thought to play more important role in case of biotrophs (rusts, downy and powdery mildews) where pathogens obtain nutrients from living host cells.

a. Enzymes

Enzymes are the major weapons employed by necrotrophs for ingress and colonization (Table 1). Biotrophs also employ such enzymes but their deployment is highly localized mainly to facilitate their penetration. They are not involved in tissue maceration. Pectin degrading enzymes, produced either constitutively or inductively are responsible for tissue maceration. Recent studies have established that pectinases determine the pathogenicity in soft rot Erwinias. *Erwinia chrysanthemi* was found to produce five pectate lyase isozymes (PLa, PLb, PLc, PLd, and PLe). Kotoujansky[12] observed that bacterial mutants lacking genes for PL isozymes a, d, and e are avirulent on *Saintpaulia ionantha*.

This indicates involvement of these isozymes in pathogenesis. Pectate lyase encoding genes (*pel* genes) and polygalacturonase-encoding genes (*peh* genes) have been cloned from four different strains of *E. carotovora* f. sp. *carotovora* and *E. chrysanthemi*.[16]

The importance of cellulases, hemicellulases, and other enzymes in pathogenesis have received scant attention. However, isolation and cloning of cellulase gene (*cel* gene) from *E. chrysanthemi* has opened up the possibilities for realistic assessment of these enzymes in pathogenesis. It is not only the production but release of wall degrading enzymes from concerned pathogens in the host tissues is essential for pathogenesis. Secretion deficient mutants of *E. chrysanthemi* and *Xanthomonas campestris* are nonpathogenic.[12]

Plant cell wall constituents may serve as effective inducers of wall degrading enzymes

TABLE 1
Cell Wall and Membrane Degrading Enzymes Produced by Certain Plant Pathogens[12,13,17-19]

Substrate	Enzymes	Pathogen
Cutin	Cutinases	*Colletotrichum gloeosporioides, Fusarium solani* f. sp. *pisi*
Cellulose	Cellulases (C_1, C_2, C_x, and β-glucosidase)	*Pseudomonas solanacearum, Erwinia carotovora, E. chrysanthemi, C. lindemuthianum, F. oxysporum* f. sp. *lycopersici, Sclerotinia sclerotiorum, Botrytis cinerea, Ascochyta pisi, Rhizoctonia solani*
Pectin pectic acid	Hydrolases/, Pectin methyl esterase (PME), Polygalacturonases (PG) (endo- or exo-), Pectin methyl galacturonases (PMG) (endo- or exo-) Lyases: Polygalacturonic acid *trans*-eliminases (exo- or endo-), (Pectin methyl *trans*-eliminases (exo- or endo-)	*E. carotovora, E. chrysanthemi, Pythium* spp.: *R. solani, Verticillium albo-atrum, Sclerotium rolfsii, S. sclerotiorum, B. cinerea*
Hemicellulose	Hemicellulases	*S. rolfsii, R. solani, S. sclerotiorum, F. roseum, E. carotovora, E. chrysanthemi*
Protein	Proteinases	*R. solani, Penicillium expansum, Pseudomomas lachrymans, E. carotovora, E. chrysanthemi*
Phospholipids	Phospholipase B, Phospholipase C	*S. rolfsii, Thielaviopsis basicola, B. cinerea, E. carotovora*
Lignin	Ligninases	*Heterobasidion annosus, B. cinerea*
Suberim	Esterases (Cutinases) to degrade aliphatic component	*F. solani* f. sp. *pisi*

in plant pathogens. Some pathogen may produce different types of wall degrading enzymes. Sequence of production and dominance of a particular enzyme may influence the nature of the symptoms. In soft rots predominant enzymes are pectinases, while cellulases and hemicellulases play important role in brown rots. Lignin degrading enzymes are reported to be employed by the white rot pathogens (Table 1).

b. Toxins

Gaumann[20] claimed that microorganisms are pathogenic only if they are toxigenic. "Toxins" can be defined as low molecular weight, non enzymatic microbial products toxic to the higher plants. These are different from other microbial products like "mycotoxins" and "antibiotics", which are toxic to the animals and microbes (except producers). Toxins are classified as phytotoxin, vivotoxin, and pathotoxin. Phytotoxins, the toxin of microbial origin, are toxic to plants but are not regarded as of primary importance during pathogenesis.[21] These are toxic to both host and nonhost plants. Vivotoxins have been defined as substance produced by the pathogen in the infected host which is involved in the production of disease but not initial incitant of disease itself.[22] Vivotoxins may also induce only a part of disease symptoms. A pathotoxin is defined as a host-specific toxin which induces all the typical symptoms of disease at reasonable concentration, the production of which is correlated with pathogenicity.[23]

In recent classification, toxins are divided into two categories. The first is "host-non-specific" which may affect many unrelated plant species in addition to main host of the pathogen producing toxin; it includes phytotoxin and vivotoxin. The second is "host-specific" which affect only the specific host of the pathogen; it includes pathotoxins. Toxins in general, interact with cell membrane or organelles (mitochondria or chloroplast) and alter their permeability. Host-specific and major host-nonspecific toxins are listed in Tables 2 and 3, respectively.

TABLE 2
Host-Specific Toxins[24]

Toxin or organism	Pathogen	Host	Chemical nature
HV-toxin	*Helminthosporium victorae*	Oats	Victoxinine (unstable and nonproteinaceous)
AK-toxin	*Alternaria kikuchiana*	Pear	Altenin (furanose ring)
PC-toxin	*Periconia circinata*	Sorghum	Proteinaceous
HC-toxin	*H. carbonum*	Corn	Cyclic peptide
AM-toxin	*A. mali*	Apple	Alternariolide
A. citri	—	Mandarin orange	Lipophilic
T-toxin	*H. maydis* race T	Corn	Linear polyketol[c]
HS-toxin[a]	*H. sacchari*	Sugarcane	α-D- galactoside of 1,2-dihydroxy-cydo propane[c]
A. alternata	*A. alternata*	Tomato	Cationic compound
P. teres[b]	*Pyrenophora teres*	Barley	Peptide
A. citri	—	Rough lemon	Lipophillic
AB-toxin	*A. brassicae*	*Brassica* sp.	Destruxin B (cyclopeptide)

[a] Report indicates host-specific toxin but data incomplete.
[b] Host selective not host specific.
[c] Not confirmed.

TABLE 3
Some Important Host-nonspecific Toxins[25,26]

Toxin	Produced by	Chemical nature
Tentoxin	*Alternaria alternata*	Cyclic tetrapeptide
Alternaric acid	*A. solani*	Hemiquinone derivative
Cercosporin	*Cercospora beticola* *C. kikuchi* *C. personata*	Benzoperylene derivative
Cerato-ulmin	*Ceratocystis ulmi*	Large M_r carbohydrate
Diaporthin	*Endothia parasitica*	Isocoumarin
Fusaric acid	*Fusarium* spp.	5-n-butylpicolinic acid
Lycomarasmin and Lycomarasmic acid	*F. oxysporum* f. sp. *lycopersici/vas-infectum*	Amino acid derivative
Tabtoxin	*Pseudomonas tabaci/ coronafaciens/ garcae*	Amino acid derivative
Pyricularin Picolinic acid	*Pyricularia oryzae* *P. oryzae*	Nitrogen containing compound picolinic acid
Malformin	*Aspergillus niger* *F. moniliforme* f. sp. *subglutinans*	Cyclic peptide

c. Growth Regulators

Alteration in concentration of growth regulators like cytokinins, auxins, gibberellins, ethylene, and abscisic acid, have been found to be associated with several plant diseases particularly in those resulting in abnormal plant/organ/cell growth like, tumors, galls, knot, stunting, epinasty, curling, hypertrophy, hyperplasia, etc. (Table 4). Increase in level of a growth regulator may be either due to its induced production (by the pathogen or host) or inhibited degradation. Decrease in concentration is due to the enhanced degradation by the host's or pathogen's enzymes. Biotrophs like powdery mildews employ these hormones to draw their nutrients from the host cell.

Alterations in the level of growth regulators seems to be a consequence of pathogenesis.

However, at least in three diseases hormones have been demonstrated to be determinant of pathogenicity. These are crown gall caused by *Agrobacterium tumefaciens,* (cytokinin and suxin), oleander, and olive knot caused by *Pseudomonas syringae* f. sp. *savastanoi* (auxin), fasciation disease caused by *Corynebacterium fascians* (cytokinin). In all these three bacteria genes have been identified and cloned which are responsible for the hormonal production.[27] Deletion of these genes result in conversion of virulent strains into avirulent.

III. FACTORS AFFECTING DISEASE DEVELOPMENT

Plant infection is influenced by many factors not only in the initial invasion but also in the subsequent spread of disease; hence these factors are of critical importance in epidemiology. These factors are summarized in the chart given below. Details of these factors are given in chapters dealing with management practices.

TABLE 4
Certain Plant Diseases Involving Altered Concentration of Growth Regulators

Growth Regulators	Diseases (pathogen)
Cytokinins	Increased concentration: crown gall (*Agrobacterium tumefaciens*), fasciation disease (*Corynebacterium fascians*), white rust of crucifers (*Albugo candida*), bean rust (*Uromyces phaseoli*), club roots of crucifers (*Plasmodiophora brassicae*), western pine blister rust (*Cronartium fusiforme*), white pine blister rust (*C. ribicola*), root knot of tobacco (*Meloidogyne incognita*) Reduced concentration: *Verticillium* wilt of tomato and cotton, root knot of tomato (*M. incognita*)
Auxins	Increased concentration: crown gall (*A. tumefaciens*), peach leaf curl (*Taphrina deformans*), wheat stem rust (*Puccinia graminis* f. sp. tritici), wheat powdery mildew (*Erysiphe graminis*), safflower rust (*Puccinia carthami*), white rust of crucifers (*A. candida*), downy mildew of crucifers (*Peronospora parasitica*), tomato wilt (*Verticillium albo-atrum*), banana wilt (*F. oxysporum* f. sp. *cubense*) bacterial wilt of solanaceous crops (*Pseudomonas solanacearum*), Oleander knot (*P. syringae* pv. *savastanoi*) Reduced concentration: Mango malformation, tobacco mosaic (TMV), potato leafroll (PLRV), curly top of sugarbeet (SBCTV)
Ethylene	Increased concentration: fruit rot of citrus (*Pencillium digitatum*), tomato wilt (*F. oxysporum* f. sp. *lycopersici*), wilt of tulip (*F. oxysporum* f. sp. *tulipae*), *Verticillium* wilt of cotton (*V. albo-atrum*)
Gibberellins	Increased concentration: 'Bakanae' disease of rice (*Gibberella fujikuroi*), creeping thistle rust (*Puccinia punctiformis*) Decreased concentration: Anther smut of sea campion (*Ustilago violacea*)

REFERENCES

1. **Mehrotra, R. S.,** *Plant Pathology,* Tata McGraw-Hill, New Delhi, 1980, 771.
2. **Klink, J. W., Dropkin, V. R., and Mitchell, J. E.,** Studies on the host-finding mechanisms of *Neotylenchus linfordi, J. Nematol.,* 2, 106, 1970.
3. **Macdonald, D.,** Some interactions of plant parasitic nematodes and higher plants, in *Ecology of Root Pathogens,* Krupa, S. V. and Dommergues, Y. R., Eds., Elsevier 1979, 281.
4. **Klingler, J.,** On the orientation of plant nematodes and of some other soil animals, *Nematologica,* 11, 4, 1965.
5. **Edmunds, J. E. and Mai, W. F.,** Effect of *Fusarium oxysporum* on the movement of *Pratylenchus penetrans* towards alfalfa roots, *Phytopathology,* 57, 468, 1967.

6. **Bird, A. F.,** The attractiveness of roots to the plant parasitic nematodes, *Meloidogyne javanica* and *M. hapla, Nematologica,* 4, 322, 1959.

7. **Aist, J. R.,** Cytology of penetration and infection—fungi, in *Physiological Plant Pathology,* Heitefuse, R. and Wieliam, P. H., Eds., Springer-Verlag, Berlin, 1976, 890.

9. **Hanchey, P. and Wheeler, H.,** Pathological changes in ultrastructure: tobacco roots infected with *Phytophthora parasitica* var. *nicotianae, Phytopathology,* 61, 33, 1971.

10. **Kraft, J. M., Endo, R. M., and Erwin, D. C.,** Infection of primary roots of bentgrass by zoospores of *Pythium aphanidermatum, Phytopathology,* 57, 86, 1967.

11. **Sisler, H. D. and Ragsdale, N. N.,** Diseases controlled by nonfungitoxic compounds, in *Modern Selective Fungicides—Properties Applications, Mechanisms of Action,* Lyr, H., Ed., Longman Group U.K., London, and VEB Gustav Fischer Verlag, Jena, 1987, chap. 23.

12. **Kotoujanski, A.,** Molecular genetics of pathogenesis by soft-rot Erwinias, *Annu. Rev. Phytopathol.,* 25, 405, 1987.

13. **Dickman, M. and Patil, S. S.,** The role of cutinase from *colletotrichum gloeosporioides* in the penetration of papaya, in *Experimental and Conceptual Plant Pathology Vol. II,* Singh, R. S., Singh, U. S., Hess, W. M., and Weber, D. J., Eds., Grodon and Breach New York, and Oxford IBH Publishing, New Delhi, 1988, 175.

14. **Soliday, C. L., Flurkey, W. H., Okita, T. W., and Kolattukudy, P. E.,** Cloning and structure determination of cDNA for cutinase, an enzyme involve in fungal penetration of plants, *Proc. Natl. Acad. Sci., U.S.A.,* 81, 3989, 1984.

15. **Sisler, H. D.** Control of fungal diseases by compounds acting as antisporulants, *Crop Prot.,* 5, 306, 1986.

16. **Kolattukudy, P. E.,** Enzymatic penetration of the plant cuticle by fungal pathogens, *Annu. Rev. Phytopathol.,* 23, 223, 1985.

17. **Bateman, D. F. and Basham, H. G.,** Degradation of plant cell walls and membranes by microbial enzymes, in *Physiological Plant Pathology,* Heitefuss, R. and Williams, P. H., Eds., Springer-Verlag, Berlin, 1976, 316.

18. **Hiittermann, A. and Haars, A.,** Biochemical control of forest pathogens inside the tree, in *Innovative Approaches to Plant Disease Control,* Chet, I., Ed., John Wiley & Sons, New York, 1987, 285.

19. **Hahlbrock, K. and Scheel, D.,** Biochemical responses of plant pathogens, in *Innovative Approaches to Plant Disease Control,* Chet, I, Ed., John Wiley & Sons, New York, 1987, 229.

20. **Gaumann, E.,** Toxins and plant disease, *Endeavour,* 13, 198, 1954.

21. **Luke, H. H. and Gracen, V. E., Jr.,** Phytopathogenic toxins, in *Microbial Toxins,* Vol. III, Kadis, S., Ciegler, A., and Ajl, S. J., Eds., Academic Press, New York, 1972, 131.

22. **Dimond, A. E. and Waggoner, P. E.,** On the nature and role of vivotoxins in plant disease, *Phytopathology,* 43, 229, 1953.

23. **Wheeler, H. and Luke, H. H.,** Microbial toxins in plant disease, *Annu. Rev. Microbiol.,* 17, 223, 1963.

24. **Scheffer, R. P.,** Host-specific toxins in relation to pathogenesis and disease development, in *Phystological Plant Pathology,* Heitefuss, R. and Williams, P. H., Eds., Springer Verlag, Berlin, 1976, 247.

25. **Rudolph, K.,** Non-specific toxins, in *Physiological Plant Pathology,* Heitefuss, R. and Williams, P. H., Eds., Springer-Verlag, Berlin, 1976, 270.

26. **Kumar, J., Singh, U. S., and Beniwal, S. P. S.,** Presence of root malformation factor in culture filtrate of *Fusarium oxysporum* f. sp. *subglutinans, J. Phytopathol.,* 119, 7, 1987.

27. **Shaw, P. D.,** Plasmids in phytopathogenic bacteria, in *Experimental and Conceptual Plant Pathology,* Vol. II, Singh, R. S., Singh, U. S., Hess W. M., and Weber, D. J., Eds., Gordon and Breach, New York, and Oxford IBH Publishing, New Delhi, 1988, 221.

Chapter 4

HOST-PARASITE INTERACTION

I. INTRODUCTION

Plants are constantly exposed to millions of microorganisms and many of them possess the faculty to attack plants. However, in nature disease is more an exception than a rule. Most of the microbes surrounding the plant are saprophytic, incapable of attacking a healthy plant. Only a few are pathogenic, but each one of these can only attack specific plant species. Furthermore, all the individuals of a host species are not equally susceptible to a given pathogen isolate. How are most of the microbes in contact with the plant surface rendered harmless? How does a plant defend itself against non-pathogens (pathogens of other species) and a pathogen? In order to properly understand these phenomenon it is essential to first understand the nature and evolution of parasitism in plant pathogens and philosophy of defense in plants.

II. EVOLUTION OF PARASITISM

Unlike their animal counterparts, plants cannot eliminate their parasites; they try to live with them by limiting the latter's deleterious effects. At the same time a successful parasite also can cause less strain to the plant. It is from these considerations that symbiosis is considered to be the most advanced form of parasitism. In the hypothetical hierarchy (Figure 1) of parasitism it is natural to consider the saprotrophs as the bottom members—they colonize only dead organic matter. The intense competition among the saprotrophs for the same source of organic matter forced some of them to develop parasitic abilities, for example *Phytium* sp. appears to have the tendencies of a true saprophyte, but it has developed the faculty to become a parasite even though restricted to only attacking the juvenile and succulent plant tissues. Pathogens such as *Rhizoctonia* sp. and *Sclerotium* sp. can be considered as more hardy because, in addition to attacking tender tissues, they can also exploit comparatively mature tissues. All these pathogens have a very wide host range. They are necrotrophs, they cause extensive damage to the host tissue by employing enzymes and secondary toxic metabolites as their major weapons. On moving up the hypothetical hierarchical ladder of parasitism (Figure 1), necrotrophs appear showing more dependence on host and less saprophytic survival ability—they rely more on toxins than on wall-degrading enzymes. Although they cause extensive tissue necrosis, little or no tissue maceration occurs. Further advancement toward semibiotrophs and biotrophs has probably been guided by: (1) lesser dependence of the pathogen on toxins and enzymes, (2) more involvement of phytohormones (3) decreased deleterious effects of the parasite and increased dependence of the parasite on living host cells, (4) decreased host range, and (5) more and more synchronization of the physiological processes of host and parasite. Increased physiological synchronization guided by genetic synchronization, probably lead to the evolution of symbionts such as lichens and mycorrhizae where both host and parasite start deriving benefits from each other. Unlike biotrophic pathogens, biotrophic symbionts exhibit a comparatively wide host range. This can be explained on the basis of (1) the ability of the parasite to breach the general defense barriers of the plant, (2) the biotrophic relationship with less strain on the host plant, and (3) the benefits derived by the host due to association with these biotrophs.

A pathogen is said to have developed basic compatibility with the host if it is able to synchronize its own physiological processes with those of the host after breaching the latter's

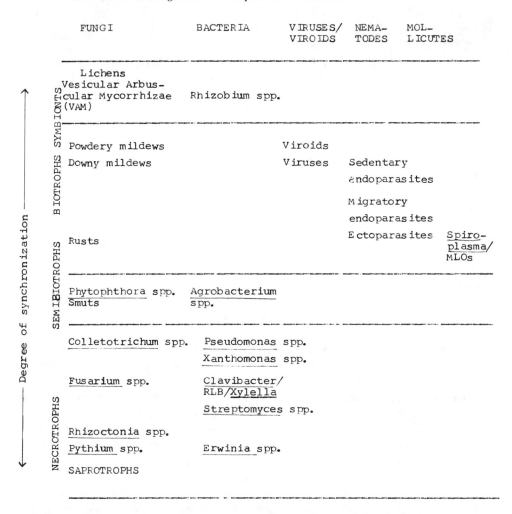

	FUNGI	BACTERIA	VIRUSES/ VIROIDS	NEMA- TODES	MOL- LICUTES
SYMBIONTS	Lichens Vesicular Arbus- cular Mycorrhizae (VAM)	Rhizobium spp.			
BIOTROPHS	Powdery mildews Downy mildews		Viroids Viruses	Sedentary endoparasites Migratory endoparasites	
BIOTROPHS	Rusts			Ectoparasites	Spiro- plasma/ MLOs
SEMIBIOTROPHS	Phytophthora spp. Smuts	Agrobacterium spp.			
NECROTROPHS	Colletotrichum spp. Fusarium spp. Rhizoctonia spp. Pythium spp. SAPROTROPHS	Pseudomonas spp. Xanthomonas spp. Clavibacter/ RLB/Xylella Streptomyces spp. Erwinia spp.			

FIGURE 1.   Hypothetical hierarchical positioning of the different plant pathogens based on their parasitic advancement and theoretically measured as their ability to establish genetic and physiological synchrony with the host species.

general defense barriers. However, mere establishment of basic compatibility does not guarantee that all the individuals of a pathogen would be able to infect all the individuals of a particular host species to the same extent under similar environmental conditions. This second level of interaction between host and parasite is termed as race-cultivar compatibility or specificity.

III. PHILOSOPHY OF DEFENSE IN PLANTS

In humans and higher animals, there is a central defense system which protects the entire body against any microbial or viral attack. The role of a localized defense system is very limited. Plants, by virtue of their peculiar morphology, anatomy and physiology, have evolved their defense systems on entirely different lines from animals. Unlike animals, plants cannot move; they cannot side step or run away from an attacking pathogen. They are bound to stand, wait, and then face the attacker. Horsfall and Cowling[2] have compared them to mediaeval castles. Whether such an analogy is correct or not, it is a fact that plants have evolved a very strong preformed defense system in the form of structural and toxic chemical barriers. Probably all the saprophytes and most, but not all, of the nonpathogens of the species are rendered ineffective because of their inability to overcome these barriers. However, preformed barriers seem to play only a limited role in providing resistance against pathogens since some of the latter have developed the ability to overcome these barriers. Only in a few cases, due to qualitative and quantitative differences, might preformed barriers play a significant role in varietal resistance.

Metabolic activities of plants, and therefore their active defense, are very much dependent upon environmental conditions. Unlike animals, plants cannot regulate the internal environment (temperature, pH, etc.) of their body, independent of the surrounding atmosphere. This factor might also have contributed toward the greater reliance of plants on constitutive rather than active defense machinery.

Being the primary consumer of solar energy, plants, in contrast to animals, can tolerate a much greater drain of their nutrients and can manage to live with limited activity of the pathogen. Often they sacrifice some of their healthy cells, even photosynthetic ones, in order to restrict the pathogen. Mechanical defoliation experiments have demonstrated that a plant suffers more due to the deleterious effects of the pathogen (e.g., toxins produced by the pathogens may be harmful to the host metabolism) rather than to the mere loss of photosynthetic area.

The most important characteristic of the plant, which has played a vital role in designing its defense machinery, is its entirely different and far less efficient vascular system. Higher animals possess an excellent blood vascular system which not only circulates the nutrients, hormones, and oxygen but also billions of different types of living cells from one part to another part of the body. It provides excellent coordination between the different tissues and organs of the body, each one of which performs specific functions. Division of labor is so pronounced in the different cells of the body that red blood cells (RBC) have even shed their nuclei. The entire defense system of the body is regulated by specific cells (lymphocytes, monocytes, etc.) present in the blood vascular system. Different functions such as recognition of foreign bodies (microbes or viruses), phagocytosis, activation of humoral (B-lymphocyte-mediated), and cellular (T-lymphocyte-mediated) immune responses, including release of helper or suppressor factors, interferons and killer factor, etc., are performed by specialized cells, present in blood but produced by different organs (such as bone marrow, thymus, spleen, lymph nodes, etc.). Activities of all these cells are well coordinated and together they constitute a central defense system. Unlike plants, in animals only infected cells are selectively killed. They are removed from the body and their chemical consitutents are reutilized.

The vascular system of plants is far less efficient than the circulatory system of animals. Upward translocation of water and solutes is passive and through nonliving parts of the plant (apoplast). Downward translocation of sugar and other metabolites is through living cells (symplastic through phloem). For downward translocation, phloem loading of a substance is essential which is selective and active process. Any compound toxic to plant metabolism cannot be loaded inside the phloem and translocated symplastically. Unfortunately, most of the chemicals (e.g., phytoalexins involved in the induced defense system of the plant are toxic, hence they do not qualify for symplastic translocation particularly in their active form. The nontoxic form may not be loaded selectively inside the phloem. Even if a substance enters the phloem by virtue of its lipid solutibility, it cannot select its own direction of translocation. It is bound to follow a source to sink pathway with respect to sugar, which may not be desirable as far as defense requirements are concerned. Movement through the plasmodesmata is only for a very short distance. Thus, although the plant behaves as a single organism, coordination between different cells/tissues/organs is not so good and division of labor is not so pronounced as in the case of higher animals. Individual cells of the plant are capable of performing multiple functions. Hence, a larger proportion of the plant cell genome is functional than of the cells in an animal's body. This may be the reason why most plant cells have retained their totipotency.

In order to suit the above situation, plants have evolved a primarily localized type of defense system where each individual cell/tissue/organ is required to defend itself without much help from distant plant parts. A pathogen (or nonpathogen which has breached its constitutive barriers) is recognized and restricted by the individual or few plant cells. Once an organism is recognized at the level of surface-to-surface interaction as incompatible or "nonself", the plant's metabolism is activated resulting in the formation of toxic substances (phytoalexins, etc.) which restrict the further advance of the pathogen and, in turn, may cause death of the few surrounding noninfected cells (hypersensitive reaction). Thanks to the poor conducting system of the plant these toxic substances are not translocated to further parts of the plant. An efficient conducting system may be the reason why the animal defense system does not involve chemicals toxic to its healthy cells.

Mere surface-to-surface interaction of the plant host and parasite might restrict the latter's activity. In several host-parasite systems, death or cessation of the pathogen's growth have been reported to precede hypersensitive host cell death and accumulation of inhibitory compounds. Cell-wall components of potato are reported to cause the bursting of the protoplasts of *Phytophthora infestans,* the wall component of resistant cultivars being more effective than susceptible ones. Recent evidence has demonstrated that an induced defense system in plants can be systemic in nature. However, in such cases it is mainly the inducer molecule (chemical nature unknown), nontoxic to the host plant, that is translocated rather than the effector. Moreover, they are more effective in providing resistance against secondary infection rather than primary infection.

Another unique feature of the plant defense system is that it does not (or cannot) eliminate the pathogens. It only tries to limit their activity within its tolerance limit. In other words, the host tries to manage the pathogen. A successful pathogen also tends to cause minimum damage to its host. Probably this tendency has lead to the evolution of symbiotic relationships where both partners become interdependent.[3]

IV. HOST-PARASITE RECOGNITION

A. PHENOMENON
In order to establish basic compatibility with the host, which is a prerequisite for the pathogenicity, a parasite must be able to breach the general defense barriers of the host and be recognized by the host as compatible partner. Host bothers about only those parasites which are able to establish basic compatibility with it. However, as discussed earlier, the establishment of basic compatibility only gives the potential, it does not guarantee that a pathogen would be able to exploit the host as it is quite evident from race-cultivar type of interactions. For that a pathogen must be able to evade the recognition by the host as a "non-self" or if it is recognized it must be able to suppress (using suppressors) or breach (using degradative enzymes) the subsequent induced reactions of the host like hypersensitive cell death, phytoalexin accumulation, cell-wall barriers (HRGP, lignin, etc.), synthesis of hydrolases (chitinase, glucanase, etc.), and/or release of systemic elicitors to induce resistant reaction in uninvaded distant cells/tissues[4] (Figure 2). So recognition between a biotrophic parasite and host plant operates at two levels, (1) for the establishment of basic compatibility and (2) for the establishment of specificity (race-cultivar compatibility). Recognition at basic compatibility level and nonrecognition as a "non-self" at race-cultivar compatibility level would result in development of disease while recognition as "non-self" at second level would trigger the defense reactions of the host (Figure 3).

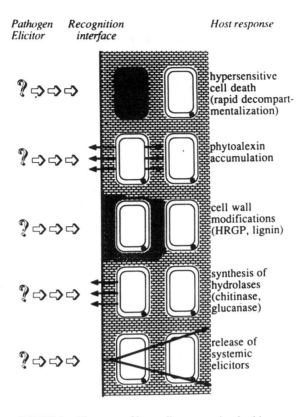

| Pathogen | Recognition | | Host response |
| Elicitor | interface | | |

FIGURE 2. The range of host-cell responses involved in active (induced) host defense (from Callow et al.[4]).

Extensive studies conducted during last 15 years in the field of nature of binding of animal hormones (insulin and glucagon, and toxins (cholera) with animal cells, pistil-pollen interaction in angiosperms in relation to inter- and intraspecific incompatibility, interaction between fungal and algal components in lichens, and work on certain host-parasite interactions such as *Phytophthora infestans*—potato, *P. megasperma* var. *sojae*—soybean, and *Pseudomonas solanacearum*—tomato, have brought following points to light (see different chapters in books by Singh et al.[5,6]).

1. Outcome of host-parasite interaction is determined very early in the infection process—latest point being the contact of fungal haustorium to host plasmalemma (Figure 4).
2. Recognition may be either for compatibility (e.g., in *Rhizobium*-legume, lichens, etc.) or incompatibility (e.g., in pistil-pollen, *P. solanacearum*-tomato interactions).
3. Binding at specific site on cell surface may influence (enhance or alter) cell metabolism (e.g., insulin and glucagon with animal cells, *Rhizobium*-legume interaction) or it may lead to cell death (cholera toxin on animal cells, host-specific toxin on plant cells, *P. solanacearum*-host plant cells).
4. Recognition of host and parasite as a compatible or incompatible partners is probably determined by constitutive binding sites present on the surface of the host and the parasite. Phenomenon like hypersensitive cell death, phytoalexin accumulation, lignification, etc., may be consequence rather than cause of resistance.
5. Binding sites determining basic compatibility are probably different from those determining race-cultivar compatibility (i.e., resistance or susceptibility).

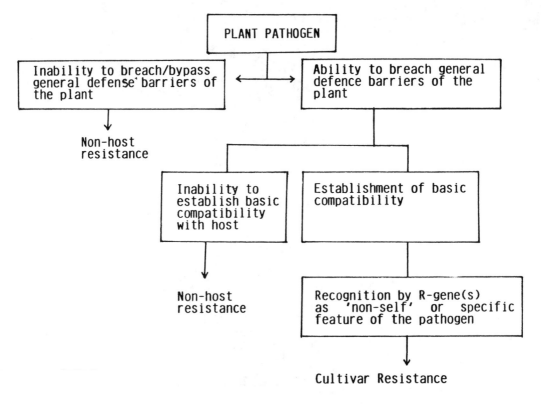

FIGURE 3. Summary of the possible origins of nonhost and cultivar resistance in terms of recognition.

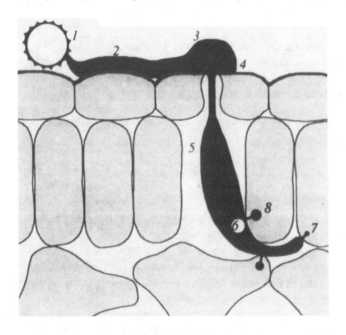

FIGURE 4. Diagrammatic representation of the interaction between a rust fungus and host-leaf tissue, illustrating the different levels of host-pathogen recognition: spore germination (1); directional and adhesive growth of germ tubes on host surface (2); appressorial differentiation over the stomata (3), and appressorial adhesion (4); expansion of the vesicle in the substomatal cavity (5); contact with mesophyll cell wall (6); initiation (7); and development (8) of haustoria in contact with host plasma membrane (Callow et al.[4]).

6. Binding of host with parasite (or its product) at site(s) determining basic compatibility will lead to an array of metabolic changes in the host resulting in increased synthesis of biomolecules and other nutrients required for the growth of the pathogen.
7. Binding at the race-cultivar compatibility site along with that of basic compatibility will result in resistant response by one or more of the following effects:
    a. Shutting off of the metabolic changes induced by the binding at the basic compatibility sites.
    b. Induction of death of the one or both the partners (hypersensitive response). Whether death of the pathogen precedes that of the host cell or vice versa would depend upon the particular host-parasite system. Even both the situations may exist within the same host-parasite system depending upon the race-cultivar composition, such as in *P. infestans*-potato system.[3]
    c. Inhibition or slowing down of further advance of the parasite with or without causing death of the host cell by inducing host's metabolic reactions leading to the synthesis of physical and/or chemical barriers.

Based on these observations and assumptions a model is proposed to explain the host-parasite interaction in *P. infestans*-potato system which could also be extended to other biotrophic interactions (Figure 5) with some modifications. Instead of direct or prior to or after, surface interaction between host and parasite, the latter may produce the chemical inducers/suppressors which are recognized by the host or vice versa. Interaction of one component with the inducer/suppressor of other component may determine the outcome of reaction. Few such chemical inducers and their effect are listed in Table 1.

B. NATURE OF RECOGNITION FACTORS

As already mentioned recognition between host and parasite as an incompatible or compatible partner, is a key factor which determines outcome of host-parasite interaction particularly in biotrophic infections. Recognition is a biochemical process and depends on the informational potential contained probably in the surfaces that come in contact and a response follows the complementary interaction of the contacting molecules. Different types of molecules which may constitute either recognition sites or may be involved in recognition are described in the following.

1. Lectins

These are sugar binding proteins or glycoproteins of nonimmune origin which are devoid of enzymatic activity toward sugars to which they bind and do not require free glycoside hydroxy group on those sugars for binding. Extra or intracellular presence of lectins have been demonstrated in almost all the groups of living organisms. Lectins present on the plant surface may constitute the recognition site and bind specifically with the sugars present on the surface of the pathogen or on elicitors produced by the pathogen and this binding may determine the outcome of plant-parasite interaction. Experimental evidences are available in support of this hypothesis from several host-parasite systems including *Rhizobium*-legume (Figure 6),[7] mycobiant-phycobiont (in lichens), *Rhizoctonia solani-Trichoderma harzianum (mycoparasite), P. infestans*-potato, *Pseudomonas solanacearum*-tomato/potato/tobacco, and *Cladosporium fulvum*-tomato. In first three examples binding leads to compatibility and in rests to incompatibility. Interference with the binding (by saturating lectin binding sites with specific sugar moieties) may alter the course of response.[6,8]

2. Common Antigens

The presence of common antigens in partners of compatible interactions involving two

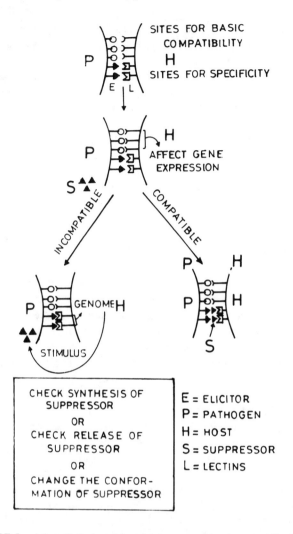

FIGURE 5. A hypothetical model explaining recognition in a specialized host (H) — parasite (P) system. Recognition, determined by complimentary binding sites at the surface of the host and parasite operates at two levels-basic compatibility leads to the genetic and subsequent physiological synchronization of the host and parasite resulting in compatible response. However, simultaneous binding at the site of specificity induces host defense system leading to an incompatible reaction. Lectins (L) at the host surface and elicitors (E) on the surface of the parasite constitute the site of specificity. Suppressors (S) produced by the parasite bind with host lectins constituting the site of specificity and convert the incompatible reaction into a compatible one: In order to counteract this situation the host, through its specificity binding sites tries to make it unsuitable for binding with suppressors but not with elicitors present on parasite's surface or it blocks the synthesis or release of suppressors in the parasite by producing some sort of inducer which ultimately leads to the incompatible response.

taxonomically distantly related organisms, higher plant and bacteria or fungi, has been demonstrated in several host-parasite systems.[9] Even plant roots are reported to share common antigens with their rhizosphere microflora.[3] In several plant-parasite interactions including maize/oat-*Ustilago maydis*[10] degree of common antigenicity was found well correlated with the degree of susceptibility. Possible role of these common antigens in establishment of basic compatibility between host and parasite, transfer of information between interacting

TABLE 1
Host Recognition "Cues" Used by Plant Pathogens and Parasites[4]

Species	Host signal	Process triggered
Bacteria		
Agrobacterium tumefaciens	Acetosyringone	Induction of *vir* genes
Rhizobium meliloti	Flavones and flavone glycosides	Induction of *nod* genes
Rhizobium leguminosarum	Isoflavonoids and flavonols	Antagonism of *nod* gene
Fungi		
Many species	Nonspecific metabolites	Spore germination on leaf surface
Ustilago violacea	α-Tocopherol	Formation of pathogenic dikaryotic mycelium
Rust fungi	Extracellular proteins	Appressorial differentiation
Zoosporic Oomycetes	Nonspecific metabolites	Chemotaxis
Phytophthora cinnamomi	Fucose-rich ligands of root mucilage	Binding to root
Phytophthora cinnamomi	Pectin, root mucilage	Zoospore encystment
Pythium aphanidermatum	Fucose-rich ligands of root mucilage	Zoospore binding Zoospore encystment

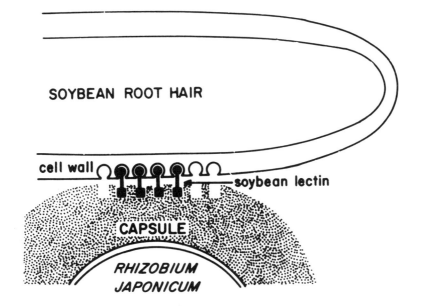

FIGURE 6.   A model depicting involvement of lectins in binding (recognition) of *Rhizobium japonicum* by soybean root hair (from Bal[7]).

partners, or in the suppression of resistance response has been postulated.[3,9] However, most of the studies are merely correlational and little attempt has been made to determine their role in plant parasite interaction. In animal system presence of common antigens helps the parasite to avoid the recognition by the immune system of the host as "non-self" but plant lacks parallel immune system.

3. Elicitors
Several chemicals of abiotic (silver, mercury, copper salts, iodoacetate, sucrose, salicylic acid, polyacrylic acid, etc.) or biotic (i.e., microbial or plant products like proteins, complex

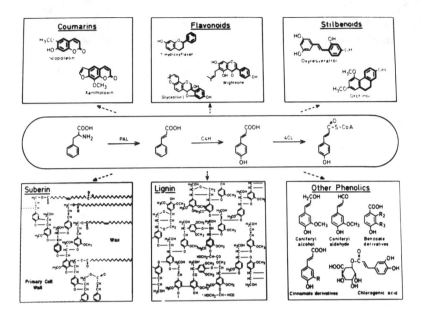

FIGURE 7.   Scheme illustrating the common biosynthetic origin of various defense-related phenolic materials in higher plants (C4H = cinnamic acid 4-hydroxylase; 4CL = 4-coumarate: CoA ligase; PAL = phenylalanine ammonia-lyase) (from Hahlbrock and Scheel[15]).

glycoproteins, fatty acids, carbohydrates) origins, physical agents (UV light, freezing injury, wounding, etc.), or microbes (fungal and bacterial cells or cell walls) can induce the host defense responses like synthesis of the phytoalexins,[11] lignification,[12] and hypersensitive response.[13] Usually they do so by inducing enzymes of phenylpropanoid metabolism which are required for the synthesis of a number of phytoalexins, lignin, suberin, etc.[14,15] (Figure 7, Table 2). Several other factors or enzymes are induced which are reported to provide resistance against microbial (chitinase, HRGP, etc.) and viral (PR proteins) infections (Table 2). Since biotic elicitors are extremely active (at hormonal levels of 10^{-10} M)[11] and they bind with the plant host lectins, they were expected to play important role in deciding race-cultivar compatibility. However, elicitors are not specific, i.e., they induce the phytoalexin synthesis in both resistant and susceptible cultivars; they alone cannot be determinant of the race-cultivar interaction.

4. Suppressors
Suppressors are the chemicals produced by the pathogen which suppress the host defense responses, like phytoalexin synthesis, hypersensitivity, etc., induced by the pathogen or elicitors produced by them.[11,13] They can convert incompatible association into compatible at race-cultivar interaction level. Suppressors probably act by blocking the binding of elicitor with the lectin, a complimentary binding site on the host cell surface (Figure 5). Ziegler and Pontzen[16] found that invertases secreted by *Phytophthora megasperma* f. sp. *glycinea,* race-specifically suppressed the glyceollin accumulation in soybean cultivar. Suppressive activity was imparted by the carbohydrate (rich in mannans) moiety of glycoprotein enzyme invertase. Since suppressors are race-specific, they are postulated to play more important role in deciding race-cultivar compatibility particularly in those host-parasite systems where their presence has been demonstrated. Whether common antigens also act as suppressors, is an interesting question for the scientists to answer.

TABLE 2
Host Gene Expression Following Interactions with Plant Pathogens and Elicitors[14]

Genes induced	Species and tissue	Pathogen or elicitor
	Phenylpropanoid metabolism	
Phenylalanine ammonia-lyase (PAL)	*Phaseolus* hypocotyls	*Colletotrichum*
	Phaseolus leaves	*Pseudomonas*
	Phaseolus culture	Elicitor
	Pisum endocarp	Elicitors
	Glycine culture	Elicitor
	Glycine hypocotyls	*Phytophthora*
	Petroselinum culture	Elicitor and UV
4-coumarate CoA-ligase (4CL)	*Petroselinum* culture	Elicitor and UV
Chalcone synthase (CHS)	*Phaseolus* hypocotyls	*Colletotrichum*
	Phaseolus culture	Elicitor
	Phaseolus leaves	*Pseudomonas*
	Petroselinum culture	Elicitor and UV
	Glycine culture	Elicitor
	Glycine hypocotyls	*Phytophthora*
Chacone isomerase (CHI)	*Phaseolus* hypocotyls	*Colletotrichum*
	Phaseolus culture	Elicitor
Cinnamyl alcohol dehydrogenase	*Phaseolus* culture	Elicitor
	Other proteins and enzymes	
Casbene synthetase	*Ricinus* seedlings	Elicitors
Chitinase	*Phaseolus* leaves	*Pseudomonas*
PR proteins	*Nicotiana* leaves	TMV
		Salicylic acid
	Petroselinum culture	Elicitor
	Phaseolus leaves	AMV
Thaumatin-like protein	*Nicotiana* leaves	TMV
		Salicylic acid
HRGP	*Phaseolus* cultures	Elicitors
Proline hydroxylase	*Phaseolus* cultures	Elicitors

V. DEFENSE MECHANISMS-PHYSICAL AND BIOCHEMICAL BASIS

Interaction between a plant and a parasite commences even before their direct contact. There are several factor/barriers, operating individually or in different combinations, which may slow down or stop the advance of a parasite/potential parasite. These factors are listed in Table 3. Since major aim of this chapter is to provide conceptual background, readers are requested to consult the specialized publications on host-parasite interaction[5,6,8,18-24] to obtain detailed information including supporting direct/indirect evidences. Usually preformed structural and chemical barriers (probably except recognition sites) play more important role in warding off the nonpathogens as pathogens have already acquired the ability to breach or bypass them. For a pathogen, after recognition, barriers induced after infection are probably more important in imparting resistances. A brief discussion of few such barriers is being presented here.

A. HYPERSENSITIVITY

The term "hypersensitivity" was coined by Stakman[25] to describe the phenomenon of rapid cell death around the sites of penetration of uredospores of *Puccinia graminis* f. sp. *tritici* in rust resistant wheat cultivars. Now it is considered as an universal phenomenon associated with resistance response irrespective of the nature of the pathogen which may be

TABLE 3
Defense Mechanisms in Plants

I.   Defense at the Perimeter
    A. Production of inhibitory compounds by the plants in root/leaf exudates
    B. Production of toxic compounds by the rhizoplane/phylloplane or rhizosphere/phyllosphere micro-flora
    C. Lack of stimulatory (e.g., inducers of resting spore germination) compounds in root/leaf exudates
    D. Nonproduction of compounds governing growth of germ tube/movement of zoospores toward plant surfaces
    E. Resistance offered by plant surface appendages (trichomes, sloughing of root hairs), surface coverings (wax, cuticle, root cap and mucilage and seed coat), surface organisms (ectomycorrhiza and epiphytic microorganisms), surface structures (epidermis, periderm, etc.) and structure or number of natural openings (stomata, ectodesmata, lenticels, hydathodes, nectaries, etc.)

II.   Preformed Internal Defenses
    A. Physical Barriers
        1. Suberized tissues
        2. Lignified tissues
        3. Walls and middle lamella
        4. Deposition of hydroxyproline-rich glycoproteins (HRGP), gums, resins, and tannin-like substances
        5. Deposition of silicic acid in cell walls and cells
    B. Chemical Barriers
        1. Lack of essential nutrients, cytoplasm (vacuolated) and recognition sites for compatibility
        2. Presence of recognition sites for incompatibility
        3. Presence of enzyme inhibitors (phenolics, tannins, proteins, etc.,
        4. Presence of hydrolytic enzymes (chitinases, β-glucanases)
        5. Presence of antimicrobial compounds (unsaturated lactones, cyanogenic glycosides, leak oils, glycosinolates, saponins, phenolics, proanthocyanidins, etc.)

III. Defenses Triggered by the Invader
    A. Physical defenses
        1. Formation of cork layers, papillae, abscission layers, or tyloses
        2. Deposition of Ca, gums, lignin, callose, suberin, etc.
    B. Hypersensitive reactions
    C. Chemical defenses
        1. Phytoalexins
        2. Pathogenesis-related proteins
        3. Lack of transport protein (in viruses)
        4. Detoxification of pathogen's toxins by host enzymes

IV.   Defense triggered by previous invaders (induced resistance/cross protection)
    A. In viral infection: different hypotheses proposed are as follows:
        1. Competition or interference between inducer (previous invader) and challenger strains for the site of attachment to the cell
        2. Blocking of uncoating of challenger virus in the previously invaded cell
        3. Encapsidation of the challenger strain nucleic acid by coat protein of the protecting strain
        4. Interference between nucleic acids resulting in inhibited transcription/translation of challenger strain. Positive strand RNA of inducer strain may anneal with negative strand RNA of challenge strain to form dsRNA which is inactive.
        5. Production of pathogenesis-related proteins and/or antiviral factors.
    B. In microbial infections
        1. Induction of host-defense system by first invader resulting in production of phytoalexins, phenolics, etc. (in locally induced resistant)
        2. Sensitization of the host by inducer produced by the first invader for quick reaction (lignification, phytoalexin synthesis, etc.) against challenger strain (in systemically induced resistance)

a fungus, bacterium, nematode, or virus.[13] Hypersensitive reactions (HR) are usually accompanied by the accumulation of antimicrobial compounds (phytoalexins) in and around the necrotic zone.[11,13] These reactions were earlier considered as the most potent active defense reaction of the plant. However, investigations carried out during last 15 years have

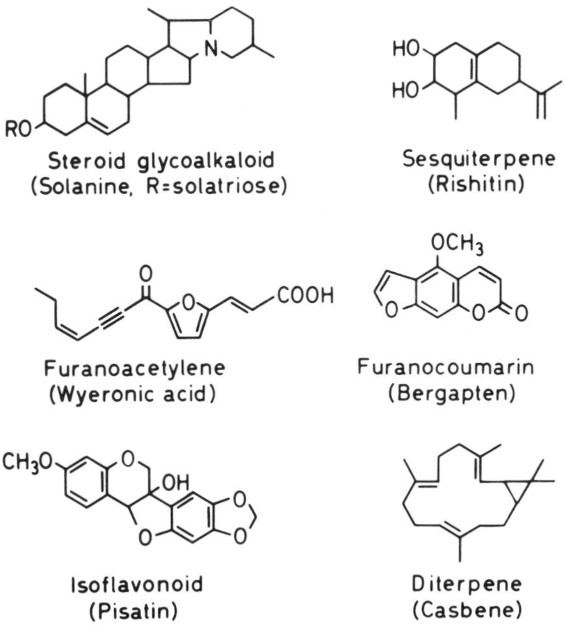

FIGURE 8. Selected phytoalexins of different biosynthetic origin (from Hahlbrock and Scheel[15]).

revealed a vast complexity and a wide variability in the phenomenon at molecular level. In certain host-parasite systems fungal cell death precedes host cell death, indicating that at least in such cases necrosis of host cell may be consequence rather than cause of resistance. So merely rapid host cell necrosis around the penetration/infection site is no more considered as determinant of resistance. Prusky[13] considers HR not merely as a rapid necrosis of host cell, but as a whole process of defense involving recognition biochemical, physiological, cytological, and morphological alterations in invaded and surrounding cells; synthesis of compounds capable of offering physical (e.g., lignin) and/or chemical (e.g., phytoalexins) barriers to the spread of the pathogen and finally the effective localization of the pathogen. Hypersensitivity is considered as not only the symptom of resistance or the nonspecific death of the host cells after the initial recognition, but as an active defense mechanism.[13]

B. PHYTOALEXINS

By definition phytoalexins "are low molecular weight antimicrobial compounds that are both synthesized by and accumulated in plants after exposure to microorganisms".[11] There are several abiotic and biotic elicitors which can induce the synthesis of phytoalexin, but not their accumulation. Over 100 phytoalexins from a wide range of plants have been recorded. They fall into several biosynthetic groups[11,15] as shown in Figure 8. Phytoalexins are host rather than pathogen specific. The legumes predominantly synthesize isoflavonoid-derived phytoalexins, solanaceous plants sesquiterpenoid-derived phytoalexins, orchids phenanthrene phytoalexins, and graminaceous plants produce alkaloid phytoalexins. Diterpene phytoalexins are synthesized by such diverse plants as rice and castor bean.[11] So far no phytoalexin have been discovered in the cucurbitaceae.[11]

Phytoalexins are degraded by host and microbial enzymes. So it is their accumulation (at proper time and site), which involves dynamic balance between synthesis and degradation in plants, rather than merely synthesis which plays more important role in plant defense.

Accumulation of phytoalexins is more following inoculation of saprophytes or nonpathogens than a pathogen indicating their more important role in nonhost than race-cultivar resistance. Although, accumulation of phytoalexins is usually associated with hypersensitivity responses, it is yet to be demonstrated conclusively whether it is cause or consequence of resistance. So role of phytoalexins in race-cultivar resistance is still an unresolved problem for most plant-pathogen interactions. Direct evidence in support of positive role of phytoalexins in host-resistance has been obtained in one system. Van Etten's group,[26] showed that pisatin, plays a decisive role in resistance of pea to *Nectria haematococca,* because only pathogen strains that are able to detoxify the phytoalexin invade the plant. Recently they have succeeded in cloning this gene, called *pda* (pisatin-demethylating ability).[27] Soon, it might be possible to confirm that ability to detoxify pisatin is an important factor in the virulence of *N. haematococca* and, by inference, the pisatin is an important factor in plant disease resistance. This determination can be made by transforming *pda* gene into a avirulent strain of *N. haematococca* lacking this gene. If transfer of *pda* converts the strain to virulent, the hypothesis will be rigorously confirmed.

C. PATHOGENESIS-RELATED PROTEINS

Pathogenesis related proteins (PR-proteins or *b* proteins) induced in a number of plant species following incompatible interaction (i.e., hypersensitive response) with the plant pathogenic bacteria, fungi and viruses.[28] Abiotic elicitors like salicylic acid, ethephon, acrylic acid etc. which induce plant's resistance to viral infections also induce synthesis of *b* proteins. However, appearance of *b* proteins also correlates with leaf senescence and stress.[28] Their roles and function in stress and disease resistance remain unknown. Dumas and Gianinazzi[29] reported that *b* proteins do not play a central role in TMV localization in *Nicotiana rustica.* However, their presence accompanies induced resistance.

Inoculation of plant viruses on lower leaves may protect the upper leaves against further infection by the virus by inducing the production of antiviral proteins (*b* proteins) in upper leaves with the help of mobile inducers, which is yet to be characterized. Since action of these inducers is biphasic and they are host rather than virus specific, several scientists have tried to correlate them or *b* proteins with animal interferons (i.e., a protein which exerts nonspecific antiviral activity at least in homologous cells through cellular metabolic processes involving synthesis of both RNA and protein).[30] However, Gianinazzi and co-workers[31] from their extensive work on *b* proteins, concluded that "although interferons in animals and inducers of resistance in plants are very similar macroscopically, at the molecular level the mechanism of action seems to be very different:.

Since genes for tobacco and parsley PR proteins have now been cloned, it would be easy to precisely assess their role by transformation experiments in near future. However, it is mobile inducer of PR-protein synthesis, rather than protein itself which may play more important role in regulation of host defense.

VI. GENETICS OF HOST-PARASITE INTERACTION

No information is available on the genetics of basic compatibility. However, a good amount of information is available regarding genetics of specificity (race-cultivar compatibility). The studies on the genetics of resistance in the host and pathogenicity in the pathogen are important to understand the basic aspects of host and parasite. Depending on the type of interaction, the breeding/management strategies of resistance could be framed.

A. GENETICS OF RESISTANCE

The rediscovery of Mendel's law of inheritance in 1900 provided the foundation necessary for the analysis of the differential reaction of varieties to diseases. The first reported genetic study of resistance to a disease was published in 1905 by Biffen[34] in England. He obtained a 3 (susceptible):1 (resistant) ratio in the F_2 populations of crosses between wheat variety (Ribet) resistant to *Puccinia striiformis* (yellow rust) and susceptible variety (Michigan, Bronza, or Red King). The F_3 families appeared in the ratio of 1:4 true breeding susceptible lines, 1:2 segregating lines, and 1:4 true breeding resistant lines. Between 1912 to 1970 numerous research papers on the inheritance of resistance were published and most of them confirmed Biffen's findings.

Person and Sidhu[35] reviewed the work carried out on the inheritance of resistance since Biffen's time. They concluded that (1) regardless of the species that was involved in host-parasite interaction, resistance generally segregated in Mendelian ratios. Resistance was usually found to be determined by "major" rather than by "minor" genes. The alleles for resistance were dominant over those for susceptibility, (2) in a relatively small number of studies the two factor genetic interaction was found, (3) in a relatively small number of studies evidence has been found for linkage of resistant genes, and (4) alternate resistance alleles were distinguished according to specific types.

The informations published after 1971 on the mode of inheritance of resistance of diseases against some of the important crop plants against important pathogens (host-parasite system) have been compiled by Singh.[36] The mode of inheritance could be monogenic, oligogenic, or polygenic.

B. GENETICS OF PATHOGENICITY

Pearson and Sidhu[35] reviewed the literature on the genetics of pathogenicity and made the following generalizations (1) the virulence/avirulence was usually under Mendelian control, (2) the genes which induced susceptible reaction were usually inherited as recessive, and (3) linkage of genes for virulence was reported only occasionally and there was no report of allelism. Van der Plank[37] observed that virulence and avirulence in the pathogen are the counterparts of vertical susceptibility and resistance in the host. Aggressiveness and unaggressiveness in the pathogen are the counterparts of horizontal susceptibility and resistance in the host, respectively.

Genetics of pathogenicity can be best explained by citing Flor's findings.[38] The two varieties Ottawa 770B and Bombay, each having one dominant gene for resistance, and their hybrid segregated in the F_2 generation to give a digenic ratio when tested with two rust races 22 and 24. Similarly, the F_2 segregates of the hybrid between race 22 and 24 showed digenic ratio when tested on the varieties Ottawa 770B and Bombay.

Nelson and Kline[39] studied the pathogenicity of 291 ascospore isolates obtained from different crosses between isolates of *Cochliobolus heterostrophus* to 9 differential gramineous species. A minimum of 13 different genes for pathogenicity was identified. Segregations and comparisons of responses of paired differential species indicated that pathogenicity to 4 host species is controlled by 5 different genes, and the pathogenicity to 4 species is conditioned by 5 different sets of 2 genes each. All pathogenic capacities are inherited independently.

C. THE GENE-FOR-GENE HYPOTHESIS

If either the virulence of the pathogen or the resistance of the host increased unopposed, it would have led to the elimination of either the host or the pathogen, respectively, which obviously has not happened. This shows that evolution of virulence and resistance are stepwise which can be explained by "gene-for gene" concept according to which for each gene that confers resistance to the host there is a corresponding gene that confers virulence to the pathogen or vice versa.

The gene-for-gene hypothesis was proposed by Flor[40-42] as the simplest explanation of

TABLE 4
Gene-for-Gene Interaction

	Genes in the plant	
	R	**r**
Genes in the pathogen		
A	A-R (resistant)	A-r (susceptible)
a	a-R (susceptible)	a-r (susceptible)

the results of studies on the inheritance of pathogenicity in the flax rust fungus, *Melampsora lini*. On the varieties of flax *(Linum usitatissimum)* that had one gene for resistance to the parent race, F_2 cultures of the pathogen segregated into monofactorial ratios. On varieties which had 2, 3, or 4 genes for resistance to the parent race, the F_2 cultures segregated into bi-, tri-, or tetrafactorial ratio.[40-42] This suggested that for each gene that conditions resistance in the host there is a corresponding gene in the parasite that conditions pathogenicity.

The gene-for-gene hypothesis, in its simplest form, is illustrated in Table 4, in which A represents the dominant gene for avirulence and a, the recessive gene for virulence in the pathogen and where R represents the dominant gene for resistance and r, the recessive gene for susceptibility in the attacked plant.

According to gene-for-gene hypothesis, only one possible gene combination, A-R, would result in resistance (Table 4). In all other gene combinations, the susceptible reaction would result because the host plant is susceptible (r), the parasite is virulent (a), or both conditions (a-r) are fulfilled in the same host-parasite interaction.

The significant aspect of the gene-for-gene concept can be further illustrated by considering interactions in which the corresponding pairs of genes occur at two or more different loci on the chromosomes. With two loci, four different gene combinations are possible. Notations for genes at two loci in the parasite would be A_1A_2, A_1a_2, a_1A_2, and a_1a_2. For corresponding genes at two loci in the host, the notations would be R_1R_2, R_1r_2, r_1R_2, r_1r_2. All possible interactions between corresponding pairs of genes, along with the disease reaction that would occur from each interaction, are given in Table 5.

Because of the specificity of interaction, resistance is expressed only when combinations A_1R_1 or A_2R_2 occur in the same host-parasite system. That is, A_1 recognizes only R_1 and A_2 recognizes only R_2.

''Gene-for-gene'' type of associations have been demonstrated in several host-parasite systems (Table 6).

VII. INDUCED RESISTANCE

Sometimes invasion of plant by a pathogen (virus, fungi, bacteria, or nematode) may protect the plant against subsequent infection by same, related (different strain/race) or even entirely different pathogens. This phenomenon in termed as induced resistance, acquired resistance, or cross protection.

Protection of plant against a severe strain by previous inoculation with mild strain is a well-documented phenomenon in plant viruses.[43] The term ''cross protection'' in viruses has been restricted to explain the interaction between two strains of the same virus. Interaction between two unrelated viruses is termed as interference. However, in microbiol interactions, cross protection has been used in a very wide sense to describe all sorts of cross protective interactions including those involving entirely different groups of the pathogens like fungi

TABLE 5
Disease Reactions Following Interactions Between
Corresponding Genes at Two Different Loci

Genes of the parasite	Genes of the host	Disease reaction
A_1A_2	R_1R_2	Resistant
A_1A_2	R_1r_2	Resistant
A_1A_2	r_1R_2	Resistant
A_1A_2	r_1r_2	Susceptible
A_1a_2	R_1R_2	Resistant
A_1a_2	R_1r_2	Resistant
A_1a_2	r_1R_2	Susceptible
A_1a_2	r_1r_2	Susceptible
a_1A_2	R_1R_2	Resistant
a_1A_2	R_1r_2	Susceptible
a_1A_2	r_1R_2	Resistant
a_1A_2	r_1r_2	Susceptible
a_1a_2	R_1R_2	Susceptible
a_1a_2	R_1r_2	Susceptible
a_1a_2	r_1R_2	Susceptible
a_1a_2	r_1r_2	Susceptible

TABLE 6
Host Parasite Systems for Which Gene-for-Gene
Relationship Has Been Demonstrated or Suggested

Disease	System
Apple scab	*Malus-Venturia inaequalis*
Bacterial disease	*Gossypium — Xanthomonas Campestris* pv. *malvacearum*
	Leguminosae—*Rhizobium* (symbiosis)
Blight	*Zea—Drechslera turcica*
Bunts	*Triticum — Tilletia caries T. contraversa*
Late blight	*Solanum—Phytophthora infestans*
Leaf mold	*Lycopersicon—Cladosporium fulvum*
Mildews	*Hordeum—Erysiphe graminis*
	Triticum—E. graminis
Potato wart	*Solanum—Synchytrium endobioticum*
Rust	*Zea—Puccinia sorghi*
	Linum—Melampsora lini
	Triticum—Puccinia graminis
	Triticum—P. striiformis
	Triticum—P. recondita
	Avena—P. graminis
	Coffee—Hemileia vestatrix
	Helianthus—P. helianthi
Smuts	*Avena—Ustilago avenae*
	Triticum—*U. tritici*
	Hordeum—U. hordei
Viruses	*Lycopersicon*—Tobacco mosaic virus
	Lycopersicon—Tomato spotted wilt virus
	Solanum—Potato viruses
	Phaseolus—Bean common mosaic virus

and viruses. Kuc and co-workers reported that cucumbers, watermelons, and muskmelons can be systemically immunized against diseases caused by viruses, bacteria, and fungi by restricted infection with viruses, bacteria, or fungi.[44] Apart from viruses, during last two decades cross-protective responses have been demonstrated in diverse plant-pathogen interactions including pear and apple-*Erwinia amylovora*, peach-*Cytospora cinta*, tomato-*Fusarium oxysporum*, cotton-*Tetranchus uritaceae*, bean-*Uromyces phaseoli*, sunflower-*Puccinia helianthi*, tobacco-*Phytophthora parasitica* var. *nicotiana*, tobacco-diverse pathogens and an aphid (see Kuc[44] for references).

Immunization may be localized or systemic in nature. In some cases it is graft transmissible and persists for the life of annual plants. Mechanism is not yet exactly understood. In microbial infections it has been proposed that inoculation with immunizing strain results in production of mobile inducer which may, locally or systemically, sensitize the plant to respond quickly to infection by challanger strain by inducing the rapid lignification and/or synthesis and accumulation of phytoalexins at the site of attack. Different hypotheses proposed to explain the cross protective response in virus[43] are listed in Table 3 and are also described briefly in Chapter 16.

REFERENCES

1. **Singh, R. S. and Singh, U. S.**, Pathologenesis and host-parasite specificity in plants, in *Experimental and Conceptual Plant Pathology*, Vol. II, Singh, R. S., Singh, U. S., Hess, W. M., and Weber, D. J., Eds., Gordon and Breach, New York, and IBH Publishing, New Delhi, 1988, 139.
2. **Horsfall, J. G. and Cowling, E. B.**, Prologue, in *Plant Diseases—An Advanced Treatise*, Vol. 5, Horsfall, J. G. and Cowling, E. B., Eds., Academic Press, New York, 1980, 1.
3. **Singh, U. S. and Singh, R. S.**, Philosophy of defence in plants, in *Experimental and Conceptual Plant Pathology*, Vol. III, Singh, R. S., Singh, U. S., Hess, W. M., and Weber, D. J., Eds., Gordon and Breach, New York, and IBH Publishing, New Delhi, 1988, 459.
4. **Callow, J. A., Estrada-Garcia, M. T., and Green, J. R.**, Recognition of non-self: the causation and avoidance of disease, *Ann. Bot.*, 60, Suppl. 4, 3, 1987.
5. **Singh, R. S., Singh, U. S., Hess, W. M., and Weber, D. J.**, *Experimental and Conceptual Plant Pathology*, Vol. II, Gordon and Breach, New York, and IBH Publishing, New Delhi, 1988.
6. **Singh, R. S., Singh, U. S., Hess, W. M., and Weber, D. J.**, *Experimental and Conceptual Plant Pathology*, Vol. III, Gordon and Breach, New York, and IBH Publishing, New Delhi, 1988.
7. **Bal, A. K.**, Pathogenesis and host-parasite specificity in *Rhizobium* species, in *Experimental and Conceptual Plant Pathology*, Vol. II, Singh, R. S., Singh, U. S., Hess, W. M., and Weber, D. J., Eds., Gordon and Breach, New York, and IBH Publishing, New Delhi, 1988, 247.
8. **Chet, I.**, *Innovative Approaches to Plant Disease Control*, John Wiley & Sons, New York, 1987.
9. **Chakraborty, B. N.**, Antigenic disparity, in *Experimental and Conceptual Plant Pathology*, Vol. III, Singh, R. S., Singh, U. S., Hess, W. M., and Weber, D. J., Eds., Gordon and Breach, New York, and IBH Publishing, New Delhi, 1988, 477.
10. **Wimalajeewa, D. L. S. and Devay, J. E.**, The occurrence and characterization of common antigen relationship between *Ustilago maydis and Zea mays*, *Physiol. Plant Pathol.*, 1, 523, 1971.
11. **Paxton, J. D.**, Phytoalexins in Plant parasite interaction, in *Experimental and Conceptual Plant Pathology*, Vol. III, Singh, R. S., Singh, U. S., Hess, W. M., and Weber, D. J., Eds., Gordon and Breach, New York, and IBH Publishing, New Delhi, 1988, 537.
12. **Bird, P. M.**, The role of lignification in plant disease, in *Experimental and Conceptual Plant Pathology*, Vol. III, Singh, R. S., Singh, U. S., Hess, W. M., and Weber, D. J., Eds., Gordon and Breach, New York, and IBH Publishing, New Delhi, 1988, 523.
13. **Prusky, D.**, Hypersensitivity: an overview, in *Experimental and Conceptual Plant Pathology*, Vol. III, Singh, R. S., Singh, U. S., Hess, W. M., and Weber, D. J., Eds., Gordon and Breach, New York, and IBH Publishing, New Delhi, 1988, 485.
14. **Collinge, D. B. and Slusarenko, A. J.**, Plant gene expression in response to pathogens, *Plant Mol. Biol.*, 9, 389, 1987.
15. **Hahlbrock, K. and Scheel, D.**, Biochemical responses of plants to pathogens, in *Innovative Approaches to Plant Disease Control*, Chet, I., Ed., John Wiley & Sons, New York, 1987, 229.

16. **Ziegler, E. and Pontzen, R.**, Specific inhibition of glucan-elicited glyceollin accumulation in soybeans by an extracellular mannan-glycoprotein of *Phytophthora megasperma* f. sp. *glycinea*, *Physiol. Plant Pathol.*, 20, 321, 1982.

17. **Agrios, G. N.**, *Plant Pathology*, 3rd ed., Academic Press, New York, 1988.

18. **Bailey, J. A.**, *Biology and Molecular Biology of Plant-Pathogen Interactions*, Springer-Verlag, Berlin, 1986.

19. **Bailey, J. A. and Deverall, B. J.**, *The Dynamics of Host Defence*, Academic Press, New York, 1983.

20. **Callow, J. A.**, *Biochemical Plant Pathology*, Wiley, New York, 1983.

21. **Fraser, R. S. S.**, *Mechanisms of Resistance to Plant Diseases*, Nijhoff/Junk, Dordecht, 1985.

22. **Groth, J. V. and Bushnell, W. R.**, *Genetic Basis of Biochemical Mechanisms of Plant Disease*, American Phytopathological Society, St. Paul, Minnesota, 1985.

23. **Horsfall, J. G. and Cowling, E. B.**, *Plant Disease—An Advanced Treatise*, Vol. V, Academic Press, New York.

24. **Wood, R. K. S.**, *Active Defense Mechanisms in Plants*, Plenum, New York, 1982.

25. **Stakman, E. C.**, Relation between *Puccinia graminis* and plants highly resistant to its attack, *J. Agric. Res.*, 4, 139, 1915.

26. **Kistler, H. C. and Van Etten, H. D.**, Regulation of pisatin demethylation in *Nectria haematococca* and its influence on pisatin tolerance and virulence, *J. Gen. Microbiol.*, 130, 2605, 1984.

27. **Yoder, O. C., Weltring, K., Turgeon, B. G., Garber, R. C., and Van Etten, H. D.**, Technology for molecular cloning of fungal virulence genes, in *Biology and Molecular Biology of Plant Pathogen Interactions*, Bailey, J. A., Ed., Springer-Verlag, Berlin, 1986, 371.

28. **Van Loon, L. C.**, Pathogenesis-related proteins. *Plant. Mol. Biol.*, 4, 111, 1985.

29. **Dumas, E. and Gianinazzi, S.**, Pathogenesis related proteins *(b)* do not play central role in TMV localization in *Nicotiana rustica*, *Physiol. Mol. Plant Pathol.*, 28, 243, 1986.

30. **Chessin, M.**, Is there a plant interferon?, *Bot. Rev.*, 49, 1, 1983.

31. **Gianinazzi, S., Dumaj, E., Antoniw, J. F., and White, R. F.**, Plant interferons: a wish or reality?, in *Experimental and Conceptual Plant Pathology*, Vol. III, Singh, R. S., Singh, U. S., Hess, W. M., and Weber, D. J., Eds., Gordon and Breach, New York, and IBH Publishing, New Delhi, 1988, 569.

32. **Hooft van Huijsduijen, R. A. M., Van Loon, L. C., and Bol, J. F.**, cDNA cloning of six mRNA induced by TMV infection of tobacco and characterization of their translation products, *EMBO J.*, 5, 2057, 1986.

33. **Somssich, I. E., Schmelzer, E., Bollman, J., and Hahlbrock, K.**, Rapid activation by fungal elicitor of genes encoding "pathogenesis related" proteins in cultured parsley cells, *Proc. Natl. Acad. Sci. U.S.A.*, 83, 2427, 1986.

34. **Biffen, R. H.**, Mendel's laws of inheritance and wheat breeding, *J. Agric. Sci.*, 1, 4, 1905.

35. **Person, C. and Sidhu, G.**, Genetics of host-parasite relationships, in *Proceedings of the Symposium in Mutation Breeding for Disease Resistance*, Int. Atomic Energy, Vienna, 1971, 31.

36. **Singh, D. P.**, *Breeding for Resistance to Diseases and Insect Pests*, Narosa Publishing House, New Delhi and Springer Verlag, Berlin, 1986.

37. **Van der Plank, J. E.**, *Principles of Plant Infection*, Academic Press, New York, 1975, 216.

38. **Flor, H. H.**, The complementary genetic system in flax rust, *Adv. Genet.*, 8, 29, 1956.

39. **Nelson, R. R. and Kline, D. M.**, Genes for pathogenicity in *Cochliobolus heterostrophus*, *Can. J. Bot.*, 47, 1311, 1969.

40. **Flor, H. H.**, Inheritance of pathogenicity in *Melampsora lini*, *Phytopathology*, 32, 653, 1942.

41. **Flor, H. H.**, Inheritance of reaction to rust in flux, *J. Agric. Res.*, 74, 241, 1947.

42. **Flor, H. H.**, Host-parasite interaction in flux rust—its genetic and other implications, *Phytopathology*, 45, 680, 1955.

43. **Bozarth, R. F. and Ford, R. E..**, Viral interactions: induced resistance (cross protection) and viral interference among plant viruses, in *Experimental and Conceptual Plant Pathology*, Vol. III, Singh, R. S., Singh, U. S., Hess, W. M., and Weber, D. J., Eds., Gordon and Breach, New York, and IBH Publishing, New Delhi, 1988, 551.

44. **Kuc, J.**, Plant immunization and its applicability for disease control, in *Innovative Approaches to Plant Disease Control*, Chet, I., Ed., John Wiley & Sons, 1987, 255.

Chapter 5

PRINCIPLES AND PRACTICES OF PLANT DISEASE MANAGEMENT

I. INTRODUCTION

Informations on etiology, symptoms, pathogenesis, and epidemiology of plant diseases are intellectually interesting and scientifically justified, but most important of all, they are useful as they help in formulation of methods for successful management of diseases and thereby increasing the quantity and improving the quality of plant and plant products. Practices employed for disease management vary considerably from one disease to another depending upon the type of pathogen, the host, and the interaction of the two under overall influence of the environment. Contrary to management of human and animal diseases, where every individual is attended, the plant diseases are generally treated as populations except for trees, ornamentals, and some times virus infected plants.

II. DISEASE CONTROL VS. DISEASE MANAGEMENT

In the past plant pathologists always aimed at the impossible task of destroying pathogens to control diseases. However, diseases really controlled are very few. No plant pathogen has ever been wiped out from the face of the earth. So long as the pathogen survives and its host is cultivated, the chances of disease incidence will persist. It is a different matter that due to some direct or indirect efforts on the part of man the disease causing agent has been subdued or has been reduced to an innocuous level. These efforts could better be called as "management practices" whereby the population of the pathogen or its disease causing potentialities have been kept under check so that it failed to cause noticeable loss to the crop.

Although disease control is an established and widely understood term, there is convincing rationale supporting the substitution of "management" for "control". The word "control" evokes the notion of finality, the final disposal of problem.[1] How many plant diseases have been finally disposed off? "Management" conveys the concept of a continuous process rather than an event accomplished through application of an intrinsic factor. The fact that in almost all examples of recommendation for disease control we recommend schedules that are to be followed almost every year in the crop suggests that these disease control methods are part of a continuous process, and the disease has not been disposed off. The diseases are inherent components of the agroecosystem that must be dealt with on a continuous, knowledgeable basis.[1] Management is based on the principle of maintaining the damage or loss below an economic injury level or at least minimizing occurrence above that level. If this concept is conveyed to the farmer he will not be frustrated if he finds a few diseased plants in his field even after adhering to recommendations made by the plant pathologists. In this book the two terms "control" and "management" have been used to convey the idea of plant disease management.

III. ESSENTIAL CONSIDERATIONS IN DISEASE MANAGEMENT

A. ECONOMIC POTENTIAL OF DISEASE

The amount of injury done to a crop which will justify the cost of control measures (economic damage), the lowest population density that will cause economic damage (economic injury-level), and the level of disease intensity that produces an incremental reduction in crop value greater than the cost of implementing a disease management strategy (economic threshold) must be known clearly to growers. In case these parameters are not known, the grower may waste efforts and resources by suppressing even those diseases of little destructive potential. In contrast, others may allow some destructive diseases to develop to intolerable levels before attempting to suppress them. Errors of both types, wasted efforts and destruction by unhindered disease, occur repeatedly.

B. COST-BENEFIT RATIO

The aim of disease control is to check reduction in economic gain from a crop and if the control measures fail to increase economic gain, even if disease incidence is reduced, no grower is likely to accept the recommendations for plant disease control. In fact, one unit of cost of control should not exceed one unit of profit derived from the control.

C. DISEASE MANAGEMENT—AN INTEGRAL COMPONENT OF CROP PRODUCTION

It has been discussed in Chapter 2 that most crop production practices influence disease development. In modern agriculture we manage ecosystem to favour growth of a single plant. The resulting simplified ecosystem (agroecosystem) that is unstable, persist only because of management efforts by growers.[2] Decision such as choice of crop, crop variety, planting date, planting method, fertilization rates, pesticides, tillage type and frequency, irrigation method and frequency, harvesting method, and crop storage all influence plant diseases.[2]

According to Fry[2] two important precepts result from the integral relation between crop production and disease development. The first is that disease management will be most successful if it is considered during all stages of crop production. Effective disease management may require several approaches, at several times during a crop cycle. For example, if a grower relies solely on weekly applications of fungicide to suppress potato late blight, the resulting disease suppression may be inefficient or inadequate. If irrigation practices, microenvironment of the field, and plant susceptibility all favor disease development and if there is a large pathogen population even weekly fungicide application may not suppress disease adequately. Conversely, if these factors do not favor disease development weekly fungicide application will be unnecessary. Disease management will be most successful if it is integrated into the crop production system and if it employs diverse approaches.

The second precept is that changes in crop production will affect disease management. For example, replacement of tillage by herbicide application is likely to alter activities of several pathogens: some may become more prevalent, others less so. Disease management must adjust to these changes.

D. DISEASE MANAGEMENT AND INTEGRATION OF PRACTICES

Selection and integrated use of appropriate practices to suppress disease is a logical approach for successful management of plant diseases. For example, changes in a pathogen population that allow it to overcome plant resistance and also to overcome the toxic effects of a fungicide are less likely in combination than singly. If both these are combined with manipulation of cultural practices, which reduce initial inoculum of the pathogen and also which do not favor growth of the pathogen, disease management should be both effective and durable.

E. DISEASE MANAGEMENT AND AREA COVERAGE

For success of any control planning, its adoption over a large continuous area is necessary. Under field conditions dissemination of pathogens takes place through the agency of air, water, insects and shifting of soil, etc. There is no method for raising an effective and permanent barrier against these agencies of transmission of plant diseases. If only one grower in the area rigidly follows principles of disease control in his crop but neighboring growers neglect them and disease develops in their crop, its transmission to the neighboring field where control procedures were followed may easily take place as soon as effect of a particular treatment has disappeared. To avoid this situation the grower will have to spend more money, time, and energy in repeating the control methods more frequently. When control methods such as spraying of fungicide on standing crop are followed on cooperative basis over large adjoining areas under a crop the frequency of application of treatments is less, chances of success of treatment are better and the cost of control is reduced.

F. DISEASE MANAGEMENT AND TIMING OF TREATMENT

A special feature of plant diseases, especially of the annual field crops, is that the disease is recognized when damage to the entire plant or its organ has already taken place. Plants do not forewarn of their sickness. They have no system of tissue rejuvenation. Object of disease control in plants is only to prevent spread of the disease from a diseased plant to healthy plant. Only in perennial trees diseased organs can be cut and removed thus, saving the life of the tree. Therefore, in plant disease most of the procedures are preventive rather curative.

IV. DISEASE MANAGEMENT PLANNING AGAINST A DISEASE OR FOR A CROP

Management planning against a disease is directed against a specific disease without taking into consideration other diseases of the same crop or plant population. Management of crop health involves a well-knit plan in which all diseases of any significance are taken into consideration, although major stress may be against the most common and severe disease. Obviously, the second approach, though difficult, is of more practical value for the grower because he is interested in increasing productivity of the crop and, therefore, prefers a plan that can provide safeguard against all possible diseases occurring in the area.

A single crop is attacked by several diseases in the same season. Potato is attacked by several viruses, late and early blight, black scurf, common scab, root knot, etc. There are specific control measures for each disease. However, it is not essential that control measures developed for one disease will control other diseases also. While deciding management practices for a specific disease the nature of causal agent is the basic consideration. Therefore, in control planning for a crop the schedule of procedures is prepared on the basis of causal agents of main diseases of the crop and their methods of control. The procedures are arranged in such a manner that maximum number of diseases of the crop are controlled by a minimum number of operations and repetition of operations are avoided. Certain principles of plant disease control help in reducing incidence of several diseases. The multiple disease control methods can reduce cost of plant protection in the crop and the grower has the ease in applying them.

V. RELATIONSHIP BETWEEN FEATURES OF PLANT PATHOGENS, DISEASE CYCLE, AND DISEASE CONTROL

The knowledge of cause of the disease, mode of perennation, dissemination, cycles of inoculum production, host-parasite relationship and influence of biotic and abiotic factors is essential for development and use of effective and economic control measures. The correct knowledge of the cause of disease avoids new problems and reduces the expenses. Abnormal conditions of the plant may be due to many causes. Yellowing of leaves in rice or stunted growth and yellow leaves in wheat may not necessarily be due to viral, bacterial, or fungal infection. It may just be manifestation of nutritional deficiency. If fungicides are used without ascertaining the cause it may be only a waste of resources. The knowledge of medium of survival of the pathogen helps in selection of correct control procedure and attacking the pathogen in its most vulnerable stage. The disease cycle of loose smut of wheat indicates that the fungus is internally seed borne. Therefore, only such seed treatments which can destroy the internally present fungus will eliminate the primary inoculum. The knowledge of environmental relations of the disease strengthen the management measures by modification of environment and also helps in ascertaining the approximate time of appearance and severity of disease so that such preventive measures as spraying of the crop can be decided with more accuracy.

A study of disease cycle reveals that primary inoculum is seed and/or soil borne or is brought by wind from external sources of survival into the main crop. Prevention of primary infection or initiation of disease depends on the management of this primary inoculum. If primary inoculum is stopped from becoming effective, there should be no disease in the standing crop. These ideal situations hardly exist anywhere. Only in such diseases where there is a single source of survival and entry of the pathogen in the field, complete prevention of the disease is possible by creating any one of the above ideal situations.

VI. PRINCIPLES OF DISEASE MANAGEMENT

A. BASED ON PRACTICES

Whetzel[3] was the first to classify methods for the control of plant diseases. His list included exclusion, eradication, protection, and immunization. As a consequence of advances in plant pathology and to accomodate control measures developed, two more principles— avoidance and therapy were included.[4]

1. Avoidance of the Pathogen

Avoiding disease by planting at times when, or in areas where, inoculum is ineffective due to environmental conditions, or is rare or absent.

Many diseases can be prevented by proper selection of land or field, choice of sowing time, selection of cultivars, seed and planting stocks, and by modification of cultural practices. The aim of these measures is to enable the host to avoid contact with the pathogen or the susceptible stage of the plant and the favorable conditions should not coincide. The main principles under this group are selection of geographic area, selection of field, choice of sowing time, selection of planting materials, disease escaping varieties, modification of cultural practices, etc.

2. Exclusion of Inoculum

Preventing the inoculum from entering or establishing in the field or area where it does not exist.

Seed certification, crop inspection, growing crop in regions unfavorable for pathogen, and quarantine measures are some of the means of preventing the spread of pathogens.

3. Eradication of the Pathogen

Reducing, inactivating, eliminating, or destroying inoculum at the source, either from a region or from an individual plant in which it is already established.

The pathogen can enter an area or crop inspite of above listed precautions. It is also possible that inoculum is already present in the field or its neighborhood. These situations necessitate eradication of inoculum of the pathogen from the field or the crop. This is attempted through the methods such as biological control, crop rotation, destruction of diseased plants or plant organs, etc.

4. Protection

The inoculum of many fast spreading infectious diseases is brought by wind from neighboring fields or any other distant place of survival. Principles of avoidance, exclusion, and eradication may not be sufficient to prevent development of the disease in such cases. Protective measures are necessary to destroy or inactivate this inoculum. It is just possible by creating a chemical toxic barrier between the plant surface and the pathogen. Methods employed to achieve such results are chemical sprays, dusts, modification of environment, modification of host nutrition, etc.

5. Disease Resistance

Preventing infection or reducing the effect of infection by managing the host through improvement of resistance in it by genetic manipulation or by chemotherapy. In any crop resistance against a specific disease can be developed by selection or hybridization. This type of resistance is genetic. Biochemical resistance of nongenetic nature can be developed in plants by chemotherapy or modification of nutrition. This type of resistance is induced and temporary, lasting until the chemical or nutrient is effective in the plant.

6. Therapy

Reducing severity of disease in an infected individual.

The first five of these principles are mainly preventive (prophylactic) and constitute the major procedures of plant disease management. They are applied to the population of plants, i.e., the crop. The last, therapy, is a curative procedure and is applied to individuals. Under the concept of "disease management" these principles have been classified into following five categories.[5]

1. Management of physical environment including cultural control
2. Management of associated microbiota which includes antagonism
3. Management of host genes
4. Management with chemicals
5. Management with therapy, radiation, and meristem culture

B. BASED ON EPIDEMIOLOGY

Epidemiology is the science of epidemics, which are widespread outbreaks of disease. As Van der Plank[6,7] pointed out, most control measures reduce either the initial inoculum from which an epidemic starts or reduce the rate at which infection builds up during the epidemic. One may, of course, also do both.[7] Crop sanitation, the growing of vertically resistant crop varieties, the sowing of disease-free planting materials, the destruction of pathogens on or in the planting material by chemical or heat treatment, and soil fumigation all reduce the initial inoculum. Measures which reduce the rate at which pathogens spread include the growing of horizontally resistant crop varieties and the application of protective fungicides; these measures reduce the number of diseased plants and the severity of infection so that the amount of inoculum at the end of the season is also reduced. Monocyclic diseases are most efficiently suppressed by reducing the amount or efficacy of initial inoculum while polycyclic diseases are most efficiently suppressed by reducing large amounts of initial inoculum and/or by limiting potentially rapid rates of disease increase.

The six principles that characterize the modern concept of plant-disease management should be viewed from three stand points: (1) whether they involve reduction in the initial

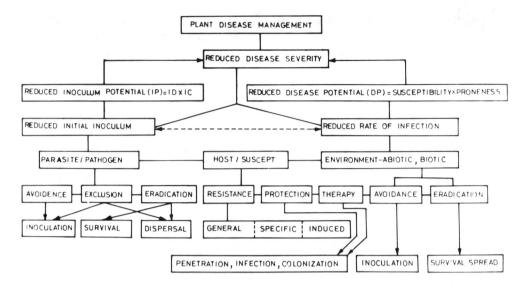

FIGURE 1.   The relationship between principles, practices, components of disease pyramid, and elements of pathogenesis.

inoculum or the rate of disease development, (2) whether they primarily involved control of the population of the pathogen, the cure or defense of the suscept, or involve the environment as it relates to disease, and (3) whether they involve interruption of dispersal, survival, inoculation, penetration, infection, or the actual course of disease.[8] These interactions originally proposed by Baker[9] and Roberts and Boothroyd[8] and subsequently modified by us are illustrated in Figure 1.

VII. INTEGRATED DISEASE MANAGEMENT

On the preceding pages major principles and practices of disease management have been discussed briefly. The vastness of methods listed suggest that we possess a big arsenal of weapons to win the war against the pathogens. The victory in any war, however, does not depend only on the quantity of weapons available. Success comes from a well-planned strategy involving the best use of weapons in a coordinated manner at the right time and with proper methods. As we know that a disease is the outcome of interaction between host, pathogen, and environment, and, therefore, in order to achieve a meaningful dominance over the pathogen and a substantial degree of control, all the three components of disease triangle are to be managed. The term "integrated pest management" (IPM) was originally designed for management of insect pests but it equally applies to plant diseases.

Since, the extent and intensity of disease development in plant population is the function of interaction between three forces, the host or suscept, the parasite or pathogen, the environment, and also the time as these interactions and their effects are not spontaneous and time of interaction influences the result of interaction, the manipulation of disease triangle will be an ideal approach for management of plant disease. In order to achieve this goal, there will have to be integration of methods directed against the causal agent, in favor of the host and for modification of the environment. This is what is now known as IPM (Figure 2). Management of pathogen involves the practices directed to exclude the inoculum, reduce inoculum, and eradicate inoculum. Management of the host involves the practices directed to improve plant vigor and induce resistance through nutrition, introduction of genetic resistance through breeding, and providing protection against attack by chemical means. Management of environment involves water management, soil management, and crop management.

75

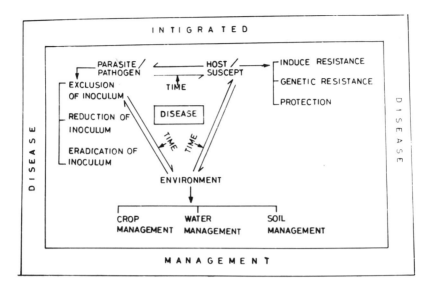

FIGURE 2. Integrated disease or pest management

REFERENCES

1. **Apple, J. L.,** The theory of disease management, in *Plant Pathology: An Advanced Treatise,* Vol. I, Horsfall, J. G. and Cowling, E. B., Eds., Academic Press, New York, 1977, 79.
2. **Fry, W. E.,** *Principles and Practices of Plant Disease Management,* Academic Press, New York, 1987, 378.
3. **Whetzel, H. H.,** The terminology of phytopathology, *Int. Congr. Plant Sci. Proc.,* 2, 1204, 1929.
4. National Academy of Sciences, *Plant Disease Development and Control,* National Academy of Sciences, Washington, D.C., 1968.
5. **Horsfall, J. G. and Cowling, E. B.,** *Plant Disease: An Advanced Treatise,* Vol. I, Academic Press, New York, 1977, 405.
6. **Van der Plank, J. E.,** *Plant Disease Epidemics and Control,* Academic Press, New York, 1963, 349.
7. **Van der Plank, J. E.,** *Disease Resistance in Plants,* Academic Press, New York, 1968, 206.
8. **Roberts, D. A. and Boothroyd, C. W.,** *Fundamentals of Plant Pathology,* W. H. Freeman & Co., Toppan Company Ltd., Tokyo, 1972, 402.
9. **Baker, R.,** Mechanism of biological control of soil-borne plant pathogens, *Annu. Rev. Phytopathol.,* 6, 263, 1968.

Chapter 6

DIAGNOSIS OF PLANT DISEASES

I. DIAGNOSIS—ITS CONCEPT AND SCOPE

Diagnosis is derived from a Greek word "diagignoskein" which means to distinguish (from "dia" through, and "gignoskein", to know). The conceptual validity of diagnosis is often a topic of discussion amongst the scientists. It is argued, whether diagnosis is an art or science or both. According to McIntyre and Sands,[1] diagnosis is an art. They have very rightly argued that diagnosis is done by precept and experience. Even today whenever a diseased sample is brought to the diagnostician, he examines the symptoms and makes an intuitive judgement as to its nature. Thus, visual observations based on experience, precept, and intuitive judgement is still the most widely used method employed for identification of diseases.

II. KOCH'S POSTULATES

In the 19th century, when criteria for determining causality of disease were hotly debated, a set of useful rules was developed in 1882 by Robert Koch and further modified by E. F. Smith[2] to demonstrate pathogenicity of microorganisms. These rules became known as Koch's postulates, and they are widely applied in plant pathology. These rules are (1) the microorganism must be constantly associated with the disease, i.e., a macroscopic as well as microscopic observations of host symptoms and if present, signs of the pathogen, (2) the microorganism must be isolated from diseased host and grown in pure culture, i.e., isolation and purification of the pathogen, (3) the specific disease must be reproduced when the microorganism from the pure culture is inoculated into the host, and (4) the same microorganism must be recovered from the inoculated diseased host.

These rules have proved very useful to identify most of the diseases caused by biotic agents. However, the modification of rules have been necessary for mesobiotic agents and/ or for pathogens that do not reproduce independently of the living cells. For example, the Koch's postulates for viruses[3] are (1) the virus must be concomitant with the disease, (2) it must be isolated from the diseased plant (separated from contaminants, multiplied in a propagated host, purified physicochemically, and identified for its intrinsic properties), (3) when inoculated into a healthy host plant, it must reproduce the disease, and (4) the same virus must be demonstrated to occur in and it must be isolated from the experimental host.

III. OBSERVATIONS AND RECORDS

Diagnosis of disorders and diseases in plants still is largely based on macroscopic and/ or microscopic observations. The symptoms that have developed on the plant or plant organs, are helpful in identification of diseases. Since, majority of the symptoms produced on the plants by abiotic, mesobiotic, and biotic agents are similar if not identical, for scientific interpretation of the symptom, cropping history, cultural practices, location of symptom on plants, distribution of diseased plants in population, etc., are essential. A form (diagnosis check list) for collection of such data is proposed in Table 1. The diagnostician must acknowledge the receipt of the sample stating its conditions, suitability for diagnosis, etc. A model acknowledgement form and diagnosis work sheet is proposed in Tables 2 and 3.

TABLE 1
Plant Disease Diagnosis Service
Diagnosis Check-List

1. Name and address of the person/agency who brought the sample for diagnosis

   _____

   _____

2. Plants

   A. Source of plants _____

   B. Crop and variety _____

   C. Date of sowing/transplanting _____

   D. Plant age and stage of growth _____

3. Fertilization

   A. Names of fertilizers used _____

   B. Doses of fertilizers used _____

   C. Methods of application — Basal/Broadcast/Spray

   D. Has soil test been done? —   Yes / No

   E. If yes   (a)   When _____

              (b)   Where _____

              (c)   Results _____

4. Soil

   A. Soil texurial class _____

   B. Soil reaction — Alkaline/Acidic

   C. Previous history of nutrient toxicity/deficiency _____

5. Irrigation

   A. Source _____

   B. Methods _____

   C. Frequency _____

   D. Drainage — Adequate/Inadequate

6. Problems in your view

   (a) Seed rot   (b) root rot   (c) Stem rot

   (d) soft rot   (e) chlorosis   (f) necrosis

   (g) wilt   (h) blight   (i) spots

   (j) any other _____

TABLE 1 (continued)
Plant Disease Diagnosis Service
Diagnosis Check-List

    B. Degree of symptom expression (✓)

        (a) whole plant   (b) only root

        (c) only foliage   (d) any other _____

    C. Symptoms, general over crop or isolated to a

        few plants _____

    D. Percentage of plants showing symptom and/or loss _____

  7. Pesticide application

    A. Name(s)

    B. Doses used

    C. Date of application

    D. Purpose

  8. Have plants been diagnosed elsewhere

    A. Yes / No   (B) If yes, where _____

    C. Results _____

  9. Observations on environmental factors

    A. Temperature _____ B. Rains _____ C. Any thing abnormal _____

  10. Any other information _____

_____

_____
Name and Signature

IV. PROCEDURES FOR DIAGNOSIS OF DISEASES CAUSED BY INFECTIOUS AGENTS

A. GENERAL PROCEDURES
1. Fungi

Diseases caused by fungi are identified primarily by symptoms produced on host and the morphological features of the causal fungus. Important symptoms (Figures 1 and 2) produced by fungal pathogens are summarized in Table 4.

Besides symptoms, the morphological features such as spores, fructifications, or sporophores, which differ from one fungus species to another, form important part of diagnosis program. If desired structures of the fungus are not readily visible on the infected host organs, the parasite may be induced to sporulate by proper incubation of the infected tissue.[4]

TABLE 2
Plant Disease Diagnosis Service
Acknowledgement Form

I am in receipt of _____

submitted to the laboratory for diagnosis and advice.

1. Condition of the sample — Excellent

   Fair

   Poor

2. Diagnosis check-list attached — Yes / No

3. Diagnosis list properly filled — Yes / No

4. Is it necessary to do clinical
   and etiological study?   — Yes / No

5. If no, results of diagnosis _____

6. If yes, approximate time interval required _____

7. Please submit another specimen

   and include   —   Roots/stems/leaves/soil

8. Any other comment _____

                                         _____
                                              (Name and Signature)

Incubation of tissue for 24 to 72 h in a moist chamber (environment of 100% RH) at moderate temperatures (20 to 25°C) induces sporulation. Growth of extraneous fungi can be restricted if the tissue is surface sterilized by soaking in the aqueous sodium hypochlorite (0.5%) for 30 to 120 s. The diagnosticians must have proper understanding of the physiology of sporulation.[1] Modification of physical factors may be more conducive for sporulation. Certain regulatory compounds such as cyclic $3',5'$adenosine monophosphate which induces both sporulation in slime molds[5] and the production of *Coprinus* fruiting bodies,[6] may promote sporulation of other fungi and allow for their identification or isolation.

It is necessary to isolate the fungus pathogen especially if it is not of a routine type, in pure culture in appropriate natural, seminatural, or synthetic medium such as potato dextrose agar (PDA) or Czapek's agar or V-8 juice agar.[4,7]

2. Bacteria

The simplest technique which one uses to ascertain bacteria as cause of a disease is the examination of ooze. The ooze can directly be seen as droplets on the infected parts if viewed in the early morning. Another method is to cut the infected part and dip the cut part in a glass of clear water. The water turns murky due to introduction of ooze along with bacterial cells into the water.

Plant pathogenic bacteria cause the development of almost as many kinds of symptoms

TABLE 3
Plant Disease Diagnosis Service
Diagnosis Work-Sheet

1. Specimen No. _____

2. Observations

    A. Symptom(s) observed _____

    B. Has pathogen been observed — Yes / No

    C. If yes, name of the pathogen and
       structure(s) observed _____

    D. If no, possibility or possible reasons

       _____

3. Has pathogen been isolated and purified — Yes / No

    A. If yes, result _____

    B. If no, possibility _____

4. If above tests negative —

    A. Is it necessary to observe the problem directly
       under field conditions — Yes / No

    B. If yes, result of the field observations

       _____

5. Final diagnosis result — Disease _____

                           Caused by _____

6. Recommendations _____

7. Any other comment/suggestion _____

8. References used _____

                    _____
                              (Name and Signature)

on the plants they infect as do fungi. Like fungi, the morphological features such as cell shape, flagellation, Gram reaction etc., are essentially employed for diagnosis of diseases caused by this group of organisms. In the Table 5, of the Chapter 1, the characteristics of bacterial genera, that are used for diagnosis are summarized.

3. Nematodes

Nematodes infect both root and above ground portion of plants.[8] Root symptoms (Figure 3) may appear as root knots or root galls, root lesions, excessive root branching, injured

FIGURE 1.   Important symptoms produced by fungal pathogens (A & B) leaf spots, (C) leaf blight, (D) canker, (E) white-rust, (F) downy mildew, (G) powdery mildew, and (H & I) rots.

root tips, and root rots when nematode infections are accompanied by parasitic or saprophytic fungi and bacteria. These root symptoms are usually accompanied by noncharacteristic symptoms in the above ground parts of plant appearing primarily as reduced growth, symptoms of nutrient deficiencies such as yellowing of foliage, excessive wilting in hot or dry weather, reduced yield, and poor quality of products.[9] Certain species of nematodes infect the above ground portions of plants. They cause galls, necrotic lesions and rots, twisting

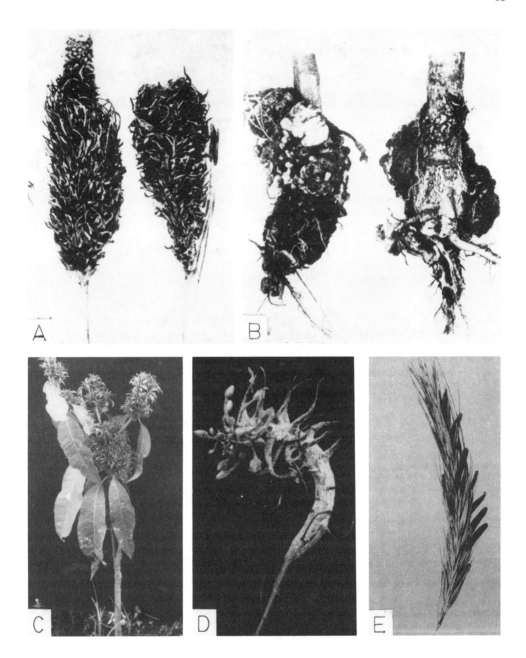

FIGURE 2. Important symptoms produced by fungal pathogens (A) green-ear of pearl millet, (B) club root of crucifers, (C) mango malformation, (D) downy mildew of crucifers, (E) ergot (sclerotial stage) disease.

and distortion of leaves and stem, and abnormal development of floral parts.[9] Certain nematodes attack grains of grasses forming galls full of nematodes in place of seed.

Diagnosis of diseases induced by sedentary nematodes is generally easier than that for migratory nematodes. Sedentary nematodes usually induce galls or enlarge sufficiently to be macroscopically visible on roots. For instance, *Meloidogyne* spp. induce variously shaped galls (Figure 3), and females of *Heterodera* and *Globodera* spp. (cyst nematodes) become so large that they are easily visible (Figure 3) on roots. One can visualize vermiform sedentary nematodes by microscopic examination of roots or in the rhizosphere.[10] Nematodes extracted

TABLE 4
Symptoms of Diseases Caused by Fungal Parasites

Symptom	Fungus	Disease
Pathogen seen as white, gray, brownish, or purple growth on host surface; the superficial growth tangled cotton or downy growth	Downy mildew fungi; (Members of family Peronosporaceae)	Downy mildew
Enormous numbers of spores formed on superficial growth giving host surface a dusty or powdery appearance; black fruiting bodies (cleistothecia) may also develop	Powdery mildew fungi; (Members of order Erysiphales)	Powdery mildew
Pustules of spores, usually breaking through host epidermis, dusty or compact, red, brown, yellow, or black in color	Rust fungi (order-Uredinales)	Rust diseases
Black or purplish black dusty mass formed on floral organs particularly the ovulary	Smut fungi (order Ustilaginales)	Smut diseases
White blister-like pustules breaking open the epidermis and expose powdery mass of spores	*Albugo*	White blister or rust
Excessive growth of host tissues; abnormal increase in size due to abnormally increased cell size (hypertrophy) or increased cell divisions (hyperplasia)	*Albugo,* Downy mildew fungi, Root knot nematodes, MLO	Gall, curl, pocket, bladder, hairy root, knot, witche's broom, clubbed roots, tumefaction, wart intume'scence
Reduced growth of host tissues, abnormally reduced size (atrophy)	Several fungi	Stunting, dwarfing, curling, puckering
Localized lesions on host leaves consisting of dead and collapsed cells	Several fungi	Leaf spots
Uniform, general and very rapid browning of foliage (leaves, branches, twigs, floral organs)	Several fungi	Blights
Necrotic and sunken ulcer like lesions on stem, leaf, flower, or fruits	*Colletotrichum* spp. *Glomerella* spp.	Anthracnose
Necrosis localized, usually surrounded by callus	Several fungi	Canker
Disintegration or decay of part or all the root system	Many fungi	Root rot
Loss of turgidity, flaccid, dropping of leaves, or shoot due to disturbance in the vascular system of root or the stem	*Fusarium oxysporum* group. *Verticillium* spp.	Wilt

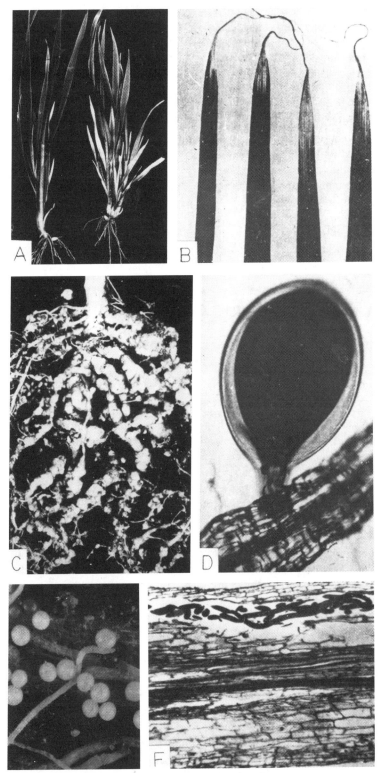

FIGURE 3.   Diagnosis of diseases caused by nematodes (A) influence of *Ditylenchus dipsaci* on oats (left hand plant uninfested, right hand plant infested showing swelling of the shoot bases and increased tillering), (B) symptoms of white tip disease on leaves, (C) root knot disease, (D & E) cysts of *Heterodera* attached to roots (F) longitudinal section of cortical cavity caused by *P. coffee* in rough lemon root (Dropkin, V. H., *Introduction to Plant Nematology,* John Wiley and Sons, New York, 1980, 293; and Wallace, H. R., *Nematode Ecology and Plant Disease,* Edward Arnold, London, 1973, 228. With permission.

from plant tissue or soil can be identified using various pictures, descriptions, and keys as aids.[11] Some knowledge of host range may also be useful for nematode identification.[8]

4. Viruses and Viroids

Methods for identification of diseases caused by viruses have been described and reviewed.[3,12-15] Methods useful in identification include symptomatology, mode of transmission, host-range, particle morphology, antigenicity and electrophoretic mobility in gels. Because some of these techniques require special methodology and equipment, symptoms and distribution patterns frequently suffice for diagnosis.[16] Characteristics of viruses and description of symptomatology are available in several sources.[17,18]

Frequently the combination of symptoms and knowledge of host-range are enough for diagnosis. For example, four sap transmissible viruses of cucurbits (cucumber mosaic virus, watermelon mosaic virus one, watermelon mosaic virus two, and squash mosaic virus) are distinguished from each other on the basis of the reaction of five indicator plants to each of them.[19] For identification, host range of unknowns are compared with those described for known viruses in literature[17,18,20] and other publications. When symptoms and related observations do not provide desired diagnosis, the samples must be referred to virology laboratories for further analysis.

V. DIAGNOSIS OF DISORDERS CAUSED BY ABIOTIC CAUSES

Disorders induced by abiotic causes are so frequently encountered in the field that a practitioner must be knowledgeable and congnizant of them.

A. PHYSICAL AGENTS
1. Temperature

Both low and high temperatures impair plant growth; sometimes the symptoms and knowledge of previous year's weather enable us to diagnose the disorder easily and readily. Low temperatures injure plants primarily by inducing ice formation between and/or within the cell. Parts of plants loose hardiness. Winter injury provides avenues for wound pathogens to initiate infection. Temperatures below freezing cause a variety of injuries to plants.[9] Such injuries include killing of buds, flowers, young fruits, and succulent twigs. Frost bands consisting of discolored, corkey tissues in a band or large area of fruit surface, are often produced on apples, pears, etc., following late frost. Low temperatures may kill young roots of trees such as apple and may cause bark splitting and canker development on trunks and large branches; other symptoms include tip necrosis, leaf margin necrosis, etc.

Injuries due to high temperatures are frequently accompanied by moisture stress. Sunscald of plant tissues occurs on many plants exposed to intense solar radiation at high temperatures.

2. Moisture

Soil saturated with water for considerable period becomes anaerobic and affect plant growth and development in several ways. Roots can die of anoxia. They become more sensitive to soil toxicants such as nitrites which accumulate due to anaerobic microbial activity.[21] Activities of several aerobic biotic pathogens such as *Phytophthora* and *Pythium* are enhanced in wet soils before oxygen depletion.

Diseases due to insufficient water are common, where water is chronically insufficient, plants grow poorly, leaves may be small, abnormally pigmented, and marginally necrotic. Water deficiency retards net photosynthesis.

TABLE 5
Nutrient Deficiencies in Plants

Nutrients	Symptoms
Boron	The bases of young leaves of terminal buds become light green and finally breakdown. Stems and leaves become distorted. Plants are stunted. Fruits, fleshy roots or stems, etc., may crack on the surface and/or rot in the center, e.g., heart rot of sugarbeets, etc.
Calcium	Young leaves become distorted, with their tips hooked back and the margins curled, leaves may be irregular in shape and ragged with brown scorching or spotting, terminal buds may die and plants have poor root system, e.g., blossom end rot of many fruits
Copper	Tips of young leaves of cereals wither and their margins become chlorotic, leaves may fail to enroll and appear wilted, heading reduced and the heads are dwarfed and distorted. Citrus, pome, and stone fruits show dieback of twigs in summer, burning of leaf margins, chlorosis, rosetting, etc.,
Iron	Young leaves become severely chlorotic, but main veins remain green, sometimes brown spots develop, part or entire leaf may dry, and leaves may be shed.
Magnesium	First the older, then the younger leaves become mottled or chlorotic, then reddish, sometimes necrotic spots appear. The tips and margins of leaves may turn upward and the leaves may appear cupped, leaves may drop off.
Manganese	Leaves become chlorotic but their smallest veins remain green and produce a checked effect, necrotic spots may appear scattered on the leaf, severely affected leaves turn brown and wither.
Nitrogen	Plants grow poorly and are light green in color. The lower leaves turn yellow or light brown and the stems are short and slender.
Phosphorus	Plants grow poorly and the leaves are bluish-green with purple tints. Lower leaves sometimes turn light bronze with purple or brown spots, shoots are short, thin, upright and spindly.
Potassium	Plants have thin shoots which in severe cases show dieback, older leaves show chlorosis with browning of tips, scorching of the margins, and many brown spots usually near the margins, fleshy tissues show end necrosis
Sulfur	Young leaves are pale green or light yellow without any spots. The symptoms resemble those of nitrogen deficiency
Zinc	Leaves show interveinal chlorosis. Later they become necrotic and show purple or redish pigmentation. Leaves are few and small, internodes are short and shoots form rosettes, and fruit production is low. Leaves are shed progressively from base to tip.

3. Light

Variation in the amount or duration of electromagnetic radiation can affect normal host physiology.[21] Lack of sufficient light retards chlorophyll formation and promotes slender growth with long internodes, thus leading to pale green leaves, spindly growth, and premature drop of leaves and flowers. This condition is called *etiolation*.

B. CHEMICAL AGENTS
1. Nutrient Deficiency in Plants

Plants require several elements for growth. When they are present in the plant in amounts smaller than the minimum levels, the plants exhibit various external and internal disorders. The kinds of symptoms produced by deficiency of a particular nutrient depend mainly on the functions of that element in plants. Several authors[22-25] have documented the subject. Major disorders/symptoms due to deficiency of elements and which could be a valid source for diagnosis are given in Table 5.

2. Toxicity of Minerals

Soils often contain excessive amounts of elements, which at higher concentration may be injurious to plants, either directly (injury) or indirectly (interference with absorption or function of other elements) or both. For instance, excessive nitrogen induces deficiency of calcium in the plant, while the toxicity of Cu, Mn, or Zn is both, direct on the plant and by inducing a deficiency of iron in the plant.

FIGURE 4.   Symptoms of herbicide injury.

Excessive amounts of sodium salts, especially sodium chloride, sodium sulfate and sodium carbonate, increase soil pH and affect plant growth directly or indirectly. The injuries vary in different plants and may range from chlorosis to stunting, leaf burning, wilting to outright death of seedlings and young plants. Some plants, e.g., wheat and apples, are very sensitive to alkali injury while sugarbeet, alfalfa, and several grasses are quite tolerant. On the other hand, when soil is too acidic, the growth of some kinds of plants is impaired and various symptoms may appear. pH influences plant growth indirectly by affecting nutrients availability. For example, blue berries grow well in acidic soil (pH 4 to 5.2) primarily because of the effect of pH on soil microflora and on the form of nitrogen.[21] Blue berries use nitrogen as NH_4^+ ions more efficiently than NO_3^- ions. At low pH, nitrifying bacteria (*Nitrosomonas* and *Nitrobacter*), which convert NH_4^+ to NO_3^- are less active, thus nitrogen remains as NH_4^+.

Hydrogen ions also affect the ionic form and solubility of several elements and thus influence the availability of the elements. At low pH Zn, Mn, Fe exist in relatively soluble forms, but at higher pH they exist in less soluble forms. Consequently, at lower pH excessive amounts of these elements may become available to plants. In contrast, at higher pH they may be so insoluble that they are unavailable. Toxicity symptoms consist of root discoloration, distortion, chlorosis, and browning of older leaves.[21,24]

3. Herbicide Injury

Pesticides, especially herbicides cause injury to plants when they are used incorrectly, i.e., on the sensitive plants, at wrong time, under improper environmental condition, or at wrong dosage. Pesticide injury is often associated with cultural practices as spraying, planting, or cultivation. However, some pesticides may be transported by air or water. For example, 2,4-D (2,4-dichlorophenoxy acetic acid), a common broad leaf herbicide, is volatile and may drift to some distance. Sensitive plants such as grapes, sunflower, etc., have been injured when growing far away from the site of application. Among the common symptoms (Figure 4) observed are smaller leaves, narrowed interveinal areas, rolled leaves, distorted petioles, yellowing of veins, leaf distortions, hypertrophied inflorescence, etc. Some symptoms of herbicide injuries on various plants have been described by Streets,[25] and Lockerman et al.[26] and should be referred and consulted if herbicide injury is expected to be involved.

Use of preplant or preemergence herbicides through applications to the soil before or

TABLE 6
Air Pollution Injury to Plants

Pollutant	Source	Susceptible plants	Symptoms
Chlorine and hydrogen chloride	Refineries, glass factories, incineration of plastic	Many plants, usually near the source; toxic at 0.1 ppm	Leaves bleached, show interveinal necrosis, margins scorched, dropping prematurely
Ethylene	Automobile exhausts, burning of gas, fuel oil and coal	Many plants; toxic at 0.05 ppm	Stunted plants, abnormal leaves, premature senescence and reduced blossoms and fruits
Hydrogen fluoride	Factories processing ore or oil	Many plants including corn, peach, tulip; toxic at 0.1 to 0.2 ppm	Leaf margins (dicot) and leaf tips (monocot) turn tan to dark, die, and fall
Nitrogen dioxide	From nitrogen and oxygen in the air by hot combustion sources like furnaces, internal cumbustion engines	Many plants including beans, tomatoes; toxic at 2 to 3 ppm	Growth reduction, bleaching and bronzing of leaves
Particulate matter (dusts)	Dusts from roads, cement factories, burning of coal, etc.,	All plants	Leaves crusty, chlorotic and grow poorly, sometimes necrosis and death
Peroxyacyl nitrates (PAN)	Automobile exhausts or internal combustion engines	Many plants including spinach, petunia, tomato, lettuce, dahlia	Causes "silver leaf", i.e., bleached white to bronze spots on lower surface of leaves
Sulfur dioxide	Factories, automobile exhausts, coal burning and internal combustion engines	Many plants including alfalfa, violet, conifers, pea, cotton, bean; toxic at 0.3 to 0.5 ppm	Leaf chlorosis, bleaching of interveinal tissues of leaf

at planting time often affect seed germination and growth of seedlings if too much or the wrong herbicide has been used. Some herbicides persist in soil for pretty long and sensitive plants grown in such fields may grow poorly and produce various symptoms.

4. Air Pollutants

Air pollution damage can be ascertained by symptoms[9,27-29] and by proximity of diseased plants to an observed source of an air pollutant. Pollutants, their source, susceptible plants and symptoms[9] are summarized in Table 6.

VI. SPECIFIC DIAGNOSIS PROCEDURES

A. SELECTIVE MEDIA

A selective medium selects the desired organism from a group by eliminating or suppressing the other members of the group. The subject has been extensively reviewed by Tsao.[30,31]

Singh and Mitchell[33] were the first to develop a medium for selective isolation of *Pythium* sp. It was based on peptone-dextrose-rose bengal-agar supplemented with pimaricin, PCNB, and agrimycin. Since then, several semiselective or selective media for *Pythium*,[30,32] *Phytophthora*,[30,32] *Fusarium*,[30,32,34] *Phoma*,[35] *Macrophomina phaseolina*,[36] *Rhizoctonia solani*,[37] *Sclerotium rolfsii*,[38] *S. cepivorum*,[39] *Neovossia indica*,[40] etc., have been described.

Similarly, media are available for the selective isolation of bacterial pathogens. The three principal selective media for isolating *Agrobacterium* are those by Schroth et al.,[41] Clark,[42] and New and Kerr.[43] Some work has been done in the isolation of *Corynebacterium fascians* from plant tissues. Mohanty[44] developed a medium by incorporating potassium dichromate to suppress various saprophytes. Crosse and Pitcher[45] used a medium consisting of 5% sucrose, 0.02% potassium tellurate, and 0.0033% potassium dichromate. Several selective media have been developed for isolating species of *Xanthomonas*.[46,47] Isolation of *Pseudomonas* spp. by using selective media especially for the group 2 pseudomonads is possible. The media developed by Brown and Lowbury,[48] Sands and Rovira,[49] and Smith and Dayton[50] could be employed successfully. A variety of selective media, many using pectin or sodium polypectate have been used for detecting soft rot bacterium, *Erwinia*.[51] The medium of Cuppels and Kelman[52] and the Miller-Schroth medium[53] amended with 167 mg l^{-1} of cobaltous chloride have been used with success.

B. SEROLOGICAL METHODS

Serology is often the best method for virus group identification and helps rapid identification of unknown viruses.[16] Antibodies can be obtained that react specifically to the pathogen[54] or to the diseased plant that may be used for plant disease diagnosis. Exposing tissue sections to pathogen specific antibodies which are covalently labeled with fluorescent compounds,[55] radioactive compounds,[56] or X-ray opaque compounds could permit the location and identification of the organism.[1] The location of the antibodies, and hence the organisms, is indicated by the label. For instance, the organism causing Pierce's disease of grapes has been located in leaf tissues and insect excreta with a fluorescent-labeled antibody.[55] Fluorescent-labeled antibodies have also been used to observe fungi[57] or bacteria[58] in the soil, and to detect viruses in mammalian tissues.[56] Monoclonal antibodies (i.e., antibodies specific to single antigenic determinant site or epitope) coupled with enzyme-linked immunosorbent assay (ELISA) have been proved to be very powerful tool for the identification and strain differentiation in plant pathogenic viruses, bacteria, MLOs, fungi and nematodes.[66]

C. MICROSCOPY

Use of transmission or scanning electron microscopy (TEM/SEM) in diagnosis of plant diseases is immense. SEM micromanipulation is a means of studying ultrastructural relationship between the host and parasites,[59] and may be useful for disease diagnosis. Virus particle morphology is useful because viruses are arranged into groups based in part on morphology. An electron microscope (TEM) is essential to visualize the particles. Nematode taxa have been separated by observing their "facial" features with SEM.[60]

D. TISSUE CULTURE TECHNIQUE

Cultures of callus, single cells, and the plantlets regenerated from them are being used extensively in plant pathology. Most commonly they are used to study the behavior of pathogens, particularly biotrophs, facultative parasitic, and/or saprophytic fungi and bacteria. This technique, though still with limited scope, could be employed for isolation, maintenance, and identification of infectious pathogens, particularly by viruses and biotrophs.

E. CHROMATOGRAPHY

Chromatographic techniques, especially gas chromatography (GC) could be successfully employed for identification of plant diseases.[1] It could be used for analyzing culture filterate, cellular constituents, extracts from diseased tissues, and volatiles like ethylene, acetone, etc., which indirectly indicate about the presence of organisms and the disease in a plant.[61,62]

F. ENZYMES

Pathogens produce enzymes either in the normal course of their activities or upon growth on certain substrates. There are several culture media which allow the detection of extracellular hydrolytic enzyme activity and other biochemical characteristics of pure cultures of fungi and bacteria.[63-65] Arginine dihydrolase,[65] oxidase,[64] or the poly-β-hydroxybutyrate produced by some *Pseudomonas* spp. and their detection in mixed bacterial cells or diseased tissues could be used for disease diagnosis.[1]

G. OTHER METHODS

Several biochemical diagnostic techniques and techniques based on genetic engineering are being employed or likely to be employed very soon for identification of phytopathogens. These techniques, which have made a mark in the field of medical science, include lectins, aminopeptidase profiles, electrophoresis, group specific polysaccharides, dot-blot hybridization, dual culture, etc. During the last 5 years blot hybridization which uses labeled complimentary DNA (cDNA) to detect the viral or viroid RNAs, employing the principle of hybridization of complimentary single strand DNA and RNA to form double stranded DNA-RNA complex, has become very popular for rapid identification of viroids and viruses even in latent infections where symptoms are not apparent.[66,67]

REFERENCES

1. **McIntyre, J. L. and Sands, D. C.,** How disease is diagnosed, in *Plant Disease: An Advanced Treatise,* Vol. 1, Horsfall, J. G. and Cowling, E. B., Eds., Academic Press, New York, 1977, 35.
2. **Smith, E. F.,** Bacteria in relation to plant diseases, *Carnegie Inst. Washington Publ.,* 27, 1, 1905.
3. **Boss, L.,** *Introduction to Plant Virology,* Longman, New York, 1983, 160.
4. **Tuite, J. F.,** *Plant Pathological Methods, Fungi and Bacteria,* Burgess, Minneapolis, Minnesota, 1969, 128.
5. **Konijn, T. M., Van de Meene, J. G. C., Bonner, M. T., and Barckley, D. S.,** The acrasin activity of adenosine 3'-5'-cyclic phosphate, *Proc. Natl. Acad. Sci. U.S.A.,* 58, 1152, 1967.
6. **Uno, I. and Ishikawa, T.,** Purification and identification of the fruiting inducing substances in *Coprinus macrorhizus, J. Bacteriol.,* 113, 1240, 1973.
7. **Sinclair, J. B. and Dhingra, O.,** *Basic Plant Pathology Methods,* CRC Press, Boca Raton, FL, 1985, 355.
8. **Christie, J. R.,** *Plant Nematodes: Their Bionomics and Control,* Agricultural Experiment Station, Gainesville, Florida, 1959.
9. **Agrios, G. N.,** *Plant Pathology,* Academic Press, New York, 1978, 703.
10. **Bergeson, C.,** Identifying methods and diagnosing the diseases which they cause, in *Plant Pathological Methods,* Tuite, J. F., Ed., Burges, Minneapolis, Minnesota, 1969.
11. **Mai, W. F. and Lyou, H. H.,** *Pictorial Key to Genera of Plant Parasitic Nematodes,* 4th Edition, Cornell Univ. Press, Ithaca, New York, 1975.
12. **Bawden, F. C.,** *Plant Viruses and Virus Diseases,* Ronald Press, New York, 1964, 361.
13. **Corbett, M. K. and Sisler, H. D.,** *Plant Virology,* University of Florida Press, Gainesville, 1964, 527.
14. **Matthews, R. W. F.,** *Plant Virology,* Academic Press, New York, 1970, 778.
15. **Ball, E. M.,** *Serological Tests for the Identification of Plant Viruses,* Am. Phytopathol. Soc., St. Paul, Minnesota, 1974, 31.
16. **Ross, A. F.,** Identification of Plant Viruses, in *Plant Virology,* Corbett, M. K. and Sisler, H. D., Eds., Univ. of Florida Press, Gainesville, 1964, 68.
17. Commonwealth Mycological Institute, *Descriptions of Plant Viruses,* CMI, Kew, Surrey, England, 1970.
18. **Smith, K. M.,** *A Text Book of Plant Virus Diseases,* 3rd eds., Longmans Group, New York, 1972, 652.
19. **Nelson, M. R. and Tuite, D. M.,** Epidemiology of cucumber mosaic and watermelon mosaic 2 of cantelopes in an arid climate, *Phytopathology,* 59, 849, 1969.
20. **Smith, K. M.,** *Plant Viruses,* 6th ed., Champman and Hall, London, 1977, 309.
21. **Fry, W. E.,** *Principles of Plant Disease Management,* Academic Press, New York, 1987, 378.
22. **Wallace, T.,** *The Diagnosis of Mineral Deficiencies in Plants by Visual Symptoms,* Chem. Publ. Co., New York, 1961, 125.
23. **Sprague, H. B.,** *Hunger Signs in Plants,* 3rd ed., David McKay Co., New York, 1964.
24. **Treshow, M.,** *Environment and Plant Response,* McGraw Hill, New York, 1970.
25. **Streets, R. B., Sr.,** *The Diagnosis of Plant Diseases,* CES, AES, University of Arizona, Tucson, 1972.

26. **Lockerman, R. H., Putnam, A. R., Rice, R. P., Jr., and Meggitt, W. F.,** *Diagnosis and Prevention of Herbicide Injury,* CES Bull, E 809, Michigan State University, East Lansing, 1975.
27. **Brandt, C. S.,** Effect of air polution on plants, in *Air Pollution,* Vol. I, Stern, A. C., Ed., Academic Press, New York, 1962, 255.
28. **Jacobson, J. S. and Hill, A. C.,** *Recognition of Air Pollution Injury to Vegetation: A Pictorial Atlas,* Air Pollution Control Assoc., Pittsburgh, Pennsylvania, 1970.
29. **Compton, O. C.,** Plant tissue monitoring for fluorides, *Hort. Science,* 5, 244, 1970.
30. **Tsao, P. H.,** Selective media for isolation of pathogenic fungi, *Annu. Rev. Phytopathol.,* 8, 157, 1970.
31. **Tsao, P. H.,** Factors affecting isolation and quantitation of *Phytophthora* from soil, in *Phytophthora: Its Biology, Ecology, Taxonomy and Pathology,* Erwin, D. C., Bartnicki-Garcia, S. and Tsao, P. H., Eds., *Am. Phytopathol. Soc.,* St. Paul, Minnesota, 1983, 392.
32. **Chaube, H. S., Singh, R. S., and Singh, U. S.,** Techniques for isolation of some soil-borne fungal pathogens, in *Perspective of Mycopathology,* Agnihotri, V. P., Ed., Mehrotra Publishing House, New Delhi, 1988, in press.
33. **Singh, R. S. and Mitchell, J. E.,** A selective medium for isolation and measuring the population of *Pythium* in soil, *Phytopathology,* 51, 440, 1961.
34. **Papavizas, G. C.,** Evaluation of various media and antimicrobial agents for isolation of *Fusarium* from soil, *Phytopathology,* 57, 848, 1967.
35. **Bugbeen, W. M.,** A selective medium for the enumeration and isolation of *Phoma betae* from soil and seed, *Phytopathology,* 64, 704, 1974.
36. **Papavizas, G. C. and Klag, N. G.,** Isolation and quantitative determination of *Macrophomina phaseolina* from soil, *Phytopathology,* 65, 182, 1975.
37. **Ko, W. and Hora, I. K.,** A selective medium for determining populations of *Rhizoctonia solani* in soil, *Phytopathology,* 61, 707, 1971.
38. **Backman, P. A. and Rodriguez-Kabana, R.,** Development of medium for selective isolation of *Sclerotium rolfsii, Phytopathology,* 66, 234, 1976.
39. **Papavizas, G. C.,** Isolation and enumeration of propagules of *Sclerotium cepivorum* from soil, *Phytopathology,* 62, 545, 1972.
40. **Singh, M.,** Studies on *Neovossia indica,* the Causal Organism of Karnal Bunt of Wheat, M.Sc. Thesis submitted to G. B. Pant University, Pantnagar, India, 1984, 70.
41. **Schroth, M. N., Thompson, J. P., and Hilderbrand, D. C.,** Isolation of *Agrobacterium tumefaciens - A. radiobacter* group from soil, *Phytopathology,* 55, 645, 1965.
42. **Clark, A. G.,** A selective medium for the isolation of *Agrobacterium* species, *J. Appl. Bacteriol.,* 32, 348, 1969.
43. **New, P. B. and Kerr, A.,** A selective medium for *Agrobacterium radiobacter* biotype 2, *J. Appl. Bacteriol.,* 34, 233, 1971.
44. **Mohanty, U.,** *Corynebacterium fascians* (Til.) Dowson; its morphology, physiology, nutrition, and taxonomic position, *Trans. Br. Mycol. Soc.,* 34, 23, 1951.
45. **Crosse, J. E. and Pitcher, R. S.,** Studies on the relationship of eelworm and bacteria to certain plant disease. I. The etiology of strawberry and cauliflower diseases, *Ann. Appl. Biol.,* 39, 475, 1952.
46. **Peterson, G. H.,** Survival of *Xanthomonas vesicatoria* in soil and diseased tomato plants, *Phytopathology,* 53, 765, 1963.
47. **Schaad, N. D. and White, W. C.,** A selective medium for soil isolation and enumeration of *Xanthomonas campestris, Phytopathology,* 64, 876, 1971.
48. **Brown, V. I. and Lowbury, E. J. L.,** Use of an improved cetrimide agar medium and other culture methods for *Pseudomonas aeruginosa, J. Clin. Pathol.,* 18, 752, 1965.
49. **Sands, D. C. and Rovira, A. D.,** Isolation of fluorescent pseudomonads with a selective medium, *Appl. Microbiol.,* 20, 513, 1970.
50. **Smith, R. F. and Dayton, S. L.,** Use of acetamide broth in the isolation of *Pseudomonas aeruginosa* from rectal swabs, *Appl. Microbiol.,* 24, 143, 1972.
51. **Schroth, M. N., Thomson, S. V., and Weinhold, A. R.,** Behaviour of plant pathogenic bacteria in rhizosphere and non-rhizosphere soils, in *Ecology of Root Pathogens,* Krupa, S. V. and Dommergues, Y. R., Eds., Elsevier, Amsterdam, 1979, 281.
52. **Cuppels, D. and Kelman, A.,** Evaluation of selective media for isolation of soft rot bacteria from soil and plant tissue, *Phytopathology,* 64, 468, 1974.
53. **Miller, T. D. and Schroth, M. N.,** Monitoring the epiphytic population of *Erwinia amylovora* with a selective medium, *Phytopathology,* 62, 1175, 1972.
54. **Rochow, W. F. and Duffus, J. E.,** Serological blocking of virus transmission by insects, in *Serological Tests for Identification of Plant Viruses,* Ball, E. M., Ed., *Am. Phytopathol. Soc.,* St. Paul, Minnesota, 1974, 29.
55. **Auger, J. G. and Shalla, T. A.,** The use of fluorescent antibodies for detection of Pierce's disease bacteria in grapevines and insect vectors, *Phytopathology,* 65, 490, 1975.

56. **Benjamin, D. R.,** Use of immunoperoxidase for rapid viral diagnosis, in *Microbiology,* Schlesenger, D., Ed., *Am. Soc. Microbiol.,* Washington, D.C., 1975.

57. **Eren, J. and Premer, D.,** Application of immunofluorescent staining to studies of the ecology of soil microorganisms, *Soil Sci.,* 110, 39, 1966.

58. **Hill, I. R. and Gray, T. R. G.,** Application of the fluorescent-antibody technique to an ecological study of bacteria in soil, *J. Bacteriol.,* 93, 1888, 1967.

59. **Kunoh, H., Ishizaki, H., Watanabe, T., Yamada, M., and Nagatani, T.,** A micromanipulating method to observe the inner structure of diseased leaves by scanning electron microscopy, *Plant Dis. Rep.,* 60, 95, 1976.

60. **Sher, S. A. and Bell, A. H.,** Scanning electron micrographs of the anterior region of some species of *Tylenchoidea, J. Nematol.,* 7, 69, 1975.

61. **Brooks, J. B.,** Identification of disease and disease causing agents by analysis of spent culture media and body fluids with electron capture gas liquid chromatography, in *Microbiology,* Schlessinger, D., Ed., Am. Soc. Microbiol., Washington, D.C., 1975.

62. **Mitruka, B. M.,** *Gas Chromatographic Applications in Microbiology and Medicine,* Wiley, New York, 1975, 472.

63. **Hankin, L. and Anagnostakis, S. L.,** The use of solid media for detection of enzyme production by fungi, *Mycologia,* 67, 597, 1975.

64. **Kovacs, N.,** Identification of *Pseudomonas pyocyanea* by the oxidase reaction, *Nature,* 178, 703, 1956.

65. **Thornley, M. J.,** The differentiation of *Pseudomonas* from other gram negative bacteria on the basis of arginine metabolism, *J. Appl. Bacteriol.,* 23, 37, 1960.

66. **Miller, S. A. and Martin, R. R.,** Molecular diagnosis of plant diseases, *Annu. Rev. Phytopathol.,* 26, 408, 1988.

67. **Singh, R. P.,** Molecular hybridization with complementary DNA for plant viruses and viroids detection, in *Perspectives in Phytopathology,* Agnihotri, V. P., Singh, N., Chaube, H. S. Singh, U. S. and Dwivedi, T. S., Eds., Today & Tomorrow's Printers and Publishers, New Delhi, 1989, 51.

Chapter 7

SURVIVAL OF PLANT PATHOGENS

I. INTRODUCTION

A successful infectious pathogen must be able to bridge the discontinuities in infection chain or disease cycle due to gaps between successive host crops and cyclic unfavorable season. It must be able to survive during unfavorable environments. Human activities aimed at interfering with the ability of the pathogen to bridge the discontinuities in the infection chain constitute the most important approach to disease management. Discontinuity in growth causes reduction in the amount of inoculum. Several workers[1-6] have described the importance and mode of survival of plant pathogens.

II. SOURCES OF SURVIVAL

The sources of survival of plant pathogens are outlined in the following chart:

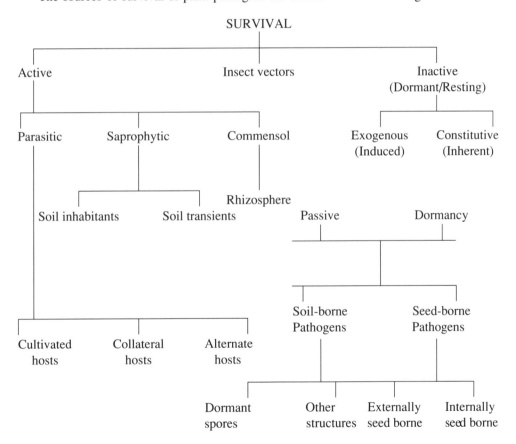

A. INFECTED PLANTS AS RESERVOIR OF INOCULUM
1. The Cultivated Host

Perennial hosts are easy source of survival of plant pathogens which attack them. In temperate regions, bacterial pathogens remain in the living but dormant tissues of the perennial hosts during the off season. Certain pathogens, such as *X. campestris* pv. *citri*, *X.*

campestris pv. *juglandis, X. campestris* pv. *pruni, Ps. syringae* pv. *morsprunorum,* and *Erwinia amylovora* are known to be carried over perennially in holdover cankers or blighted twigs which may produce bacterial ooze in favourable weather to furnish fresh inoculum.

Similarly, several fungal pathogens like *Colletotrichum gloeosporioides (Glomerella cingulate)* (anthracnose of mango), *Oidium mangiferae* (powdery mildew of mango), *Podosphaera leucotricha* (powdery mildew of apple), *Botryosphaeria ribis* (canker of apple), *Phyllosticta* sp. (apple blotch), etc., survive on infected organs of the plants.

2. Collateral and Other Weed Hosts

Along with cultivated crops, several undesirable plants (weeds, etc.), both annual and perennial, grow independently in the nature. Such plants which belong to the same botanical family to which cultivated hosts belong, are known as collateral hosts. Some important examples of pathogens surviving on collateral and weed hosts are given in Table 1.

3. Alternate Hosts

One of the two kinds of plants on which a parasitic fungus (e.g., rust) must develop to complete its life cycle is known as "alternate host". The role of such hosts in perpetuation is not as important as of collateral hosts. In temperate regions the alternate host of *Puccinia graminis* (black or stem rust of wheat), the barberry bush, grows side by side with the cultivated host. In such areas this wild host belonging to a different family may be of some importance for survival of the fungus. It helps in completion of the heterogenous infection chain of the rust fungus. A list of some weeds and wild plants that serve as alternate hosts for rusts affecting field or fruit crops is given in Table 2.

B. SAPROPHYTIC SURVIVAL

Garrett[23] has characterized many of the root pathogens ecologically as root inhabitants or soil inhabitants. Root inhabitants are considered to be ecologically obligate parasites, whereas soil inhabitants grow well saprophytically and can survive longer in soil in the absence of a susceptible host plant.

Fungal pathogens like *Pythium, Rhizoctonia, Sclerotium,* etc., survive as soil inhabitants for considerable period by the virtue of their ability to attack and colonize dead plant materials and thus, remain active as saprophytes. However, the chances of very prolonged active survival of even these pathogens has been doubted by many workers.[24] Antagonism by other soil microflora reduces their ability to continue saprophytic activity unless they have strong competitive saprophytic ability.

Another category of saprophytic survival is of those fungi that are slightly more tolerant to antagonism and usually predominate in the rhizosphere. In this region these pathogens are restricted to pioneer colonization of dead substrates and are tolerant to competition by other soil microflora.

The third category of saprophytic survival in soil is of those pathogens that have low competitive saprophytic survival ability and survive saprophytically for only a short time. These are described as "root inhabiting" fungi. Many vascular wilt causing species of *Fusarium (F. oxysporum* f. sp. *udum, F. oxysporum* f. sp. *vasinfectum), Verticillium,* the root rot pathogen *Phymatotrichum omnivorum,* etc. fall in this category.

Besides fungi, the bacteria form another important group of plant pathogens to survive in soil. Plant pathogenic bacteria have been grouped in relation to survival in soil: (1) transient visitor, (2) resident visitor, and (3) saprophytes.[25] Crosse[26] subsequently proposed a slightly modified scheme consisting of four groups: (1) with permanent soil phase, (2) with protracted soil phase, (3) with transitory soil phase, and (4) with no soil phase. Phytopathogenic bacteria characterized as transient visitors[25] consist of species whose populations

TABLE 1
Collateral and Other Weed Hosts

Cultivated host	Disease	Pathogen	Wild or weed hosts	Ref.
Barley	Barley yellow dwarf disease	Virus	*Lolium perenne*	7
	Powdery mildew	*Erysiphe graminis*	Wild species of *Hordeum*	8
Carrot	Carrot thin leaf motley dwarf	Virus	*Daucus* spp.	9
Cotton	Root knot	*Meloidogyne incognita*	*Cyperus esculentus*	10
		Hoplolaimus columbus	*Cyperus rotundus*	
Cowpea	Web blight	*Rhizoctonia solani*	*Amaranthus spinosus*	11
			Aspilia africana	
			Fleurya estruans	
			Newbauldia laevis	
	Cowpea mosaic	Virus	*Phaseolus lathyroides*	12
	Cowpea aphid borne mosaic	Virus	*Chenopodium amaranticolor*	7
Cucurbits	Powdery mildew	*Erysiphe cichoracearum*	*Sonchus oleraceus*	13
Maize	Downy mildew	*Sclerophthora rayssiae* var. *zeae*	*Digitaria sanguinalis*	14
		Sclerospora sorghii	*Heteropogon contortus*	15
	Maize dwarf mosaic	Virus	*Sorghum* sp.	16
	Maize chlorotic dwarf	Virus	*Halepense* sp.	7
	Maize rough dwarf	Virus	*Digitaria sanguinalis*	7
Oat	Cyst nematode	*Heterodera avenae*	*Avena* spp. (wild oats)	7
Pea	Pea seed borne mosaic	Virus	*Vicia villosa*	17
Pear	Fire blight	*Erwinia amylovora*	*Crataegus* spp.	18
Rice	Bacterial leaf blight	*X. campestris* pv. *oryzae*	*Leersia oryzoides*	19
	Hoja blanca	Virus	*Echinocloa* spp.	7
			Panicum spp.	
	Tungro	Virus	*Oryza* spp.	7
			Echinocloa spp.	
			Leersia hexandra	
Soybean	Rhizoctonia blight	*Rhizoctonia solani*	*Cynodon dactylon*	20
Sugarbeet	Curly top	Virus	*Sophia* spp.	7
Sugarcane	Red rot	*Colletotrichum falcatum*	*Sorghum vulgare*	21
			S. halepense	
			Saccharum spontaneum	
	Downy mildew	*Peronosclerospora sacchari*	*Zea mays*	21
Watermelon	Watermelon mosaic	Virus	*Momordia dioica*	22
			M. charantia	
			Coccinia grandis	
Wheat	Stem rust	*Puccinia graminis*	Wild species of *Triticum; Aegilops; Hordeum;*	7
	Stripe rust	*P. striiformis*	*Agropyron repens*	7

are developed almost exclusively in the host plant where maximum number of generations are produced. When such bacteria reach soil, their populations decline rapidly and do not remain important source of primary inoculum for the next season. Being poor competitors, most phytopathogenic bacteria fall under this category. Examples are most species and pathovars of *Xanthomonas*, nonsoft rot species of *Erwinia*, and pathovars of *Corynebacterium michiganense*. However, in temperate regions and under conditions not favorable for intense soil microbial activities, these bacteria may perpetuate in soil in association with crop debris.

TABLE 2
Alternate Hosts on Which Rust Fungi Complete Their Life Cycle[7]

Cultivated hosts	Rust fungus	Alternate host
Barley	*Puccinia hordei*	*Ornithogalum* spp.
Cherry	*Puccinia cerasi*	*Eranthis* spp.
Cotton	*Puccinia stakmanii*	*Bouteloua* spp.
Gooseberry	*P. caricis* var. *grossulariata*	*Carex* spp.
Oat	*Puccinia coronata*	*Rhamnus* spp.
Pear	*Gymnosporangium* spp.	*Juniperus* spp.
Plum	*Tranzschelia discolor*	*Anemone coronaria*
Red currant	*Cronartium ribicola*	*Pinus* spp.
Sorghum	*Puccinia sorghi*	*Oxalis stricta*
Wheat	*Puccinia graminis*	*Berberis* spp.
	Puccinia recondita	*Thalictrum* spp.
		Isopyrum spp.
		Anchusa spp.

Investigations have shown that *Xanthomonas* spp. do not multiply or survive in soil in the free state. *X. campestris* pv. *phaseoli*, pv. *translucens*, pv. *malvacearum*, pv. *citri*, pv. *campestris*, pv. *vesicatoris*, and pv. *oryzae* are known to decline rapidly and reach extinction within few days or weeks after their introduction into soil.[4] Certain species of *Pseudomonas* are incapable of persisting in free state in natural soil for extended periods. The corynebacteria though not soil borne, can survive in soil in safely placed crop debris.

The resident visitors also have their maximum generations in the host but their numbers gradually decline in soil. If populations enter the soil at a sufficiently high rate, as happens in case of bacterial diseases of underground parts, the slow decline would permit net increase of such bacteria from season to season. The long term persistence of such bacteria in soil is host dependent; increase, decrease or total extinction depending on the cropping practices. Bacterial pathogens with an extended soil phase include crown gall bacterium *(Agrobacterium tumefaciens)* and *Psudomonas solanacearum* race 1 (bacterial wilt of solanaceous crops). These can be considered true soil-borne pathogens or soil inhabitants.[4,5] *Ps. solanacearum* race 1 can survive in soil in free state. This pathogen survives under bare fallow for 4 to 6 years and up to 10 years in soil cropped with non host or nonsusceptible crops. Susceptible weed hosts prolong its soil survival. *Agrobacterium tumefaciens* can be reisolated from artificially infested soil without crop cover but gradually declines in natural soil.[3] *Streptomyces scabies* and *S. ipomoea* are other examples of soil inhabiting bacteria. Their primary inoculum comes from soil.

There are very few phytopathogenic bacteria which are truly saprophytes with permanent soil phase. The group of Buddenhagen's[2] classification is typified by bacteria whose populations are largely produced in the soil including the rhizoplane and whose relation to plant disease is only ephimeral. This group includes the true soil saprophytes, the rhizoplane bacteria, the green fluorescent *Pseudomonas* causing soft rots, the species of *Bacillus* and the pectolytic soft rotting *Clostridium* species which are opportunistic pathogens.

Survival of soft rot causing bacterium *Erwinia* species has always been a matter of controversy. *E. carotovora* considered a soil inhabitant, was found in subsequent studies that, after being introduced into soil, its populations reach nondetectable level rapidly. Populations of *E. carotovora* sub sp. *carotovora* in potato root zone soil increase as seed tubers decay but subsequently decline to very low levels after the plants are removed.[27]

C. SURVIVAL THROUGH DORMANT/RESTING STRUCTURES (ORGANS)

Among infectious plant pathogens fungi, nematodes and phanerogams survive through their resting or dormant structures. Phanerogams produce seeds just like any other flowering plant and through these seeds they can live in dormant stage, sometimes for years. Majority of phytophagous nematodes, survive through their dormant structures (eggs, cysts, galls formed from host tissues). The eggs are mainly present in soil. Cysts or similar structures such as galls are also present in soil or sometimes in seed lots. Plant parasitic fungi are the only organisms that produce spores, analogous to eggs of nematodes, and other structures for their inactive survival. In most fungal pathogens these dormant stages are the major sources of survival.

1. Dormancy

It is "reversible interruption of phenotypic development of an organism".[28] Viewed in this way, dormancy can be of two types: "exogenous or induced" dormancy and "constitutive or inherent" dormancy.[28] Exogenous dormancy or induced dormancy is a condition wherein development is delayed because of unfavorable physical or chemical conditions of the environment. On the other hand, constitutive or inherent dormancy is a condition wherein development is delayed due to an innate property of the dormant stage, such as a barrier to the penetration of nutrients, a metabolic block or self inhibitory substances produced by the spores.

In soil the presence of widespread "fungistasis" has been studied in detail since 1953.[29] This phenomenon is described as failure of propagule to grow even if necessary physical conditions for germination are favorable. It is responsible for induction of dormancy in fungal spores especially those that depend on some external source of nutrients for germination. Fungistasis is of various types[30-32] but the most important mechanism is one which has a biotic origin.

Thus, the asexual spores (conidia, chlamydospores), sexually produced spores (oospores, ascospores), fruiting bodies (acervuli, pycnidia, sporodochia, cleistothecia, perithecia, etc.), and other dormant structures like thickened hyphae, sclerotia, rhizomorphs, etc., are the main structures for dormant survival. When the hemibiotrophs/facultative saprophytes fail to continue as saprophytes in host plant debris or in soil, they produce resting structures. Wilt causing *Fusarium* spp. (vascular pathogens) perpetuate in the form of chlamydospores formed due to conversion of conidia and vegetative hyphae. In addition to nutritional factors and competition, in *Fusarium oxysporum* chlamydospores are reported to be formed under the influence of metabolites from certain bacteria.[33-35] The formation of such structures ensures survival of these fungi under the conditions of intense antagonism in soil because dormant structures are not easily affected by antagonistic activities of other soil microflora.

D. SURVIVAL WITH SEEDS

Seed can harbor a wide range of microflora, viruses and other causal agents of plant diseases. When a pathogen is carried with or within seed without necessarily producing disease in the off-spring, it is "transport" of the pathogen by the seed. If the seed-borne inoculum necessarily produces disease in the off-spring it is "transmission". A pathogen is "externally seed borne" when it is external to the functional seed or fruit parts essential to production of a new plant. When embedded in the tissues of the seed, the pathogen is "internally seed-borne".[36] (Figure 1).

1. Fungal Pathogens

Neergaard[37] reviewed the data available on maximum longevity of seed borne fungi. The longevity of some important seed-borne fungal pathogens are given in Table 3.

2. Bacterial Pathogens

A large number of bacterial pathogens have been demonstrated as seed-borne. Some important phytopathogenic bacteria and their longevity are listed in Table 4.

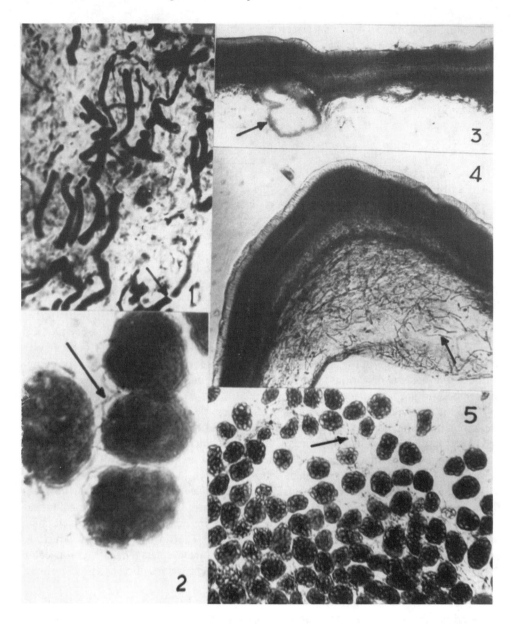

FIGURE 1. Location of *Ascochyta rabiei* in chickpea seed. (1) Mycelium in seed coat, (2) mycelium in cotyledons, (3) pycnidium in seed-coat, (4) mycelium in seed-coat, and (5) mycelium in cotyledons. (Courtesy: Dr. K. Vishnuwat).

3. Viruses

About 100 viruses have been reported to be transmitted by seed.[76] However, only a small portion (1 to 30%) of the seeds derived from virus infected plants transmit the virus and the frequency varies with the host-virus combination.[76] In most seed transmitted viruses, the virus seems to come primarily from the ovule of infected plants, but several cases are known in which the virus in the seed seems to be just as often derived from the pollen that fertilized the flower.[76] A list of seed-borne viruses and their longevity is presented in Table 5.

TABLE 3
Longevity of Some Important Seedborne Fungi

Fungus	Crop	Viability in years	Ref.
Alternaria sp.	Hordeum vulgare	10,6	38,39
A. brassicicola	Brassica oleracea	7	6
Ascochyta phaseolorum	Phaseolus vulgaris	2.5	6
A. pisi	Vicia faba	9	40
	Pisum sativum	7	41
A. rabiei	Cicer arietinum	2—3	42
Botrytis allii	Allium cepa	3.5	43
B. cinerea	Linum usitatissimum	3.33	44
B. fabae	Phaseolus vulgaris	0.75	45
Cercospora beticola	Beta vulgaris	2.5	46
C. kikuchii	Glycine max	2	47
C. gossypii	Gossypium sp.	13.5	48
Drechslera avenae	Avena sativa	10	38
		7	49
D. graminea	Hordeum vulgare	5	6
D. oryzae	Oryza sativa	10	6
F. moniliforme	Gossypium sp.	13.5	48
F. oxysporum f.sp. udum	Cajanus cajan	2	50
Macrophomina phaseolina	Phaseolus spp.	2.42	51
Phoma betae	Beta vulgaris	5	52
Sclerospora graminicola	Pennisetum typhoides	2	53
S. nodorum	Triticum aestivum	7	38
Tilletia caries	T. aestivum	9	6
Ustilago nuda	Hordeum vulgare	11	54,55
U. tritici	Triticum aestivum	7	56

4. Plant Parasitic Nematodes

Survival and seed transmission of plant parasitic nematodes (Figure 2) have been found in the species of genera: *Anguina, Aphelenchoides, Ditylenchus, Heterodera*, etc. Seedborne nematodes known to cause some important plant diseases along with their viability are given in Table 6.

E. SURVIVAL IN ASSOCIATION WITH INSECTS

Survival of plant pathogens in association with insects is not uncommon in the nature. Carter[94] has reviewed the relationship between bacterial pathogens and the insects. The bacterium associate with corn wilt *(E. stewartii)* is present in the intestinal tract of its vector, *Diabrotica undecimpunctata* (spotted cucumber beetle) and *Chaetocnema pulicaria* (the corn flea beetle). It not only survives in these beetles but is distributed over long distances also. The cucurbit wilt bacterium *(E. tracheiphila)* is totally dependent on cucumber beetles for its survival between seasons. The hibernating adult striped cucumber beetle *(Acalymma vittatum)* and the spotted cucumber beetle *(D. undecimpunctata)* harbor the pathogen during off-season in their intestinal tract and transmit it during feeding in the crop season.

Similarly, the potato black leg organism, *E. carotovora* sub sp. *atroseptica* can live in all stages of the seed corn maggot *(Hylemya platura)* and may persist in the intestinal tract of both adult flies and larvae. Since the pathogen survives pupation, the emerged adult may contaminate eggs as they are laid.

F. SURVIVAL IN VEGETATIVELY PROPAGATED PLANT PARTS

Several horticultural and cash crops, and most of those crops where modified roots or

TABLE 4
Longevity of Some Important Seed Borne Bacteria

Bacterium	Crop	Viability in years	Ref.
Corynebacterium flaccumfa-ciens var. aurantiacum	Phaseolus vulgaris	15	57
C. flaccumfaciens pv. flac-cumfaciens	P. vulgaris	24	58
C. flaccumfaciens var. viola-ceum	P. vulgaris	8	57
C. michiganense pv. insi-diosum	Medicago sativa	3	59
C. michiganense pv. tritici	T. aestivum	5	60
E. carotovora pv. carotovora	Nicotiana sp.	0.66	61
P. syringea pv. apii	Apium grave-olens	1	6
P. syringae pv. glycinea	Glycine max	1.33	6
P. syringae pv. lachrymans	Cucumis sativus	2	62
P. syringae pv. pisi	Pisum sativum	0.84	63
P. syringae pv. sesami	Sesamum indicum	0.9	64
P. syringae pv. tabaci	Glycine max	1.5	65
	Nicotiana sp.	2	66
Xanthomonas campestris pv. campestris	Brassica oleracea	3	6
X. campestris pv. malva-cearum	Gossypium sp.	4.75	67
		7	68
X. campestris pv. manihotis	Manihot esculenta	1.5	69
X. campestris pv. oryzae	Oryza sativa	0.16	70
		0.9	71
X. campestris pv. phaseoli	Phaseolus vulgaris	15	57
X. campestris pv. sesami	Sesamum indicum	1.33	72
X. campestris pv. glycinea	Glycine max	2.5	65
X. campestris pv. tomato	Lycopersicon esculen-tum	20	73
X. campestris pv. vesicatoria	Capsicum frutescens	10	73
X. campestris pv. zinniae	Zinnia elegans	4	74
X. campestris pv. phaseoli var. fuscans	Phaseolus vulgaris	3	75

stems happen to be the actual edible parts, are raised from vegetatively propagated organs such as tubers, suckers, cuttings, runners, etc. In majority of such cases the plant organs used for raising new crops, are the major sources of survival for plant pathogens. Out of many examples, the survival of pathogens in two such crops — potato and sugarcane where tubers and sugarcane pieces are used for raising new crops, are given in Table 7.

TABLE 5
Longevity of Some Important Seed Borne Viruses

Virus	Crop	Viability in years	Ref.
Alfalfa mosaic	*Medicago sativa*	3,5	77,78
Barley stripe mosaic	*Triticum aestivum*	3	79
		19	80
Bean common mosaic	*Phaseolus vulgaris*	3	6
Bean southern mosaic	*Phaseolus vulgaris*	0.6	6
Bean western mosaic	*P. vulgaris*	3	6
Cowpea aphid-borne mosaic	*Vigna unguiculata*	2	6
Cucumber mosaic	*Phaseolus vulgaris*	2.25	81
	Stellaria media	1.75	82
Muskmelon mosaic	*Cucumis melo*	3	83
Prune dwarf	*Prunus* sp.	3.5	84
Raspberry ringspot	*Capsella bursa-pastoris*	6	85
	Stellaria media	6	85
Sowbane mosaic	*Chenopodium murale*	6.5	86
		14	87
Squash mosaic	*Cucurbita pepo*	3	88
Tobacco mosaic	*Lycopersicon esculentum*	3,9	89,90
Tobacco ring spot	*Petunia violacea*	5	91
	Glycine max.		

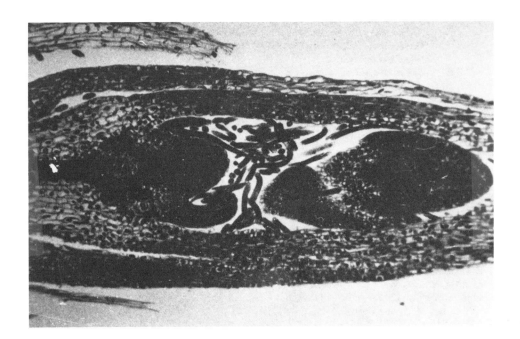

FIGURE 2. *Anguina agropyronifloris* into germinating seed (From Norton, D.C. and Sass, J. E., *Phytopathology,* 56, 769, 1966. With permission).

TABLE 6
Longevity of Some Important Seed Borne Nematodes

Nematode	Crop	Viability in years	Ref.
Anguina tritici	*Triticum aestivum*	28	6
		14	92
Aphelenchoides besseyi	*Oryza sativa*	3	6
Ditylenchus dipsaci	*Avena sativa*	8	6
	Medicago sativa	5	6
Heterodera glycines	*Glycine max*	1.8	93

TABLE 7
Survival of Plant Pathogens in Potato Tubers and
Sugarcane Pieces[21,95]

Pathogens (Sugarcane)	Pathogens (Potato)
Peronosclerospora sacchari	*Colletotrichum atramentarium*
Colletotrichum falcatum	*Rhizoctonia solani*
(Physalospora tucumanensis)	*Phytophthora infestans*
Ustilago scitaminea	*Spongospora subterranea*
Fusarium moniliforme	*Helminthosporium solani*
Acremonium furcation	*Synchytrium endobioticum*
Xanthomonas vasculorum	Alfafa mosaic virus,
Xanthomonas albilineans	Corky ring spot virus,
Grassy shoot (MLO)	Potato virus A
White leaf disease (MLO)	Potato virus M
Mosaic of sugarcane (virus)	Potato virus Y
Spike of sugarcane (virus)	Potato aucuba mosaic
	Potato leaf roll
	Potato mop-top
	Potato spindle tuber
	Potato witches' broom
	Tobacco black ring
	Yellow dwarf virus

REFERENCES

1. **Menzies, J. D.,** Survival of microbial plant pathogens in soil, *Bot. Rev.,* 29, 79, 1963.
2. **David, P.,** Survival of microorganisms in soil, in *Ecology of Soil-Borne Plant Pathogens,* Baker, K. F. and Snyder, W. C., Eds., Univ. of California Press, Los Angeles, 1965, 571.
3. **Schuster, M. L. and Coyne, D. P.,** Survival mechanisms of phytopathogenic bacteria, *Annu. Rev. Phytopathol.,* 12, 199, 1974.
4. **Schroth, M. N., Thomson, S. V., and Weinhold, A. R.,** Behaviour of plant parasitic bacteria in rhizosphere and non-rhizosphere soils, Krupa, S. V. and Dommergues, Y. R., Eds., Elsevier, Amsterdam, 1979, 281.
5. **Chaube, H. S. and Singh, R. A.,** Survival of plant pathogenic bacteria, in *Progress in Microbial Ecology,* Mukerjee, K. G., Agnihotri, V. P., and Singh, R. P., Eds., Print House (India), Lucknow, 1984, 653.
6. **Agrawal, V. K. and Sinclair, J. B.,** *Principles of Seed Pathology,* Vol. I, CRC Press, Boca Raton, Florida, 1987, 176.
7. **Palti, J.,** *Cultural Practices and Infectious Crop Diseases,* Springer-Verlag, Berlin, 1981, 243.
8. **Eshed, N. and Wahl, I.,** Rose of wild grasses in epidemics of powdery mildew on small grains in Israel, *Phytopathology,* 65, 57, 1975.

9. **Howell, W. E. and Mink, G. E.,** The role of wild hosts, volunteer carrots, and overlapping growing seasons in the epidemiology of carrot thin leaf and carrot motley dwarf viruses in central Washington, *Plant Dis. Rep.*, 61, 217, 1977.

10. **Bird, G. W. and Hogger, C.,** Nutsedges as hosts of plant parasitic nematodes in Georgia cotton fields, *Plant Dis. Rep.*, 57, 402, 1973.

11. **Onesirosan, P. T.,** Seed-borne and weed-borne inoculum in web blight of cowpea, *Plant Dis. Rep.*, 50, 338, 1975.

12. **Lima, J. A. A. and Nelson, M. R.,** Etiology and epidemiology of mosaic of cowpea in Ceara, Brazil, *Plant Dis. Rep.*, 61, 864, 1977.

13. **Stone, O. W.,** Alternate hosts of cucumber powdery mildew, *Ann. Appl. Biol.*, 50, 203, 1962.

14. **Bains, S. S., Jhooty, J. S., Sokhi, S. S., and Rental, H. S.,** Role of *Digitaria sanguinalis* in out breaks of brown stripe downy mildew of maize, *Plant Dis. Rep.*, 62, 143, 1978.

15. **Dange, S. R. S., Jain, K. L., Sirdhana, B. S., and Rathore, R. S.,** Perpetuation of sorghum downy mildew *(Sclerospora sorghi)* of maize on *Heteropogon contortus* in Rajasthan, India, *Plant Dis. Rep.*, 50, 285, 1974.

16. **Damsteegi, V. D.,** A naturally occuring corn virus epiphytotic, *Plant Dis. Rep.*, 60, 858, 1976.

17. **Stevenson, W. R., Hagedorn, D. J.,** Overwintering of pea seed-borne virus in hairy vetch, *Vicia villosa, Plant Dis. Rep.*, 57, 349, 1973.

18. **Eden-Green, S. J. and Billing, E.,** Fire blight, *Rev. Plant Pathol.*, 53, 353, 1974.

19. **Mizukami, T. and Wakimoto, S.,** Epidemiology and control of bacterial leaf blight of rice, *Annu. Rev. Phytopathol.*, 7, 51, 1969.

20. **O'Neill, N. R., Ruch, M. C., Horn, N. L., and Carver, R. B.,** Aerial blight of soybean caused by *Rhizoctonia solani, Plant Dis. Rep.*, 61, 713, 1977.

21. **Agnihotri, V. P.,** *Diseases of Sugarcane,* Oxford & IBH Publishing Co., New Delhi, 1983, 363.

22. **Bhargava, B., Bhargava, K. S., and Joshi, R. D.,** Perpetuation of water-melon mosaic virus in eastern Uttar Pradesh, India, *Plant Dis. Rep.*, 59, 634, 1975.

23. **Garrett, S. D.,** *Pathogenic Root-Infecting Fungi* Cambridge, Univ. Press, London, 1970, 294.

24. **Singh, R. S.,** *Introduction to Principles of Plant Pathology,* Oxford and IBH Publishing Co., New Delhi, 1984, 390.

25. **Buddenhagen, I. W.,** The relation of plant pathogenic bacteria to the soil, in *Ecology of Soil-Borne Plant Pathogens—Prelude to Biological Control,* Baker, K. F. and Snyder, W. C., Eds., Univ. of California Press, Los Angeles, 1965, 571.

26. **Crosse, J. E.,** Plant pathogenic bacteria in soil, in *Ecology of Soil-Bacteria,* Gray, T. R. C. and Parkinson, D., Eds., Liverpool Univ. Press, Liverpool, 1968, 522.

27. **De Boer, S. H.,** Frequency and distribution of *Erwinia carotovora* associated with potato in the Pemberton Valley of British Columbia, *Can. J. Plant Pathol.*, 5, 279, 1983.

28. **Sussman, A. S.,** Dormancy of soil microorganisms in relation to survival, in *Ecology of soil-Borne Plant Pathogens—Prelude to Biological Control,* Baker, K. F. and Snyder, W. C., Eds., Univ. of California Press, Los Angeles, 1965, 571.

29. **Dobbs, C. G. and Hinson, W. H.,** A widespread fungistasis in soil, *Nature,* 172, 197, 1953.

30. **Lockwood, J. L.,** Soil fungistasis, *Annu. Rev. Phytopathol.*, 2, 341, 1964.

31. **Watson, A. G. and Ford, E. J.,** Soil-fungistasis—a reappraisal, *Annu. Rev. Phytopathol.*, 10, 327, 1972.

32. **Lockwood, J. L.,** Fungistasis in soil, *Biol. Rev.*, 52, 1, 1977.

33. **Ford, E. J., Gold, A. H., and Snyder, W. C.,** Soil substances inducing chlamydospore formation by *Fusarium, Phytopathology,* 60, 124, 1970.

34. **Ford, E. J., Gold, A. H., and Snyder, W. C.,** Induction of chlamydospore formation in *Fusarium solani* by soil bacteria, *Phytopathology,* 60, 479.

35. **Singh, N.,** Growth and Antagonism of Fusarium spp. in Soil Amended with Organic Matter, Ph.D. Thesis, submitted to G. B. Pant Univ. of Agric. & Technol., Pantnagar, (India), 1976, 190.

36. **Baker, K. F. and Smith, S. H.,** Dynamics of seed transmission of plant pathogens, *Annu. Rev. Phytopathol.*, 4, 311, 1966.

37. **Neergaard, P.,** *Seed Pathology,* The Macmillan Press, London, 1979.

38. **Machacek, J. E. and Wallace, H. A. H.,** Longevity of some fungi in cereal seed, *Can. J. Bot.*, 30, 164, 1952.

39. **Christensen, J. J.,** Longevity of fungi in barley kernels, *Plant Dis. Rep.*, 47, 639, 1963.

40. **Sprague, R.,** Host range and life history studies of some leguminous Ascochytae, *Phytopathology,* 19, 917, 1929.

41. **Wallen, V. R.,** The effect of storage for several years on the viability of *Ascochyta pisi* in pea seed and on the germination of the seed and emergence, *Plant Dis. Rep.*, 39, 674, 1955.

42. **Chaube, H. S.,** Studies on Ascochyta blight of chickpea, *Expt. St. Res. Bull.* G. B. Pant Univ. of Agric. & Techn., Pantnagar, India, 1987, 109.

43. **Maude, R. B. and Presly, A. H.,** Neck rot *(Botrytis allii)* of bulb onions. I. Seed-borne infection and its relationship to the disease in the onion crop, *Ann. Appl. Biol.,* 86, 163, 1977.

44. **Colhoun, J. and Muskett, A. E.,** A study of the longevity of the seed-borne parasites of flax in relation to the storage of the seed, *Ann. Appl. Biol.,* 35, 429, 1948.

45. **Harrison, J. G.,** Role of seed-borne infection in epidemiology of *Botrytis fabae* on field beans, *Trans Br. Mycol. Soc.,* 70, 35, 1978.

46. **Wenzl, H.,** The importance of seed infection of beat seed by *Cercospora beticola* and its control, *Pflanzenschuts-berichte,* 23, 33, 1959.

47. **Lehman, S. G.,** Survival of the purple seed stain fungus in soybean seeds, *Phytopathology,* 42, 285, 1952.

48. **Arndt, C. H.,** Survival of *Colletotrichum gossypii* on cotton seeds in storage, *Phytopathology,* 43, 220, 1953.

49. **Sheridan, J. E. and Tan, P. E. T.,** Incidence and survival of *Pyrenophora avenae* in New Zealand seed oats, *N.Z. J. Agric. Res.,* 16, 251, 1973.

50. **Haware, M. P., Nene, Y. L., and Mathur, S. B.,** *Seed Borne Diseases of Chickpea,* Technical Bulletin, Danish Government Inst. of Seed Technology For Developing Countries, Copenhagen, No. 1, 1966, 32.

51. **Watanabe, T.,** *Macrophomina phaseoli* found in commercial kidney bean seed and in soil and pathogenicity to kidney bean seed, *Ann. Pathogen. Soc. Jpn.,* 30, 100, 1972.

52. **Newton, W. and Bosher, J. E.,** The longevity of *Phoma betae* in garden beet seed, *Sci. Agric.,* 26, 305, 1946.

53. **Shetty, H. S., Mathur, S. B., and Neergaard, P.,** *Sclerospora graminicola* in pearl millet seeds and its transmission, *Trans. Br. Mycol. Soc.,* 74, 127, 1980.

54. **Porter, R. H.,** Longevity of *Ustilago nuda* in barley seed, *Phytopathology,* 45, 637, 1955.

55. **Russell, R. C.,** The influence of aging of seed on the development of loose smut in barley, *Can. J. Bot.,* 39, 1741, 1961.

56. **Tapke, V. F.,** Longevity of *Ustilago nuda, Phytopathology,* 43, 407, 1953.

57. **Schuster, M. L. and Sayre, R. M.,** A coryneform bacterium induces purple-colored seed and leaf hypertrophy of *Phaseolus vulgaris* and other Leguminosae, *Phytopathology,* 57, 1064, 1967.

58. **Burkholder, W. H.,** The longevity of the pathogen causing the wilt of the common bean, *Phytopathology,* 35, 743, 1945.

59. **Cormack, M. W.,** Longevity of the bacterial wilt organism in alfalfa hay, pod, debris and seed, *Phytopathology,* 51, 260, 1961.

60. **Mathur, R. S. and Ahmad, Z. U.,** Longevity of *Corynebacterium tritici* causing tundu disease of wheat, *Proc. Natl. Acad. Sci. India. Sect. B.,* 34, 335, 1964.

61. **McIntyre, J. L., Sands, D. C., and Taylor, G. S.,** Overwintering, seed disinfestation and pathogenicity studies of the tobacco hollow stalk pathogen, *Erwinia carotovora* var. *carotovora, Phytopathology,* 68, 435, 1978.

62. **Walker, J. C., Chand, J. N., and Wade, E. K.,** Relation of seed- and soil-borne inoculum to epidemiology of angular leaf spot of cucumber in Wisconsin, *Plant Dis. Rep.,* 47, 15, 1963.

63. **Skoric, V.,** Bacterial blight of pea: overwintering, dissemination, and pathological histology, *Phytopathology,* 17, 611, 1927.

64. **Vajavat, R. M. and Chakravarti, B. P.,** Survival of *Pseudomonas sesami* and effect of an antagonistic bacterium isolated from seeds on the control of the disease in seed, *Indian Phytopathol.,* 31, 286, 1978.

65. **Graham, J. H.,** Overwintering of three bacterial pathogens of soybean, *Phytopathology,* 43, 189, 1953.

66. **Johnson, J. and Murwin, H. F.,** Experiments on the control of wildfire of tobacco, *Wisc. Agric. Exp. Stn. Res. Bull.,* 62, 1928.

67. **Hunter, R. E. and Brinkerhoff, L. A.,** Longevity of *Xanthomonas malvacearum* on and in cotton seed, *Phytopathology,* 54, 617, 1964.

68. **Schnathorst, W. C.,** Longevity of *Xanthomonas malvacearum* in dried cotton plants and its significance in dissemination of the pathogen on seed, *Phytopathology,* 54, 1009, 1964.

69. **Persley, G. J.,** Studies on the survival and transmission of *Xanthomonas manihotis* on cassava seed, *Ann. Appl. Biol.,* 93, 159, 1979.

70. **Kauffman, H. E. and Reddy, A. P. K.,** Seed transmission studies of *Xanthomonas oryzae* in rice, *Phytopathology,* 65, 663, 1975.

71. **Singh, R. A. and Rao, M. H. S.,** A simple technique for detecting *Xanthomonas oryzae* in rice seeds, *Seed Sci. Technol.,* 5, 123, 1977.

72. **Habish, H. A. and Hammad, A. H.,** Survival and chemical control of *Xanthomonas sesami, FAO Plant Prot. Bull.,* 19, 36, 1971.

73. **Bashan, Y., Okon, Y., and Henis, Y.,** Long term survival of *Pseudomonas syringae* pv. *tomato* and *Xanthomonas campestris* pv. *vesicatoria* in tomato and pepper seeds, *Phytopathology,* 72, 1143, 1982.

74. **Strider, D. L.,** Detection of *Xanthomonas nigromaculans* f. sp. *zinniae* in zinnia seed, *Plant Dis. Reptr.,* 63, 869, 1979.

75. **Basu, P. K. and Wallen, V. R.,** Influence of temperature on viability, virulence, and physiologic characteristics of *Xanthomonas phaseoli* var. *fuscans in vivo* and *in vitro, Can. J. Bot.,* 44, 1239, 1966.
76. **Agrios, G. N.,** *Plant Pathology,* 3rd ed., Academic Press, New York, 1988, 803.
77. **Frosheiser, F. I.,** Virus-infected seeds in alfalfa seed lots, *Plant Dis. Rep.,* 54, 591, 1970.
78. **Frosheiser, F. I.,** Alfalfa mosaic virus transmission to seed through alfalfa gametes and longevity in alfalfa seed, *Phytopathology,* 64, 102, 1974.
79. **McNeal, F. H., Mills, I. K., and Berg, M. A.,** Variation in barley stripe mosaic virus incidence in wheat seed due to storage and continuous propagation and the effect of the disease on yield and test weight, *Agron. J.,* 53, 128, 1961.
80. **McNeal, F. H., Berg, M. A., and Carroll, T. W.,** Barley stripe mosaic virus data from six infected spring wheat cultivars, *Plant Dis. Rep.,* 60, 730, 1976.
81. **Bos, L. and Maat, D. Z.,** A strain of cucumber mosaic virus seed transmitted in beam, *Neth. J. Plant Pathol.,* 80, 113, 1974.
82. **Tomlinson, J. A. and Walker, V. M.,** Further studies on seed transmission in the ecology of some aphid-transmitted viruses, *Ann. Appl. Biol.,* 73, 293, 1973.
83. **Rader, W. E., Fitzpatrick, H. F., and Hildebrand, E. M.,** A seed-borne virus of muskmelon, *Phytopathology,* 37, 809, 1947.
84. **Cilmer, R. M.,** Longevity of sour cherry yellows virus in infected cherry seeds, *Plant Dis. Rep.,* 48, 338, 1964.
85. **Lister, R. M. and Murant, A. F.,** Seed transmission of nematode-borne viruses, *Ann. Appl. Biol.,* 59, 49, 1967.
86. **Bennett, C. W. and Costa, A. S.,** Sowbane mosaic caused by a seed-transmitted virus, *Phytopathology,* 51, 546, 1961.
87. **Bennet, C. W.,** Seed transmission of plant viruses, *Adv. Virus Res.,* 14, 221, 1969.
88. **Middleton, J. T.,** Seed transmission of squash mosaic virus, *Phytopathology,* 34, 405, 1944.
89. **Alexander, L. J.,** Inactivation of tobacco mosaic virus from tomato seed, *Phytopathology,* 50, 627, 1960.
90. **Broadbent, L.,** The epidemiology of tomato mosaic. XI. Seed transmission of TMV, *Ann. Appl. Biol.,* 56, 177, 1965.
91. **Laviolette, F. A. and Athow, K. L.,** Longevity of tobacco ringspot virus in soybean seed, *Phytopathology,* 61, 755, 1971.
92. **Raeder, J. M.,** A note on the longevity of the wheat nematode, *Anguina tritici, Plant Dis. Rep.,* 38, 268, 1954.
93. **Epps, J. M.,** Survival of soybean cyst nematodes in seed bags, *Plant Dis. Rep.,* 52, 45, 1968.
94. **Carter, W.,** *Insects in Relation to Plant Diseases;* Interscience Publishers, New York, 1962, 705.
95. **Rich, A. E.,** *Potato Diseases,* Academic Press, New York, 1983, 238.

Chapter 8

DISSEMINATION OF PLANT PATHOGENS

I. INTRODUCTION

Knowledge of modes by which pathogens are dispersed is essential for devising suitable and effective disease management practice. Several terms such as distribution, dissemination, dispersal, spread, and transmission have been used invariably and rather loosely. Distribution implies the spread of a pathogen into new geographical areas and its establishment there. Dispersal is what happens between take-off of a spore and its deposition—it does not include its germination or infection of the plant—whereas spread implies that the pathogen reaches and infects plants. Spread can be used to describe progressive colonization of the infected organs or plants, passage of the pathogen from infected plants to others in the same field or crop area, or long distance spread of pathogens between plants which are widely separated, as in intercontinental spread.[1]

II. MODES OF DISPERSAL

A. DIRECT (ACTIVE)
1. Through Seed

The survival of plant pathogens with seeds has been explained in the preceding chapter. Since most of the cultivated crops are raised from seed the transmission of diseases and transport of pathogens by seeds has much importance for plant pathologists and farmers. Dispersal of inoculum through seed is accomplished either as mixture and contaminant dormant structures of the pathogen (e.g., sclerotia of ergot fungus, galls containing nematode larvae, cysts, smut sori, etc.), or through presence of propagules on the seed coat (externally seed-borne, covered smut of barley), or as dormant mycelium in the seed (internally seed-borne loose smut of wheat). Agrawal and Sinclair[2] have made a thorough review of seed-borne plant pathogens that cause major diseases of major crops.

2. Transmission by Vegetative Propagation Materials

Over 40% of the bacterial plant pathogens are transmitted on vegetatively propagated material. Among the pertinent examples are *X. campestris* pv. *citri* (citrus canker), *Agrobacterium tumefaciens* (crown gall) and pathogens of potato such as *Erwinia carotovora* ssp. *carotovora* and ssp. *atroseptica*, *Corynebacterium sepidonicum*, and *Pseudomonas solanacearum*. Dispersal of bacteria to healthy plants often occurs during cultural practices (cutting, pruning, grafting, etc.). Vegetative propagation is most important, often the only, method of transmission of MLO and RLB.

3. Through Soil

As discussed in the preceding chapter (survival of plant pathogens), it has been conclusively demonstrated that a large number of plant pathogens survive in soil. Thus, soil as such becomes an important means for their short or long distance dispersal. Zoosporic fungi, flagellate bacteria and nematodes are capable of active mobility. Several others are passively transported.

a. Active Motility

Hickman and Ho[3] observed that zoospores of *Phythium aphanidermatum* moved at speeds up to 14.4 cm/h over short periods of chemotactically directed movement. However, Lacey[4] noted that zoospores of *Phytophthora infestans* moved about 1.3 cm in 2 weeks in wet soils.

Autonomous dispersal of fungal pathogens have been described by Muskett[5] and Griffin.[6] Autonomous dispersal takes place by active growth of hyphae or hyphal strands. *Phytophthora cinnamomi* moved at least 4.5 m uphill in 22 months.[7] According to Shipton[8] runner hyphae of *Gaeumannomyces graminis* grew for a radius of 1.5 m in a growing season. Wehrle and Ogilvie[9] also observed that mycelial growth of *G. graminis* was about 1.5 m in a season. The rate of spread of *Rhizoctonia solani* has been estimated to be 25 cm/month;[10] 21.2 cm in 23 d[11] and 2.5 cm/d.[12] The rhizomorphs of *Armillaria mellea* grew through field soil at about 1 m/year. The rate of such spread in *Phymatotrichum omnivorum* (root rot of cotton) is estimated as 3 to 30 ft. per season and 2 to 8 ft/month in alfalfa crop.[13] *Fomes, Ganoderma, Polyporus, Sclerotium*, etc., also move independently.

Active movement of plant parasitic nematodes has also been studied. According to Wallace[14] spread rate in soil of 0.1 to 0.5 cm/d appear to be reasonable approximation for nematodes. *Pratylenchus zeae* travelled about 0.1 cm/d in sandy loam soil[15] and *Heterodera rostochiensis* at about 0.3 cm/d.[16] Tarjan[17] found that *Radopholus similis, Pratylenchus coffeae*, and *Tylenchulus semipenetrans* moved at about 0.4, 0.2, and 0.1 cm/d, respectively. The average speed of *Pratylenchus penetrans* in soil under optimum conditions was about 0.3 cm/d.[18] *Ditylenchus dipsaci* moved at 0.5 to 1.0 cm/d[19] and *Tylenchulus semipenetrans*, 0.1 cm/d.[20]

b. Passive Dispersal in or alongwith Soil

Pathogens perpetuating in soil either saprophytically on plant debris or in dormant stage are transported from one place to another within a plot or from one plot to another or even from one area to another area. During the cultural operations in a field, soil is moved from one point to another within the field through agricultural implements, workers' feet, erosion, etc. Movement of farm equipments and animals from one farm to another may likewise transport the pathogen. Irrigation or rain water, wind with high velocity, hail storms, etc., also transport soil particles and plant debris from one location to another and thus short or long distance dispersal of pathogen propagules takes place in nature.

B. INDIRECT (PASSIVE)
1. Animal Dispersal and Other Agencies
a. Insect

Since the discovery by Waite[21] that bees and wasps can transmit *Erwinia amylovora* (fire blight of apple and pear), much information has been accumulated on the role of insects and other small animals in the dissemination of plant pathogenic fungi, bacteria and viruses.

b. Fungi

Spread of fungal pathogens by insects, although accidental, is fairly widespread especially for those diseases in which pathogens produce sugary and sticky substances. According to Austwick[22] about 66 species of fungi causing plant diseases are transmitted by more than 100 species of insects belonging to at least 6 orders. Spores of *Ustilago violaceae* (smut of Caryophyllaceae) and *Botrytis anthophilla* (mold of clover) are carried from the infected anthers to healthy flowers by pollinating insects. The conidia of *Claviceps* are disseminated from infected ovaries to healthy ones by insects which feed on the sugary honeydew.

c. Bacteria

A wide variety of insects act as vectors of bacterial pathogens either through incidental association or through intimate biological association. The best example is *Erwinia amy-*

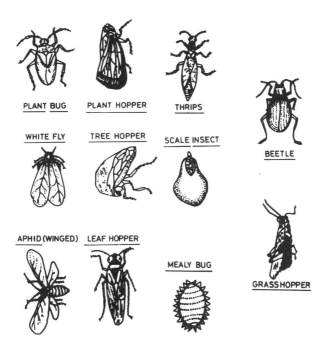

FIGURE 1.   Insect vectors of plant viruses.

lovora (fire blight of apple and pear). More than 100 insects that visit apple blossoms easily spread this bacterium to healthy blossoms. Close biological association includes transmission of bacteria with a limited host range by insects restricted to a small group of plants. Examples include transmission of *E. stewartii* (corn wilt) by corn flea beetle (*C. pulicaria*) and *E. tracheiphila* (cucurbit wilt) by cucumber beetle (*A. vittatum*).

2. Viruses

Japanese workers were the first to have established the transmission of rice dwarf disease by the leaf hopper. The virus-vector relationship is specialized phenomenon. This relationship is usually classified according to the length of time the virus can persist in and transmitted by the vector. Insects with sucking mouth parts carry plant viruses on their stylets—*stylet-borne* or *nonpersistent viruses*—or they accumulate the viruses internally, and after passage of the virus through the insect system, they introduce the virus again into plants through their mouthparts — *circulative* or *persistent viruses*. Some circulative viruses may multiply in their vectors and are then called *propagative viruses*.

Members of relatively few groups of insects as shown in Figure 1 are known to transmit viruses.

a. Mites

Agents of at least 14 diseases are transmitted by ereophyid mites. These include *Agropyron* mosaic, cherry leaf mottle, currant reversion, fig mosaic, rye grass mosaic, peach mosaic, sterility mosaic of pigeonpea, wheat spot mosaic, and wheat streak mosaic viruses.

b. Nematodes

Hewitt et al.[23] were the first to demonstrate that nematodes can transmit a virus when they proved that fanleaf disease of grapevine was transmitted by the nematode *Xiphinema index*. Nepoviruses are transmitted by species of related nematode genera *Longidorus Par-*

TABLE 1
Nematode Vectored Plant Viruses[26]

Type of virus	Vector	Hosts
Nepoviruses		
Arabis mosaic	*Xiphenema diversicaudatum, X. coxi*	Cherry, cucumber, grapevine
Cherry leaf roll	*X. diversicaudatum* *X. coxi, X. vuittenzi*	Black berry, cherry, dog-wood, elm, rhubarb, walnut
Grapevine fan leaf	*X. index* *X. italiae*	Grapevine
Mulberry ring spot	*Longidorus martini*	Mulberry
Raspberry ring spot	*L. elongatus* *L. macrosoma* *X. diversicaudatum*	Blackberry, raspberry, redcurrants, Strawberry
Strawberry latent ring spot	*X. coxi* *X. diversicaudatum*	Black current, cherry, celery rose, Strawberry
Tobacco ring spot	*X. americanum* *X. coxi*	Bean, blueberry, gladiolus, grapevine, tobacco
Tobacco black ring	*L. attenuatus* *L. elongatus*	Celery, potato, strawberry, tomato
Tomato ring spot	*X. americanum*	Blackberry, cherry, grapevine, peach, tobacco
Tobraviruses		
Pea early browning	*Paratrichodorus* spp. *Trichodorus* spp.	Pea, alfalfa
Tobacco rattle	*Paratrichodorus* spp. *Trichodorus* spp.	Potato, lettuce, tobacco
Other Viruses		
Brome mosaic	*X. diversicaudatum* *L. macrosoma*	Grasses
Carnation ring spot	*X. diversicaudatum*	Carnation
Prunus necrotic ring spot	*L. macrosoma*	Plum, sour cherry

alongidorus, and *Xiphinema,* while tobra-viruses are transmitted by species of *Trichodorus* and *Paratrichodorus.*[24,25] A list of virus diseases transmitted by these nematodes is given in Table 1.

c. Fungi

A number of soil-borne fungi of the groups of Chytridiomycetes and Plasmodiophoromycetes transmit viruses.[27,28] Viruses transmitted by soil-borne fungi are listed in Table 2.

d. Transmission by Dodder

Many plant viruses are transmitted from one plant to another through establishment of parasitic relationship between two plants by the twining stems of the parasitic plant dodder (*Cuscuta* spp.). The tobacco rattle virus is transmitted by at least 6 species of *Cuscuta*. The cucumber mosaic virus (*Cucumis virus* 1) is transmitted by at least 10 species. Some important viruses transmitted by dodder are summarized in Table 3.

3. Wind Dispersal (Anemochory)

Some plant pathogenic bacteria, seeds of some angiospermic plant parasites, eggs and

TABLE 2
Virus Diseases Transmitted by Fungi

Fungal vector	Virus disease	Ref.
Olpidium brassicae	Lettuce big vein and tobacco necrosis on tobacco, bean, potatoes and tulips	27
	Tobacco stunt	27
Olpidium cucurbitacearum	Cucumber necrosis	29
Polymyxa betae	Beet necrotic yellow vein	30
Polymyxa graminis	Barley yellow-dwarf mosaic	31
	Oat mosaic	31
	Rice necrosis	29
	Wheat spindle Streak mosaic	31
	Wheat soil-borne mosaic	31
Pythium ultimum	Pea false leaf roll	32
Spongospora subterranea	Potato mop-top	27
Synchytrium endobioticum	Potato virus X	32

TABLE 3
Some Virus Diseases Transmitted by Dodder

Virus	*Cuscuta* sp.
Arabis mosaic	*C. subinclusa* and *C. californica* frequently and *C. campestris* occasionally
Barley yellow dwarf	*C. campestris*
Citrus tristeza	*C. americana*
Citrus exocortis viroid	*C. subinclusa*
Cucumber green mottle mosaic virus (*Cucumis* virus 2)	*C. subinclusa* *C. lupuliformis* *C. campestris*
Potato leaf roll	*C. subinclusa*
Tobacco etch virus	*C. californica* *C. lupuliformis*

cysts of nematodes and majority of fungal pathogens such as downy mildew fungi, powdery mildew fungi, rusts, smuts, sooty molds and leaf spots, and/or blight (Figure 2) causing fungi which produce abundant spores of extremely varied morphology and origin (conidia, uredospores, basidiospores, ascospores, etc.) on the host surface, are disseminated by winds. Air dispersal of inoculum (Figure 3) is accomplished through rain drop splashes, sprinkler irrigation, dew or fog drip, air-borne debris, air-borne dust, aerial strands, aerosols, etc.

The number of spores produced can be astronomical. Powdery mildews (Figure 2) may produce several thousand conidia per square centimeter of infected leaf surface, a fairly modest out put compared with 10^5 or more spores in some downy mildews. A single smut sorus may contain millions of spores, a heavily infected barberry bush is said to produce up to 70×10^3 million aeciospores of *Puccinia graminia,* and a relatively small apothecium of *Sclerotinia* can produce about 30 million ascorpores.[1]

In general, the fungus spores are more abundant in the air close to the earth/plant surface than at higher altitudes. However, several investigators[2,13] have encountered clouds of spores, numerous bacteria and other minute objects several thousand meters above the earth. Uredospores of *Puccinia graminis f. sp. tritici* have been caught as high as 14,000 ft above infected fields, living spores of various fungi have been caught from aeroplanes above the

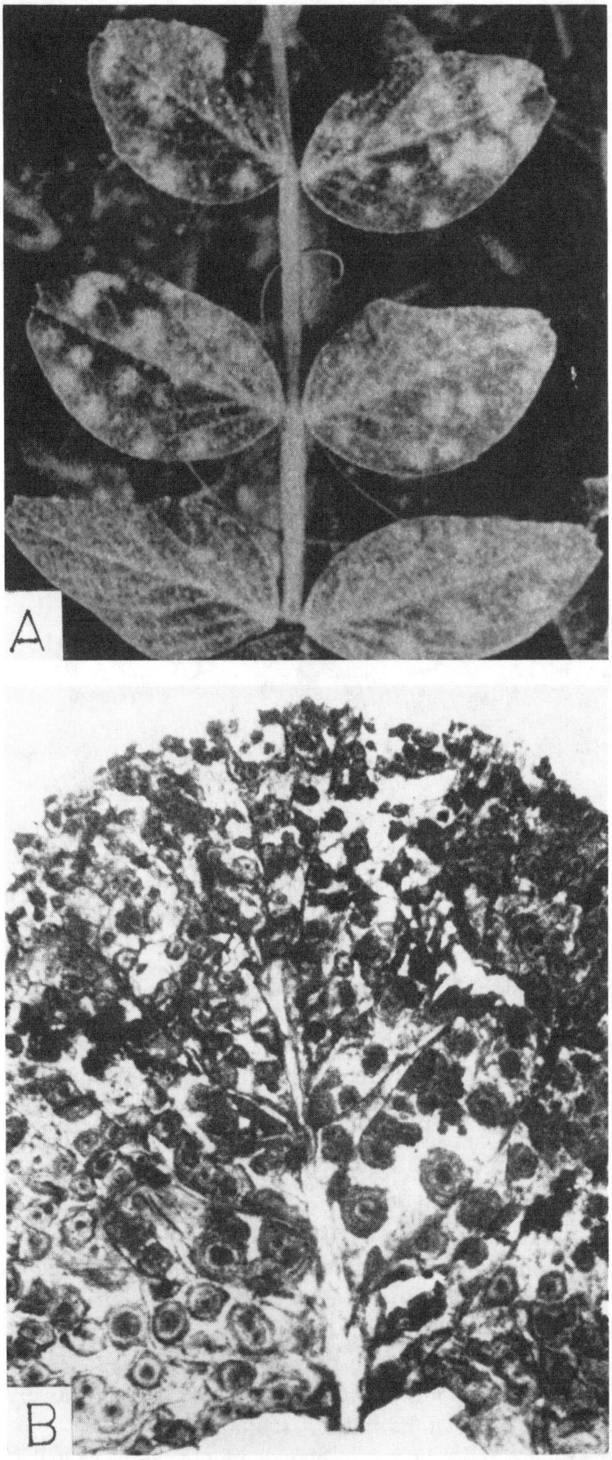

FIGURE 2.   (A) Powdery mildew disease—ectophytic growth of fungus and production of spores on host surface, (B) Leaf spot disease—spores produced esternally.

FIGURE 3. Production and dispersal of spores, (A) pycnidia and pycnidiospore of *Ascochyta rabiei,* (B) sporidia of *Neovossia indica* (germination at soil surface and dispersal by air currents, courtesy, Dr. E. J. Warhan.

carribean sea 600 mi. from their nearest possible source and living spores of several common molds were caught in spore trap released from the balloon Explorer II at 72,500 ft and set to close at 36,000 ft.[13]

4. Water Dispersal (Hydrochory)

Rain, flood and/or irrigation water may carry the propagules, especially that in or near the soil. The water separates and distributes spores in a microenvironment. Spores that are extruded in gelatinous tendrils that may harden when dry, can be separated from each other and washed down trees and other kinds of plants during rain or heavy dews. The splashing and spattering of water during heavy rains may result in distributing inoculum to plant parts near the soil and may distribute propagules to different parts of the same plant or to neighboring plants (Figure 3).

The secondary spread of some bacterial pathogens such as *X. amylovora* (fire blight bacterium), *X. campestris* pv. *campestris* (black rot of crucifers), and *C. michiganense* pv. *michiganense* (bacterial blight of tomato), results largely from splashing during rains.

Surface water resulting from rains or irrigation is known to disseminate the inoculum of several pathogens. Propagules of *X. malvacearum, Plasmodiophora brassicae, Phytophthora infestans, Colletotrichum* spp., etc., spread from plant to plant by irrigation or surface water. Eggs, larvae and cysts of plant parasitic nematodes, sclerotial bodies of *Sclerotium rolfsii, Rhizoctonia solani, Sclerotinia sclerotiorum*, and seeds of phanerogamic plant parasites also disseminate through irrigation or rain water.

5. Human Dispersal (Anthropochorus)

Man is, to a large extent, responsible for the dissemination of plant pathogens which he does in two ways—through his person and through the objects he transports from one place to another.

The long distance dissemination of plant pathogens by man has been discussed in detail in Chapter 12 under plant quarantines.

REFERENCES

1. **Tarr, S. A. J.,** *The Principles of Plant Pathology,* Winchester Press, New York, 1972, 632.
2. **Agrawal, V. K. and Sinclair, J. B.,** *Principles of Seed Pathology,* Vols. I & II, CRC Press, Boca Raton, Florida, 1987, 176.
3. **Hickman, C. J. and Ho, H. H.,** Behaviour of zoospores in plant-pathogenic phycomycetes, *Annu. Rev. Phytopathol.,* 4, 105, 1966.
4. **Lacey, J.,** The role of water in spread of *Phytophthora infestans* in the potato crop, *Ann. Appl. Biol.,* 59, 245, 1967.
5. **Muskett, A. E.,** Autonomous dispersal, in *Plant Pathology,* Vol. 3, Horsfall, J. G. and Dimond, A. E., Eds., Academic Press, New York, 1960, 57.
6. **Griffin, D. M.,** *Ecology of Soil Fungi,* Champman and Hall, London, 1972, 218.
7. **Zentmyer, G. A. and Richards, S. J.,** Pathogenicity of *Phytophthora cinnamomi* to avocado trees and the effect of irrigation on disease development, *Phytopathology,* 42, 35, 1952.
8. **Shipton, P. J.,** Take-all in spring-sown cereals under continuous cultivation. Disease progress and decline in relation to crop succession and nitrogen, *Ann. Appl. Biol.,* 71, 33, 1972.
9. **Wehrle, V. M. and Ogilvie, L.,** Spread of take-all from infected wheat plants, *Plant Pathol.,* 5, 106, 1956.
10. **Dimock, A. W.,** The *Rhizoctonia* foot-rot of annual stocks *(Mathiola incana), Phytopathology,* 31, 87, 1941.
11. **Blair, I. D.,** Behaviour of fungus *Rhizoctonia solani* Kuhn in the soil, *Ann. Appl. Biol.,* 30, 118, 1943.
12. **Shurtleff, M. C.,** Factors that influence *Rhizoctonia solani* to incite turf brown patch, *Phytopathology,* 43, 484, 1953.

13. **Mehrotra, R. S.,** *Plant Pathology,* Tata McGraw-Hill, New Delhi, 1980, 771.
14. **Wallace, H. R.,** Dispersal in time and space: soil pathogens, in *Plant Disease: An Advanced Treatise,* Vol. 2, Horsfall, J. G. and Cowling, E. B., Eds., Academic Press, New York, 1978, 181.
15. **Endo, B. Y.,** Responses of root-lesion nematodes, *Pratylenchus brachyurus* and *P. zeae,* to various plants and soil types, *Phytopathology,* 49, 417, 1959.
16. **Rode, H.,** Untersuchungen iiber das Wandervermogen Von larven des Kartoffelnematoden (*Heterodera rostochinensis* Woll.) in modellversuchen mit verschiedenen bodenarten, *Nematologica,* 7, 74, 1962.
17. **Tarjan, A. C.,** Migration of three pathogenic citrus nematodes through two Florida soils, *Soil Crop Sci. Soc. Fla. Proc.,* 31, 253, 1971.
18. **Townshend, J. L. and Webber, L. R.,** Movement of *Pratylenchus penetrans* and the moisture characteristics of three Ontario soils, *Nematologica,* 17, 47, 1971.
19. **Webster, J. M. and Greet, D. N.,** The effect of a host crop and cultivations on the rate that *Ditylenchus dipsaci* reinfested a partially sterilized area of land, *Nematologica,* 13, 295, 1967.
20. **Baines, R. C.,** The effect of soil type on movement and infection rate of larvae of *Tylenchulus semipenetrans, J. Nematol.,* 6, 60, 1974.
21. **Waite, M. B.,** Results from recent investigations in pear blight, *Bot. Gaz.,* 16, 259, 1891.
22. **Austwick, P. K. C.,** Quoted in *Biological Aspects of the Transmission of Diseases,* Smith, H., Ed., Oliver and Boyd, Edinburgh, 1957, 73.
23. **Hewitt, W. B., Raski, D. J., and Goheen, A. C.,** Nematode vectors of soil-borne fanleaf virus of grapevines, *Phytopathology,* 48, 586, 1958.
24. **Martelli, G. P.,** Some features of nematode-borne viruses and their relationship with the host plants, in *Nematode Vectors of Plant Viruses,* Lamberti, F., Taylor, C. E., and Seinhorst, J. W., Eds., Plenum, New York, 1975, 223.
25. **Taylor, C. E. and Robertson, W. M.,** Virus vector relationships and mechanics of transmission, in *Symposium on Nematode Transmission of Viruses, Proc. Am. Phytopathol. Soc.,* St. Paul, Minnesota, 1977, 20.
26. **Taylor, C. E. and Robertson, W. M.,** Acquisition, retention and transmission of viruses by nematodes, in *Nematode Vectors of Plant Viruses,* Lamberti, F., Taylor, C. E., and Seinhorst, J. W., Eds., Plenum, New York, 1975, 253.
27. **Teakle, D. S.,** Fungi as vectors and hosts of viruses, in *Viruses, Vectors and Vegetation,* Maramorosch, K., Ed., Wiley, New York, 1969, 23.
28. **Teakle, D. S.,** Fungi, in *Vectors of Plant Pathogens,* Harris, K. F. and Maramorosch, K., Eds., Academic Press, New York, 1980, 417.
29. **Fry, W. E.,** *Principles and Practices of Plant Disease Management,* Academic Press, New York, 1987, 378.
30. **Palti, J.,** *Cultural Practices and Infectious Crop Diseases,* Springer-Verlag, Berlin, 1981, 243.
31. **Slykhuis, J. T.,** Virus and virus like diseases of cereal crops, *Annu. Rev. Phytopathol.,* 14, 189, 1976.
32. **Singh, R. S.,** *Introduction to Principles of Plant Pathology,* 3rd ed., Oxford and IBH Publishing Co., New Delhi, 1984, 534.

Chapter 9

PATHOMETRY-ASSESSMENT OF DISEASE INCIDENCE AND LOSS

I. INTRODUCTION AND IMPORTANCE

Pathometry or phytopathometry, a term coined by Large,[1] can be defined as the quantitative study of the suffering plant.[1]

Knowledge of plant disease incidence,[2] severity, and spatial pattern is becoming increasingly important as the economics of agriculture require more critical decisions at all levels. Government, public, and private institutions use this informatin to evaluate their long-term research goals and resource allocations. Growers and agricultural advisors use it to make pest management decisions. It is also an important element for improving the efficiency of crop surveillance, disease monitoring and forecasting system. Disease management through breeding resistant cultivars is now at forefront in countries where resource limitations do not permit high cost plant protection technology. Pathometry is a major tool for developing resistant varieties.

It was Large[3] who brought the first comprehensive review on the subject. Many more excellent reviews on pathometry have appeared in recent years.[4-8] The FAO of the United Nations with its concern for the ''Hungry Nations'' exhibited an enthusiastic interest in assessing disease and crop loss. Chiarappa[9] edited the most comprehensive manual on crop loss assessment methods.

II. PREREQUISITES

Large[3] has pointed out that devising methods for disease assessment involves a detailed knowledge of the disease and the way in which it affects the plant. He suggested a ''strategy of investigation''.

A. THE HOST

It is necessary to have a thorough knowledge of the healthy crop for any assessment. It includes morphology and development of the healthy plant from sowing to harvest. It also includes sketches of the characteristic successive stages of crop development. The distinguishable stages are named and coded in an easily memorizable way. Feekes[10] designed the well-known ''Feekes Scale'' of wheat and other smaller grains in the verbal form. Large[11] modified the original ''Feekes scale'' and prepared illustration as standard diagram for the recording of growth stages in cereals (Figure 1). Later several others[9,12,13] published standard diagrams for growth stages, assessment methods and scales.

B. THE DISEASE

The entire sequence of host × pathogen interaction leading to development of disease must be clearly understood. Description of disease symptoms in standard diagrams, color plates and in words are essential for disease appraisal. This requires rigorous experimentation. Disease symptomatology should distinguish between response and severity.[1] Response is a measure of quality, a lesion may be chlorotic, necrotic, poorly sporulating, or heavily sporulating. Severity is a measure of quantity; a leaf may be covered with lesion on 5 or 90% of its surface area. Key should include both response and severity.[4]

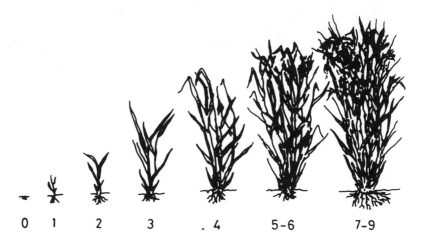

FIGURE 1.   Standard diagram for recording of growth stages in cereals (Courtesy, Dr. C.D. Mayee).

Disease offers three parameters for measurement.[1] They are

1.   Disease incidence: It is defined as the number of plant units infected such as, whole plant, leaves, fruits, tubers, twigs, stems, etc., and is expressed as proportion (0 to 1) or as percentage (0 to 100) of diseased entities within a sampling unit. It is the most popular parameter for disease measurement because it is easiest and quickest.[14] One can count accurately and reproducably the number of smutted ears of wheat, wilted plants, rotted plants, etc. This makes incidence the favored measure for detection and enumeratin of pattern of disease spread. The expression made as percentage or proportion are transformed into logs, probits, or logits as is suitable for comparative analysis.
2.   Disease prevalence: Zadoks and Schein[4] considered the term ambiguous and suggested its use with caution. It has been associated with large scale measurement of disease incidence and, therefore, has been considered a synonymous term for incidence.[15]
3.   Disease severity: It is defined as the quantity of disease affecting entities within a sampling unit. It is often precisely stated as the area[2] or area and volume[14] of plant tissues affected. Severity has also been called "intensity" by various authors.[9,14]
4.   Yield loss: Yield loss is defined as the measurable reduction in quantity and quality of yield.[2,4] Yield has been often a measure of disease. However, it is too gross a parameter as several factors account to it by the time it comes to a measurable stage. There are several treaties on crop loss assessment.[2-4,6,7]

In addition, field experiments are carried out in which development of disease is observed in detail and crop yields are estimated. Yields of plots kept free from disease by spraying or other means are also recorded for comparison, and the trials are run over a number of years. These are adequately replicated trials with a sound statistical foundation, and they permit a more detailed analysis of disease development and crop yields. From these trials the methods of disease assessment which are likely to be most useful for disease survey work in the field are devised, and the calibration of disease severity with likely crop loss is attempted.

III. PATHOMETRIC METHODS

The tactics of disease measurement vary according to the nature of the disease and, therefore, no single method is applicable to all the diseases. Various methods that are used include visual assessment, incidence-severity (I-S) relationship, inoculum-disease (I-D) relationship, remote sensing and video image analysis.

A. Visual Assessment Methods

1. Percentage of Diseased Plants, Organs, or Tissues

This is particularly applicable to diseases which kill plants rather quickly or which cause about the same amount of damage to all the infected plants. These include damping-off diseases, root rots and wilts; diseases which cause total destruction of the infected organs, such as ergot and some inflorescence smuts and many virus diseases originating from infected planting materials such as tuber-borne viruses of potato. Recording the percentage of diseased plants and organs is a direct measure of crop loss involved.

2. The Descriptive Scales

For such diseases where the amount of disease varies greatly on different plants in the population (rusts, mildews, blights, leaf spots, etc.) many arbitrary indices and ratings have been in practice.[16] These are widely used on a numerical scale to subjective estimates such as "moderate", "severe", and so on. Such scales may be meaningless to workers other than the ones who devised them, since "moderate" disease in a region or season in which the disease is very prevalent may correspond to "severe" disease in a year or location with less abundant disease. These scales can be useful if the grades are realistic, well described, usable in practice and comparable from one worker location or season to another. In these scales the grades are rather wide and it takes no account of the fact that the eyes assess diseased area in logarithmic steps.

3. Logarithmic Scale

In accordance with Weber-Fechner law, visual grades progress in logarithmic steps, and that up to 50% the eye tends to judge the percentage of the total area that is diseased (or covered by lesions) while above 50% it judges the percentage that is healthy (or free from lesions). On this basis Horsfall and Barratt[17] devised a logarithmic scale for plant disease measurement. The scale has 50% as a natural center and turning point, i.e.,the scale is balance around the 50%. For convenience the scale as published is based on ratio of two except for the upper and lower ends. Ignoring fractions the scale is 0—3, 3—6, 6—12, 12—25, 25—50, 50—75, 75—87, 87—94, 94—97, 97—100. These are percentages; i.e., the leaf area diseased and the grades are alloted as 1 = nil, 2 = 0—3%, 3 = 3—6% and so to 11 = 97—100, and 12 = 100%. Thus the scale reads the diseased tissues in logarithmic units below 50% and healthy tissue in the same unit above 50%.

A system using percentage scale, developed by British Mycological Society[18] for measuring late blight of potato and later used in Canada by James et al.[19,20] for loss appraisal in potato is given in Table 1.

The percentage scale has many advantages such as (1) the upper and lower limits of the scale are always well defined, (2) the scale is flexible in that it can be divided and subdivided conveniently, and (3) it is universally known and can be used to record both the number of plants infected (incidence) or area damaged (severity) by a foliage or root pathogen.

4. Standard Diagrams

With the help of preliminary portfolio and after deciding the method of disease mea-

TABLE 1

Key for Assessment of Disease Severity of Late Blight of Potato[18]

Blight %	Description
0	Not seen on the field
0.1	Only a few plants affected here and there, up to 1 or 2 spots in 12 yd radius
1	Up to 10 spots per plant or general light spotting
5	About 50 spots per plant or up to 1 leaflet in 10 attacked
25	Nearly every leaflet with lesions plants still retaining normal form; field may smell of blight but looks green, although every plant is affected
50	Every plant affected and about $1/2$ of leaf area destroyed by blight; field looks green flecked with brown
75	About $3/4$ of leaf area destroyed by blight; field looks neither predominantly brown or green
95	Only a few leaves left green but stems green
100	All leaves dead, stems dead or dying

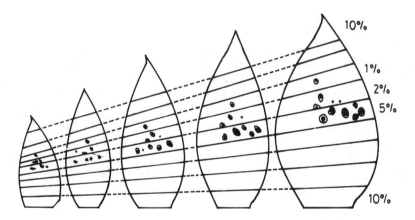

FIGURE 2.   Variation in leaf size and measurement of disease (groundnut leaves, courtesy, Dr. C. D. Mayee).

surement, a standard diagram, or research key for the assessment of the disease is prepared in order to get high degree of uniformity in rating disease intensity. In this case, use is made of visual standards including photographs, drawings, or preserved specimens, representative of each of a series of grades of disease intensity. No standard diagram can show all differing distributions of lesions that can make up any given percentage cover. What the observer really has to do is to visualize what area the lesions would cover if he could gather them all together and then estimate this area as percentage of the total area of the leaf. A useful aid can be devised to cover variation in different sizes of leaf. For example, James[21] prepared a diagram for barley leaves of four different sizes and indicated how to cover variation in leaf size of disease symptom measurement. It is illustrated in the Figure 2.

The Cobb scale for assessment of wheat rust was among the first of standard area diagrams that have been developed.[22] The idea of Cobb scale has been extended to many other diseases. Dixon and Doodson[23] have published many scales and keys for measuring diseases on crops.

5. Some Examples of Assessment Keys

These keys can be used for comparing the samples collected in the field and calculating the mean percentage area damaged by the disease. It can also be used for calculating infection index, severity index, coefficient of infection, average infection, disease intensity, etc., by

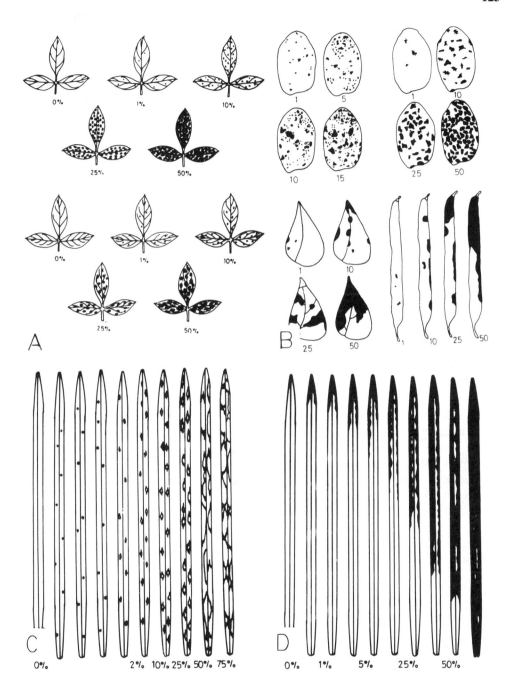

FIGURE 3.   Assessment keys, (A) leaf spot diseases (courtesy, Dr. C. D. Mayee), (B) black scurf of potato, common scab of potato, common bacterial blight of bean (leaf and pod) (courtesy, Dr. R. S. Singh), (C) brown spot of rice (courtesy, Dr. R. A. Singh), (D) bacterial leaf blight of rice (courtesy, Dr. R. A. Singh).

converting the keys into grades, i.e., giving a number to each range of percentage are affected. Thus, for example, in a foliage disease if the assessment keys give 1, 5, 25, 50% area covered, the grade numbers can be given as 0 for no infection, 1 for up to 1% area covered, 3 for up to more than 1% and less than 5%, 5 for more than 5% and less than 25% area covered, 7 for more than 25% and less than 50% area covered, and 9 to more than 50% area covered (Figure 3).

B. INCIDENCE-SEVERITY (I-S) RELATIONSHIP

The relationship between incidence and severity (I-S relationship) is an epidemiologically significant concept.[1] Some workers[1] have argued that since measurement of incidence is easier than severity, any quantifiable relationship between the two measures would permit estimation of severity based on incidence data, which are more precise. Related percentage of dead tomato leaves with an index of disease severity of early blight infection;[24] relationship between severity and incidence for coffee rust;[25] for apple scab and currant rust;[26] powdery mildew and rust of what[27] have been described. Three types of analysis have been used to describe the I-S relationship, correlation and regression, multiple infection models, and measurement of aggregation.[1,28]

C. INOCULUM-DISEASE INTENSITY RELATIONSHIP

Besides measuring diseased area, the counting of spores has been tried as an alternative method, especially for fungal diseases. Several workers used uredospore numbers, in addition to other variables, to predict rust severity on wheat. Sclerotial counts in soil to estimate and predict incidence of sugarbeet root rot caused by *Sclerotium rolfsii* has been described by Backman et al.,[29] while Zadoks[15] based on logical and technical grounds has justified this approach, others feel that it is essential to distinguish between inoculum-incidence and inoculum-severity relation.

D. REMOTE SENSING

Aerial photography can detect objects on land over a large area. It dates back to around 1860 but it was not until 1956 that Colwell[30] demonstrated its potential usefulness in plant disease work. He showed that panchromatic, colour and especially infra red aerial photography could be used to detect rusts and virus diseases of small grains and certain diseases of citrus. Later infrared photography was used in England for potato late blight disease. The remote sensing by aerial photography has now been tried for many types of assessment and detection work in agriculture including survey of plant diseases.[31-34]

The detection of diseased plant tissue on false-color infrared film is due to its greater reflection of near infrared light (λ = 700 to 950 nm) compared to healthy tissue.[35] Color infrared photographs have been analyzed with microdensitometers or other types of electronic scanning devices to quantify the disease severity.[33-34] Although disease severity of single plant has been assessed accurately by analysis of photographs of the plant,[36] the high cost and inconvenience of taking and analyzing color infrared photographs has limited the employment of this method.[1]

E. VIDEO IMAGE ANALYSIS

Several workers[37-39] have employed video image analysis for assessment of plant disease severity. Recent advances in computer technology and electronics allow video cameras to interface directly with a microcomputer. Rapid automated, nonsubjective estimates of disease severity are made possible by computer-controlled analysis of video images. Assessment of only these plant diseases with lesions having different color from healthy tissues is possible by this method.[1]

IV. SAMPLING TECHNIQUES

A. GENERAL INFORMATIONS

Certain preliminary steps are essential for better accuracy of assessment. Growth stage of the crop at the time of sampling must be recorded according to standards prescribed. For example in cereals (wheat, barley, etc.), there are ten broad growth stages such as tiller initiation, tillers formed, first node of stem visible, leaf sheath just visible, etc. (Figure 1).

Next step is the sampling procedure. Methods of sampling is an important factor in measurement of plant diseases. Depending upon disease concerned and the purpose, the emphasis in disease measurement is given on either the incidence (proportion of diseased entities) or severity (quantity of disease affecting entities) within a sampling unit. An "entity" is the plant part or plant population that is measured. A "sampling unit" is a group of entities that form a single composite or average measure.[1]

Thus, the level of spatial hierarchy at which measurements are made include cells, leaves, tillers, shoots, plants/trees, field, farm or total fields per spatial unit such as district, region, state, or country. For example, proportion of rust disease of a crop on 100 leaves constitutes only one incidence value. If 100 plants are collected and then 100 incidence values can be computed because each plant will consist of a group of entities from which incidence of disease can be determined as proportion of infected leaves per plant. Severity can be calculated as the average amount of disease per leaf or per plant.

In measuring diseases over large field area the sample size, sample point, sampling fraction, and sampling method need to be considered.[1] "Sample size" is the number of sampling units chosen for assessment. The spatial unit for which disease is measured is called "sample point". "Sampling fraction" is the proportion of sample points to the total number of units in the population. In plant disease survey, the sample point is commonly a field within some range of area. The selection of number sample points is often determined by the sampling fraction, stipulated before actual survey. Selection of sample point and sampling fractions are critical factors for satisfactory disease measurement. The procedures employed for such selections are simple random, stratified random and multistage.[6]

B. A NEW APPROACH—FIELD RUNNER: A DISEASE INCIDENCE, SEVERITY, AND SPATIAL PATTERN ASSESSMENT SYSTEM

Disease incidence, severity, and spatial pattern depend on data obtained from field samples. The accuracy of these data, as well as time and effort required to obtain them, is affected by sampling technique used. Field runner, the computer software system simplifies the task of sampling fields.[40] Field runner uses the stratified random sampling design (SRSD) with single stage cluster sampling(s), in which the field is divided into equal-sized sectors and a randomly located sample is collected within each sector. Thus, sample sites are distributed throughout the field without bias to any section of the field. Sample units are located randomly within each sector. Each sample unit is composed of a "cluster" of adjacent plants, which is referred to as a "transect" because of the linear arrangement of the plants (Figure 4).

1. Sampling System

Field runner is a computer software system developed by Delp et al.[40] The software was written in MBASIC and developed for the Espon HX-20 (Espon America, Inc., Torrance, CA). This microcomputer was chosen because it is light enough to carry through the field and it has a video screen to display prompts for the operator, a printer to provide an immediate copy of the analysis, a tape drive to store data for later use and an internal clock for elapsed time calculations.

2. Field Layout

Field runner requires field dimensions and plant spacing to divide the field into uniform sectors. The number of sectors in the field is determined by the system as a function of the total number of plants in the field (calculated from field dimension and plant spacing), the transect size, and the sample intensity.

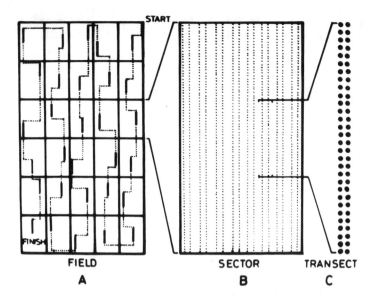

FIGURE 4.   (A) Path (broken line) generated by the computer system (Field Runner) used to direct the operator to the sampling sites (solid lines) of a stratified random design, (B) one sector composed of two-row beds of individual plants (points) and a randomly located transect selected within the sector (bracketed line), (C) two-row transect with 30 plants per row (Delp, B. R. *et al., Plant Disease,* 70, 954, 1986. With permission).

3. Data Collection

Field runner direct the operator to each randomly located sample site with a series of instructions, e.g., FACE 90 DEGREES, WALK 29 PACES, FACE 180 DEGREES, WALK 7 BEDS. Once at a sample site, the operator is prompted to enter data for each plant in the transect and is then directed to the next site. One such path is illustrated in Figure 4. Plant evaluations are entered as codes that represent the condition of the plant. These codes are defined by the operator at the beginning of each sampling session. Plants can be evaluated for incidence or severity. Multiple diseases can also be rated simultaneously for incidence; evaluations for severity are limited to one disease at a time.

4. Data Analysis

Data can be analysed immediately by Field runner, or the stored data can be transferred to a microcomputer for more detailed analysis by a related program, Data runner. Field runner and data runner were designed to answer two questions: (1) What is the incidence or severity of the disease in the field? (2) Is the spatial pattern of the disease uniform, random, or aggregated within the field and within the transects? The system provides five analyses of disease in a field: (1) estimates of the mean and variance of disease incidence, (2) the variance-to-mean ratio,[41] (3) k parameter of the negative binomial distribution,[42] (4) Lloyd's indices of mean crowding and mean patchiness,[43] and (5) an ordinary runs analysis.[44,45]

The variance-to-mean ratio, where the variance and mean are estimated from the number of diseased plants in each transect, is an indicator of disease spatial pattern. A variance-to-mean ratio of 1 indicates that disease is randomly distributed. When the disease is aggregated, the variance-to-mean ratio is greater than 1. The authors[40] have discussed certain formulae as well as generating additional informations.

V. COMPUTATION OF DISEASE INDEX

Index number is a specialized type of average and is used to measure the changes in some quantity which we can not observe directly. It is a relative measure of central tendency of a group of items. The index number can be unweighted or weighted type where the relative importance of item is not equal, weighted average gives better results than on unweighted one. As such index numbers should be weighted. Since, we must arrive at a single index number for summarizing a large amount of information, it is easy to realize that averages play an important role in computing disease index. Following are some of the ways in which disease index is computed.

A. AS A SIMPLE ARITHMETIC AVERAGE
1. When Class Rating is Expressed in Percent
Disease index for the sample is usually obtained by multiplying the number of leaves (frequency) in each category by the mean percentage of that category, adding these products, and dividing by the total number of leaves. This is simple arithmetic average and is given by the following formula:

$$\text{Disease Index}^{16} = \frac{\text{Frequency} \times \text{Mean of rating category}}{\text{Number of plants or leaves examined}} \quad \text{ue1}$$

2. When Class Rating is Expressed in Arbitrary Numbers

$$\text{Disease Index}^{46} = \frac{\text{Sum of all numerical ratings} \times 100}{\text{Total number of observations} \times \text{Maximal disease index}} \quad \text{ue2}$$

In this formula highest numerical point on a rating scale is used because that is the rating of the maximal disease category while the factor 100 converts the final rating to percent ranging from 0 (for no disease) to 100 (where every plant is diseased to the maximal extent).

B. AS WEIGHTED ARITHMETIC AVERAGE
Sometimes we associate with percent value X_1, X_2, X_3 ———— X_k, certain weights, W_1, W_2, W_3 ——— W_k, depending upon the significance or importance attached to percent value. Then the index is calculated as weighted arithmetic average and is given by the formula:

$$\text{Disease Index} = \frac{W_1X_1 + W_2X_2 + W_3X_3 \dots W_kX_k}{W_1 + W_2 + W_3 \dots W_k}$$

C. AS A LOGARITHM OF THE GEOMETRIC MEAN OF THE PERCENTAGE ASSESSMENT
This has been used by Horsfall and Barratt.[17] Each leaf is scored by a category number, and the mean score for the sample is obtained simply by adding up these scores and dividing by the number of leaves. The mean score thus obtained is a logarithm of the geometric mean of the percentage assessments and is better than arithmetical mean in that it is less distorted by extreme individual score

$$\text{Disease Index} = \frac{\Sigma \text{ Scores of all individual leaves or plants}}{\text{Number of leaves or plants examined}}$$

D. SEVERITY ESTIMATES FOR LARGER AREAS
The following formula[47] is used for such calculations:

$$\text{Disease Index} = \frac{\text{Field class rating} \times \text{Number of hectares in class}}{\text{Total number of hectares}}$$

VI. ASSESSMENT OF YIELD LOSS

A. METHODS RELATING CROP LOSS TO DISEASE
Many of the methods used in relating crop loss to disease intensity are of two main types—experimental and statistical.[48]

1. Experimental Methods
Most of the experimental methods are based on yield comparisons, (1) between infected plants or uninfected ones, or between different grades of infection; (2) between disease resistant and susceptible varieties; (3) between infected plants and plants kept free from disease by protective chemicals; or (4) between uninfected plants and plants mutilated to stimulate disease damage.[48] In all these methods it is essential that the experiments be properly designed so that the results can be analyzed statistically and, if possible, they should be repeated over a number of seasons and in different areas.

2. Statistical Methods
Reports of disease incidence, estimated crop losses, and yield figures can be used for analysing yield losses. However, the accuracy of such methods rarely approaches that obtainable in scientifically designed field experiments. Statistical methods employed[48]

1. Analysis of yields in relation to estimated disease incidence over many seasons can be analyzed to get information regarding loss in yield.[47] However, several factors such as weather, pests, farming practices, varieties, etc., which also affect productivity must be taken into account. This method is useful when a single disease is responsible.
2. Comparison of expected and actual yields[16] based on the experience of the farmers and agricultural agencies can often make reasonably accurate forecasts of yield. The difference between actual and estimated yields can be an indication of crop loss caused by a pathogen if the latter is the main cause of the reduction in yield, but many other factors are likely to be involved and this method has obvious limitations.
3. Analysis of yields before and after the application of control measures can give valuable information about the overall damage caused by a disease, over several years and over large areas. According to Chester,[16] the introduction of curly top resistant varieties of sugarbeet in the northwest of the U.S. raised yields from about 12.5 to 35.0 t/ha and averted crop failure over large areas. In this method, too, the possible effects of other factors such as improved agronomic practices, improved varieties, and the control of diseases and pests, must be considered.
4. In some countries the growers and agricultural officers are instructed to use question-naires on disease and pest incidence.[49] They are required to furnish informations on particular diseases, pests—their prevalence, severity, date of appearance, weather conditions, varietal susceptibility, estimated crop loss, etc. This method has the ad-

vantage of enabling numerous data to be collected at fairly slight expense as copies of a printed questionnaire can be sent out to thousands of growers. The success of this method depends on the ability and training of the growers and also the correctness of the entries made.

VII. MODELS FOR ESTIMATING YIELD LOSS

Mathematical models for establishing the relationship between disease and loss can be of two basic types—criticial-point models and multiple-point models.[2]

A. CRITICAL-POINT MODELS

Linear regression equations often are used to characterize critical-point models where the independent variable is disease measurement and percentage loss in yield is the dependent variable. Such models provide loss estimates for a given amount of disease at a given point in time, or for any point in time when a given amount of disease is present.[50] A major shortcoming of all critical-point models is that they fail to account for the epidemiological variables of infection rate and shape of the disease progress curve. Katsube and Koshimizu[50] estimated yield loss in rice due to rice-blast disease by using the critical-point model. They reported that for every 10% of neck blast (at 30 days after heading) there was about a 6% yield reduction. Other examples of such models are those of James et al.[51] for leaf blotch in spring barley, and for stem rust on wheat by Roming and Calpouzos.[52]

B. MULTIPLE-POINT MODELS

The second type of loss assessment model[2] is the multiple-point model which measures disease at many points during epidemic. Yield loss (dependent variable) is expressed as the sum of linear functions of other (independent) epidemiologically important variables. The general form of regression equation used is $Y = b_1 X_1 + b_2 X_2 + b_3 X_3 - - -$, where Y is percentage loss in yield and $X_1 X_2 X_3 - - -$ are disease increment for the first, second, third week, respectively. The equation passes through the origin with a multiple correlation coefficient of 0.976. The estimate of loss can be made at any point during epidemic or for whole epidemic by corresponding weekly disease increment and obtain a product. The sum of the product represents the loss at any point in time.

Examples of multiple regression analysis (MRA) are those of Sallans[53] for common root rot on wheat, Watson et al.[54] for sugarbeet yellows virus, Burleigh et al.[55] for wheat leaf rust, and James et al.[20] for late blight of potato.

C. AREA UNDER THE CURVE MODELS

Relating the area under the disease progress curve[56] to loss represents a third type of model which can be described as midway between critical and multiple point models. Loss in yield from two epidemics with different areas under the curve, but with the same disease severity at some critical point in crop development, can be distinguished.[50] However, even this model cannot account for the time factor inherent in epidemics with differing dates of disease onset.

D. LOSS APPRAISAL—TOWARD NEW APPROACH

In the recent years, interest has grown in developing simulations of crop growth driven by plant physiological process in relation to the physical environment, and which include mechanistic models of pests and diseases to enable the assessment of losses.[57] It has been suggested that the effects of pests and diseases can be modelled on a similar way to the effects of agronomic factors such as plant population density and nutrient and water supply on crop growth and yield. Sutherland and Benjamin[58] and Waggoner and Berger[59] have proposed such models for crop loss appraisal.

REFERENCES

1. **Mayee, C. D. and Datar, U. D.**, *Phytopathometry, Technical Bulletin* -1 Marathwarda Agricultural University, Parbhani, 1986, 146.
2. **James, W.**, Assessment of plant disease and losses, *Annu. Rev. Phytopathol.*, 12, 27, 1974.
3. **Large, E. C.**, Measuring plant disease, *Annu. Rev. Phytopathol.*, 4, 9, 1966.
4. **Zadoks, J. C. and Schein, R. D.**, *Epidemiology and Plant Disease Management*, Oxford Univ. Press, New York, 1979, 42.
5. **Lindow, S. E.**, Estimating disease severity of single plant, *Phytopathology*, 73, 1576, 1983.
6. **Teng, P. S.**, Estimating and interpreting disease intensity and loss in commercial fields, *Phytopathology*, 73, 1587, 1983.
7. **Madden, L. V.**, Measuring and modelling crop losses at the field level, *Phytopathology*, 73, 1591, 1983.
8. **Seem, R. C.**, Disease incidence and severity relations, *Annu., Rev. Phytopathol.*, 22, 133, 1984.
9. **Chiarappa, L., Ed.**, *Crop Loss Assessment Methods*, F.A.O., Rome, 1971.
10. **Feekes, W.**, De tarwe en hear milieu, *Versl. Techn. Tarwe Comm. Hoitsewa, Groningen*, 12, 522, 1941.
11. **Large, E. C.**, Growth stages in cereals, illustration of the Feekes scale, *Plant Pathol.*, 3, 128, 1954.
12. **Zadoks, J. C., Chang, T. T., and Konzak, C. F.**, Decimal code for the growth stages of cereals, *Euorpia Bull. No. 7*, 1974, 10.
13. **Anon.**, *Standard Evaluation System for Rice*, International Rice Testing Programme, IRRI, Los Banos, Phillipines, 1980, 44.
14. **Horsfall, J. G. and Cowling, E. B.**, Pathometry: The measurement of plant diseases, in *Plant Diseases: An Advanced Treatise*, Vol. 2, Horsfall, J. G. and Cowling, E. B., Eds., Academic Press, New York, 1978, 436.
15. **Zadoks, J. C.**, Methodology of experimental research, *Annu. Rev. Phytopathol.*, 10, 253, 1972.
16. **Chester, K. S.**, Plant disease losses: their appraisal and interpretation, *Plant Dis. Rep.*, Suppl. 193, 189, 1950.
17. **Horsfall, J. G. and Barratt, R. W.**, An improved grading system for measuring plant diseases, *Phytopathology*, 35, 655, 1945.
18. British Mycological Society, The measurement of potato blight, *Trans. Br. Mycol. Soc.*, 31, 140,1947.
19. **James, W. C., Callbeck, L. C., Hodgsan, W. A., and Shih, C. S.**, Evaluation of a method used to estimate loss in yield of potatoes caused by late blight, *Phytopathology*, 61, 1471, 1971.
20. **James, W. C., Shih, C. S., Hodgson, W. A., and Callbeck, L. C.**, The quantitative relationship between late blight of potatoes and loss in their yield, *Phytopathology*, 62, 92, 1972.
21. **James, W. C.**, Proceedings, 4th British Conference: *Insecticides and Fungicides*, 1967, 111.
22. **Cobb, N. A.**, Contribution to our economic knowledge of the Australian rust (Uredineae), *Agric. Gaz. N.S. Wales*, 3, 60, 1892.
23. **Dixon, G. R. and Doodson, J. K.**, Assessment keys for some diseases of vegetable, fodder, and forage crops, *J. Natl. Inst. Agric. Bot. (G.B.)*, 23, 293, 1971.
24. **Horsfall, J. G. and Heuberger, J. W.**, Measuring magnitude of a defoliation disease of tomato, *Phytopathology*, 32, 226, 1942.
25. **Rayner, R. W.**, Measurement of fungicidal effects in field trials, *Nature*, 190, 328, 1961.
26. **Analytis, S. and Kranz, J.**, Uber and Korrelation Zwischen Befallschaufigkeit und Befallstaerke bei Pflanzen-Krankheiten, *Phytopath. Z.*, 73, 201, 1972.
27. **James, W. C. and Shih, W.**, Relationship between incidence and severity of powdery mildew and leaf rust on winter wheat, *Phytopathology*, 63, 183, 1973.
28. **Seem, R. C.**, Disease incidence and severity relations, *Annu. Rev. Phytopathol.*, 22, 133, 1984.
29. **Backman, P. A., Rodriguez-Kabana, R., Caulin, M. C., Beltramini, E., and Zilliani, N.**, Using the soil-tray technique to predict the incidence of sclerotium rot in sugarbeet, *Plant Dis.*, 65, 419, 1981.
30. **Colwell, R. N.**, Determining the prevalence of certain diseases by means of aerial photography, *Hilgardia*, 26, 223, 1956.
31. **Jackson, R.**, Detection of plant disease symptoms by infra red photography, *J. Biol. Photograph. Assoc.*, 32, 45, 1964.
32. **Brenchley, G. M.**, Aerial photography for the study of plant diseases, *Annu. Rev. Phytopathol.*, 6, 1, 1968.
33. **Wallen, U. R., and Jackson, H. R.**, Aerial photography as a survey technique for assessment of bacterial blight of field beans, *Can. Plant Dis. Surv.*, 51, 163, 1971.
34. **Toler, R. W., Smith, B. D., and Harlan, J. C.**, Use of serial color infra red photography to evaluate crop disease, *Plant Dis.*, 65, 24, 1981.
35. **Gausman, H. W.**, Leaf reflectance of near infra red, *Photogramn. Eng.*, 40, 18, 1974.

36. **Eyal, Z. and Brown, M. B.,** A quantitative method for estimating density of *Septoria tritici* pycnidia on wheat leaves, *Phytopathology,* 66, 11, 1976.
37. **Lindow, S. E. and Webb, R. R.,** Use of digital video image analysis in plant disease assessment, *Phytopathology,* 71, 891, 1981.
38. **Lindow, S. E. and Webb, R. R.,** Measurement of foliar plant disease using microcomputer controlled digital video image analysis, *Phytopathology,* 73, 520, 1983.
39. **Blanchette, R. A.,** New techniques to measure tree defect using an image analyzer, *Plant Dis. Rep.,* 66, 394, 1982.
40. **Delp, B. R., Stowell, L. J., and Marius, J. J.,** Field runner: A disease incidence, severity, and spatial pattern assessment system, *Plant Disease,* 70, 954, 1986.
41. **Cochran, W. G.,** *Sampling Techniques,* 3rd ed., John Wiley and Sons, New York, 1977, 428.
42. **Anscombe, F. J.,** Sampling theory of the negative bionomial and logarithmic series distribution, *Biometrika,* 37, 358, 1950.
43. **Lloyd, M.,** Mean crowding, *J. Anim. Ecol.,* 36, 1, 1967.
44. **Gibbons, J. D.,** *Nonparametric Statistical Inference,* McGraw-Hill, New York, 1971, 306.
45. **Madden, L. V., Louie, R., Abt, J. J., and Knoke, J. K.,** Evaluation of tests for randomness of infected plants, *Phytopathology,* 72, 195, 1982.
46. **McKinney, R. H.,** Influence of soil temperature and moisture on infection of wheat seedlings by *H. sativum, J. Agric. Res.,* 26, 195, 1923.
47. **Large, E. C.,** Methods of plant disease measurement and forecasting in Great Britain, *Ann. Appl. Biol.,* 42, 344, 1955.
48. **Tarr, S. A. J.,.** *The Principles of Plant Pathology,* McMillan Publishers, Ltd., London and Basingstoke, 1981, 632.
49. **Moore, W. C.,** The measurement of plant disease in field, *Trans. Br. Mycol. Soc.,* 26, 28, 1943.
50. **Main, C. E.,** Crop destruction—the raison d' etre of plant pathology, in *Plant Disease—An Advanced Treatise,* Vol. 1, Horsfall, J. G. and Cowling, E. B., Eds., Academic Press, New York, 1977, 55.
51. **James, W. C., Jenkins, J. E. E., and Jammett, J. L.,** The relationship between leaf blotch caused by *R. secalis* and losses in grain yield of spring barley, *Ann. Appl. Biol.,* 62, 273, 1968.
52. **Roming, R. W. and Calpouzos, L.,** The relationship between stem rust and loss in yield of spring wheat, *Phytopathology,* 60, 1801, 1970.
53. **Sallans, B. J.,** Interrelations of common root rot and other factors with wheat yield in Saskatchewan, *Sci. Agric.,* 28, 6, 1948.
54. **Watson, M. A., Watson, D. J., and Hull, R.,** Factors affecting the loss of yield of sugarbeet caused by beet yellows virus, I. Rate and date of infection, date of sowing and harvesting, *J. Agric. Sci.,* 36, 151, 1946.
55. **Burleigh, J. R., Roelfs, A. P., and Eversmeyer, M. G.,** Estimating damage to wheat caused by *Puccinia recondita tritici, Phytopathology,* 62, 944, 1972.
56. **Van der Plank, J. E.,** *Plant Diseases: Epidemics and Control,* Academic Press, New York, 1963.
57. **Hughes, G.,** Models of crop growth, *Nature,* 332, 16, 1988.
58. **Sutherland, R. A. and Benjamin, L. R.,** A new model relating crop yield and plant arrangement, *Ann. Bot.,* 59, 399, 1987.
59. **Waggoner, P. E. and Berger, R. D.,** Defoliation, disease and growth, *Phytopathology,* 77, 383, 1987.

Chapter 10

EPIDEMIOLOGY OF PLANT DISEASES

I. INTRODUCTION

Epidemiology is primarily concerned with epidemics (epiphytotics), but the term has a wide meaning and has come to include most field aspects of plant diseases. It is the interaction of crop, pathogen and environment, populations of plants, and pathogens rather than individuals being involved. Epidemiology covers the effects of environmental factors on disease prevalence, incidence and severity, in addition, to survival and spread of plant pathogens. A proper understanding of epidemiology is necessary for prediction of plant diseases and formulation of effective control measures. In the recent years, especially after the application of statistics and mathematical analysis by Van der Plank,[1] the subject has received serious considerations.

II. CONCEPT

The word "epidemic", stems from the Greek, epi = (on) and demons = (people). Literally epidemic is defined as "affecting or tending to affect many individuals within a population, a community, or a region at the same time—an outbreak, a sudden and rapid growth, spread and development".

In recent years the term "epidemiology" has come to have a broad meaning within plant pathology.[2] The term has been variously defined as: the study of diseases in populations; the study of environmental factors that influence the amount and distribution of disease in population, and the study of rates of change (either increase or decrease) in the amount of disease in time, in space, or both.[2]

III. THE ELEMENTS OF AN EPIDEMIC

Disease results from the interaction of a pathogen with its host but the intensity and extent of this interaction is markedly affected by the environmental factors. Although these factors are not the causal agents of infectious diseases, they are the final determinants of almost all events that constitute the infection chain leading to pathogenesis and also the events that follow, viz., spread of the disease in the population. The interaction of these three components of disease have been visualized as triangle, generally referred to as the "disease triangle". However, to understand an epidemic, a fourth dimension—time—must be added to the "disease triangle" to give a "disease pyramid" or "disease cone".[2] Agrios[3] has added another dimension, i.e., the activities of humans (Figure 1). Fry[4] has explained the complex nature of disease involving interactions between host × pathogen × environment through an equation:

$$D_t = \sum_{i=0}^{t} f(p_i, h_i, e_i), \quad \text{where } D_t \text{ is disease at time t}$$

and is the sum of interaction of pathogen (p_i includes inherent ability to induce disease and population size), host (h_i includes susceptibility, distribution, and population size), and environment (e_i includes physical, biological, and chemical factors) over time (from i = 0 to t).

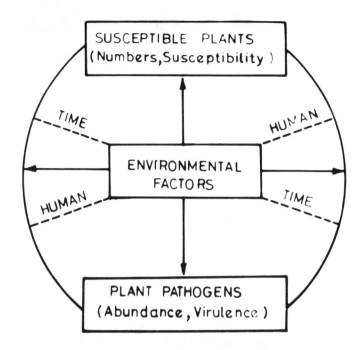

FIGURE 1.   A schematic diagram of the interrelationships of the factors involved in plant disease epidemics.

A. HOST

Levels of genetic susceptibility or resistance, degree of genetic uniformity of host plants, distance of susceptible host from the source of primary inoculum, abundance, and distribution of susceptible host, disease proneness in the host due to environment, and availability of alternate or collateral hosts are the factors which play important role in the development of epidemics involving host.

The host plant having high levels of resistance (vertical resistance) do not allow some races and/or biotypes of a pathogen to become established in them and thus, no epidemic can develop unless and until there is evolution of new races and/or biotypes that can attack that resistance and thus, making the host susceptible. Plants with horizontal or general resistance will become infected, but the rate of development will be slow. On the other hand, susceptible host plants lacking genes for resistance against existing races of the pathogen provide suitable and adequate substrate for establishment and development of new infections.

The disease in an area is initiated by primary inoculum surviving at some source. Longer the distance from the source of primary inoculum, longer will be time required for build up of epidemic in a susceptible crop. During dispersal in different directions the density of primary inoculum is diluted and as the distance increases fewer propagules are likely to reach the susceptible surface. Continuous cultivation of a susceptible variety in a given area, large areas under a similar susceptible variety, and distribution of the variety over large contiguous areas help in build up of inoculum and improve the chances of epidemic. Under these conditions, the pathogen is able to use maximum number of its propagules effectively, increase the rate of multiplication many times and repeat the disease cycle quickly. This phenomenon has been observed in case of *Helminthosporium* blight of victoria oats and Southern maize leaf blight with Texas male sterile cytoplasm.

Plants change in their susceptibility to disease with age. For example, in root rots, downy mildews, smuts, rusts, etc., the hosts or their target organs are susceptible only during the growth period and become resistant during the adult period. In some other diseases such as flower or fruit blights caused by fungal pathogens like *Botrytis, Alternaria, Glomerella, Penicillium* and *Monilinia,* and in all post harvest infections, plant parts (fruits) are resistant during growth and early adult period but become susceptible near ripening.

For pathogens spreading through heterogenous infection chain, presence of an alternate host is necessary for providing primary inoculum. The amount of inoculum thus available will determine the intensity of primary infection and subsequent spread. Presence of collateral hosts plays the same role for pathogens of homogenous infection chain. For example, grass hosts of *Sclerospora philippinensis* (downy mildew of maize and sorghum), *Ustilago scitaminea* (sugarcane smut), *Pyricularia oryzae* (rice blast), may produce abundance of inoculum aiding in build-up of epidemics of these diseases. In annual or seasonal crops, such as maize, wheat, barley, vegetables, chickpea, cotton, tobacco, etc., the epidemic generally develops much faster than do in perennial woody crops such as fruit and forest trees. For instance, epidemics of fruit and forest trees like pear decline, Dutch elm, chestnut blight, citrus tristeza, etc., take years to develop.

B. THE PATHOGEN

For any epidemic rapid cycles of infection is essential and successful infection can be caused only by virulent or aggressive isolates of the pathogen. High birth rate or fast reproductive cycles is another important contributory factor for epidemics. The pathogens that assume epidemic form invariably have the capacity to produce enormous quantity of spores that are adapted to quick and long distance dispersal in a short time so that they can take advantage of favorable weather conditions during that short period. In most cases, these spores are asexually produced usually on the exposed surfaces of the host for quick dispersal by wind, water, and insects. These are polycyclic pathogens that usually cause leaf spots, blights, rusts, mildews, and are responsible for most of the catastrophic plant disease epidemics in the world. Soil-borne pathogens like *Fusarium, Phymatotrichum, Verticillium,* etc., and most nematodes usually have 2 to 4 reproductive cycles per growing season. Since, the dispersal of such pathogens are limited both in space and time, only localized and slower developing epidemics are caused. On the other hand, monocyclic pathogens like smut fungi require an entire year to complete a life cycle. In such diseases, inoculum builds up from one year to the next, and the epidemic develops over several years. Similarly, epidemics caused by pathogens that require more than one year to complete a reproductive cycle are slow to develop. Examples are cedar apple rust (2 years), white pine blister rust (3 to 6 years), and dwarf mistletoe (5 to 6 years). Such pathogens produce inoculum and cause series of infections each year only as a result of overlapping of the polyetic generations.[3]

Among fungi the vicissitudes of dispersal by wind, minute size of unprotected spores, possible chances of falling on wrong plants, etc., are many factors that cause high death rate among the propagules. However, this weakness is offset by extremely high birth rate. Epidemics attributed to low death rate of pathogens are those in which the causal organism is systemic and protected by the plant tissues. Thus, the chances of high mortality are considerably reduced. The chief source for accumulation of inoculum for epidemics of such diseases is the diseased plant organs used for vegetative propagation and, therefore, the build-up of epidemic is comparatively slow. When a particular area becomes saturated with diseased planting materials chances of occurrence of epidemics are very high.

Adaptability of pathogen is another factor vital to development of epidemics. For pathogens having capacity to adapt to adverse conditions, the occurrence of epidemics is almost certain. The units of propagation produced by the pathogen are dispersed by external agencies which must be available if epidemics are to develop.

C. ENVIRONMENTAL FACTORS

The effect of weather on disease development has been discussed elsewhere (Chapter 3). Optimum moisture, temperature, light, etc., are necessary for activities of biotic pathogens. Assuming that a particular fungal pathogen meets all the conditions discussed above for causing epidemic; high reproductive cycles, high aggressiveness, ensured and effective dispersal, susceptible host; even then development of epidemic may not occur if weather is not favorable for germination of spores or in the absence of light stomata have not opened to permit entry of the infection thread or when the stomata open the moisture is so deficient that the germtube has dried. The weather also affects the activity of the pathogen on the host surface. It may not permit sporulation on the host surface thus reducing amount of inoculum for secondary spread.

D. ACTIVITIES OF HUMANS

Many activities of farmers, such as selection of sites, propagative materials and various disease management practices including cultural practices, have a direct or indirect effect on plant disease epidemics. The use of infected seeds, nursery stocks, and other propagative materials, increase the amount of primary inoculum within the crop and thus, greatly favor the development of epidemics. Contrary to it, the use of healthy, pathogen free or suitably treated planting materials can greatly reduce the chance of epidemic. Continuous monoculture, large acreages planted to the same or related variety of crop, high levels of nitrogen fertilization, overhead (sprinkler) irrigation, herbicide injury, and poor sanitation all increase the possibility and severity of epidemics. The use of certain chemical or planting of a certain variety may lead to selection of virulent strains that are either resistant to the chemical or can attack the resistance of the variety and thus lead to epidemics.

IV. DECLINE OF THE EPIDEMICS

In epidemics of crop plants, no epidemic remains forever in a population. In an area where a disease assumes epidemic proportion, majority of the plant contact infection, and therefore, there is saturation of the pathogen in the host population as nonavailability of more uninfected plants limits the spread of the pathogen. All these result in production of less inoculum, fewer secondary infections, and finally no new infections. The plants that escape infection are those that possessed resistance or in which resistance developed during the epidemic. It is possible that in future only these plants will be grown in that area. Therefore, one of the positive effects of an epidemic is that it eliminates susceptible individuals and permits only the resistant individuals of the population to survive and breed.

With passage of time, decline of proneness in the host also accounts for decline of the epidemics. Most pathogens attack the plant at a particular stage of its growth. Once the plant has crossed this stage, its proneness for contacting infection is reduced or completely lost. Under these conditions the epidemic will automatically decline. When the plant is receptive for infection throughout its life and its population has been affected by an epidemic, the weather conditions may not remain always congenial for disease development. This will result in reduced spread of the inoculum and the epidemic will decline. The aggressiveness of the pathogen may also be reduced. After the population of susceptible host has been destroyed, the pathogen may try to parasitize the remaining resistant individuals of the same species. In these adverse conditions it may lose its power of successful infection, its reproduction may slow down, and thus it may not remain as aggressive as when the conditions were favorable.[5]

V. STRUCTURES OF EPIDEMICS

As discussed earlier, the epidemics develop as a result of interactions of the populations of two components, hosts and pathogens under the influence of environment and human activities over time. This interaction leads to the production of the third component, disease.

Kranz[6] has described the details of components of epidemic structures. Each of these components of epidemic consists of subcomponents. The subcomponents of host include plant type (herb, tree, shrub, annual, deciduous, perennial), growth rhythm (growth stages— seedling, tillering, flowering, grain setting, etc.), propagation, population resistance (disease incidence), and reaction to disease (type of symptom).

The subcomponents of the pathogen include pathogenicity (type of parasitism, mode of infection); virulence (host specialization, races, biotypes), sporulation (kind and amount of inoculum); dispersal (autonomous, by vectors, wind, water); and survival (longevity, form).

The subcomponents of disease include infection (localized or systemic, number of lesions, disease severity); pathogenesis (incubation period); latency; lesion growth (size, rate); infectiousness (time and amount of sporulation, amount of new inoculum); spread (infection gradient); multiplication (length of infection cycle, duration and/or number of generations per season); and survival (longevity in time).

VI. PATTERNS OF EPIDEMICS

Interactions of structural elements in epidemics that are triggered by factors in the environment or by human interference result in system behavior that is expressed in patterns and rates.[6] The pattern of an epidemic in terms of disease incidence or severity is expressed in curves that show the progress of a disease over distance or over time. This curve is called "disease-progress curve". The shape of a disease-progress curve may reveal informations about the onset of the disease, amount of infective propagules, changes in host susceptibility and proneness over time, recurrent weather events and the effectiveness of disease management practices. Disease-progress curves are generally characteristic of some group of disease, though they vary somewhat with location and time due to influence of weather, crop variety, etc. For example, a saturation-type curve is characteristic for monocyclic diseases (Figure 2), a sigmoid curve is characteristic for polycyclic diseases (Figure 2), and a bimodal curve (Figure 2) is characteristic for diseases affecting different organs of the plant.[3] There is another type of disease curve called "disease-gradient curve" (Figure 3). Since the amount of disease is generally higher near or at the source of inoculum and decreases with increasing distance from the source, most disease curves are hyperbolic and quite similar, at least in the early stages of the epidemic. The incidence and severity of disease decrease steeply within short distance of the source and less steeply at greater distances until they reach zero or a low background level of occasional diseased plants.[3] Disease gradients, however, are sometimes flattened near the source as a result of multiple infections and may become flatter with time as secondary spread occurs.

The rate of growth of the epidemic can be obtained by plotting the disease-progress curve from the informations obtained at various time intervals. The rate of epidemic is the amount of increase of disease per unit of time in the plant population under study. The patterns of epidemic rates are given by curve called "rate curves" and these curves are different for various groups of diseases (Figure 4). These rate curves identify at least three major classes of epidemic patterns: symmetrical (bell-shaped, Figure 4), for example, in late blight of potato; or asymmetrical curves (Figure 4) with epidemic rate greater early in the season due to greater susceptibility of young leaves—for example, in apple scab or most downy and powdery mildews; or asymmetrical with the epidemic rate greater late in the season (Figure 4) as observed in many diseases which start slowly but accelerate markedly as host susceptibility increases late in the season.

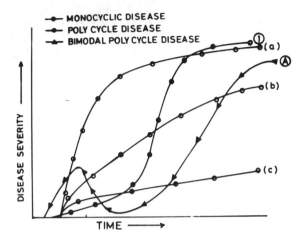

FIGURE 2. Schematic diagrams of disease — progress curves of some basic epidemic patterns (a-c). Three monocyclic diseases of different epidemic rates (1) a polycyclic disease, (A) a biomodal polycyclic disease.

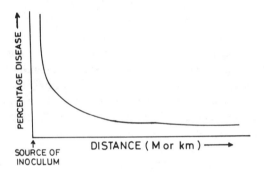

FIGURE 3. Schematic diagram of disease-gradient curve.

VII. ANALYSIS OF EPIDEMICS

Epidemic is a system. Analysis of epidemics is a modern approach in the study of epidemiology. Although the disease-progress curves and mathematical equations for rate reveal the rise and fall of an epidemic over a period of time during the growing season for the crop and about the comparison of the epidemic between varieties, fields, or regions, it does not answer how the different components of the epidemic behaved during that time.

As a system, the epidemic exists in a relatively very small compartment of a much bigger system, the "agroecosystem" which has developed over a long period of time from the "ecosystem". The epidemics interact with other subsystems of the agroecosystem such as cropping system or crop management system, pest management system, associated biological systems other than epidemic, etc., which interact among themselves also (Figure 5). These systems and their components are influenced by physical parameters of the agroecosystem and, in turn, influence the latter also. These different interacting systems, with their own interacting component give a picture of the complexity of data for analysis of epidemics.

No.

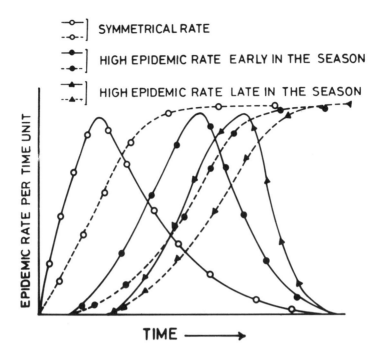

FIGURE 4. Schematic diagrams of epidemic rate curves with a symmetrical epidemic rate, with high epidemic rate early in the season, and with a high epidemic rate late in the season. Dotted curves indicate possible disease-progress curves that may be produced in each case from the accumulated epidemic rate curves.

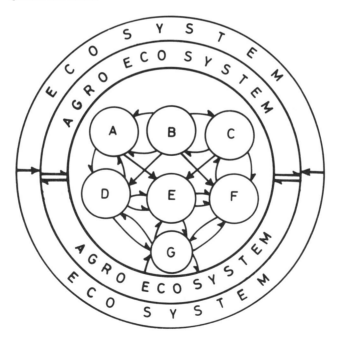

FIGURE 5. Position of epidemic in the ecosystem.

VIII. MODELS OF EPIDEMIC DEVELOPMENT

An epidemic is a dynamic process. Researchers have used mathematical models as a tool in analysis of disease dynamics. Van der Plank was instrumental in establishing mathematical models for analysing epidemics when he published his book *Plant Disease Epidemic and Control* in 1963.

A. MONOCYCLIC PATHOGENS

The amount of disease caused by a monocyclic pathogen in a season is the function of several interacting factors. The pathogen factors include size and distribution of the pathogen population and the genetic potential of the pathogen to induce disease. Host factors are genetic potential of the host as well as its size and distribution. Environmental factors include both biotic and abiotic factors. The equation x_t = QRt explains how disease develops from the interaction of host × pathogen × environment through time. X_t is the amount of disease at time t, Q is the size of initial inoculum, R is the efficacy of inoculum and t is the length of time that host and pathogen interact in their environment.[4]

This explains several important characteristics of the interaction between the host population and monocyclic pathogen. The size of initial inoculum (Q) does not increase during the season as pathogen produces no additional inoculum which could induce disease in the same season. R (pathogen efficacy) may range from zero to some positive value. If Q or R is zero, no disease will occur, t (duration of host × pathogen interaction) may influence the amount of disease.

Fry[4] with the help of Van der Plank's equation[1] d_x/d_t = QR where increase in disease (dx) during a short time period (dt) is a function of initial inoculum (Q) and its efficacy (R, a rate), explains that this equation neglects an important factor (the amount of healthy tissues) that can have a greater influence on disease increase. He explains that for a given pathogen population in a given environment, the rate of disease increase is likely to be greater when there is availability of large amount of healthy tissues than when there is a small amount. Thus, he corrects the equation and describes it as dx/dt = QR (1 − x) where (1 − x) describes the amount of available tissues or the proportion of the host population which is not yet infected. To make it more useful he rearranges this equation as dx′ (1 − x) = QR dt and then after integration another equation ln [1/ (1 −x)] = Q Rt + k is developed. The symbol ln indicates natural logarithms (to the base e). Thus, the left-hand side of this equation is natural logarithm of 1/ (1 − x). The letter k is the constant which results from integration (k = ln [1/ (1 − x_0)] where x_0 = the amount of disease at t = 0). This equation can be used to predict the required effect of disease management to achieve a desired degree of disease suppression.[4]

B. POLYCYCLIC PATHOGENS

Diseases caused by polycyclic pathogens are influenced by the size and distribution of initial pathogen population, the inherent ability of the pathogen to induce disease, host reaction, environmental factors including cultural manipulations, time during which host and pathogen interact, and rate of reproduction of the pathogen.

To explain the increase of disease induced by polycyclic pathogen Fry[4] described an equation dx/dt = xr (1 − x), where dx/dt is the instantaneous rate of disease increase at a specific time, x is the proportion of tissue diseased, r is the rate at which new infections occur (apparent infection rate), and (1 − x) is the proportion of tissue available for infection. He proposed a simplified equation if the amount of diseased tissue is very small (i.e., <0.01). Then (1 − x) is approximately 1.0 and the equation becomes dx/dt = xr. Upon rearrangement it becomes dx/x = r dt and then integrated to obtain ln x = rt + k, where t = 0, the constant of integration = x_0, the value of x at the beginning of the time period of concern. After taking antilogs it was found that x = x_0e^{rt}, where x is the amount of disease at time t, x_0 is the amount of initial disease (at t = 0), e is the base of natural logarithm, r is the exponential rate of disease increase, and t is the interval during which host and pathogen interact. This equation is the description of exponential growth. At low levels, disease induced by a polycyclic pathogen increases exponentially. At higher levels of disease, the rate of increase is limited by the diminishing supply of uninfected tissues.[4]

IX. COMPUTER SIMULATION OF EPIDEMICS

In the last two decades, the use of computers has helped pathologists to write programs that allow simulation of several plant disease epidemics. The computer is given data pertaining to various subcomponents of the epidemic and control practices at specific points in time. The computer then provides continuous information concerning not only spread and severity of the disease over time but also the economic losses likely to be caused by the disease under the conditions of the epidemic as given to the computer.

The first computer simulation program called EPIDEM, was written in 1969 to simulate *Alternaria* early blight of potato and tomato. Subsequently, computer simulators were written for *Mycosphaerella* blight of chrysanthemums (MYCOS), for southern corn leaf blight caused by *Helminthosporium maydis* (EPICORN), and for apple scab caused by *Venturia inaequalis* (EPIVEN).[3]

X. MODELS AND DISEASE MANAGEMENT STRATEGIES

Monocyclic diseases are most efficiently managed by reducing the amount or efficacy of initial inoculum, as for a monocyclic pathogen, disease is directly related to the size of the population at the start of the season because inoculum produced during the season do not cause new infections in the same season. Hence, there is a direct relationship between initial inoculum and disease at the end of the season.

Polycyclic diseases are most efficiently managed by reducing large amounts of initial inoculum and/or by limiting potentially rapid rates of disease increase because polycyclic pathogens produce inoculum effective during the same season, and disease induced by them increases exponentially or logistically.

REFERENCES

1. **Van der Plank, J. E.,** *Plant Disease Epidemic and Control,* Academic Press, New York, 1963, 349.
2. **Cowling, E. B. and Horsfall, J. G.,** Prologue: how disease develops in populations, in *Plant Disease: An Advanced Treatise,* Vol. II, Horsfall, J. G. and Cowling, E. B., Eds., Academic Press, New York, 1978, 436.
3. **Agrios, G. N.,** *Plant Pathology,* 3rd ed., Academic Press, New York, 1988, 803.
4. **Fry, W. E.,** *Principles of Plant Disease Management,* Academic Press, New York, 1987, 378.
5. **Singh, R. S.,** *Introduction to Principles of Plant Pathology,* Third Edition, Oxford & IBH Publishing Co. New Delhi, 1984, 534.
6. **Kranz, J.,** Comparative Anatomy of Epidemics, in *Plant Disease: An Advanced Treatise,* Vol. II, Horsfall, J. G. and Cowling, E. B., Eds., Academic Press, New York, 1978, 436.

Chapter 11

PLANT DISEASE FORECASTING

I. INTRODUCTION

The need for plant disease prediction in intensive agriculture hardly needs an emphasis. The practical implications of forecasting have been incorporated in the definition proposed by Miller and O'Brien[1] who state: "Forecasting involves all the activities in ascertaining and notifying the growers of community that conditions are sufficiently favorable for certain diseases, that application of control measures will result in economic gain, or on the other hand, and just as important that the amount of disease expected is unlikely to be enough to justify the expenditure of time, energy and money for control." Miller and O'Brien thus include a warning system/service for plant disease prevention in their definition of forecasting. There are excellent reviews covering this aspect and related area, such as those of Bourke,[2] Cox and Large[3] concerned with the forecasting from weather data, of potato blight, and other plant diseases. Bourke[4] has dissertated on the "use of weather information in the prediction of plant disease epiphytotics", Miller[5] on "plant disease epidemics—their analysis and forecasting", Miller and O'Brien[6] on "prediction of plant disease epidemics", Waggoner[7] on "forecasting epidemics" and "ground level climate and disease forecasting", Krause and Massie[8] on "predictive system-modern approaches to disease control", Shrum[9] on "forecasting of epidemics".

II. PURPOSE AND REQUISITES

Shrum[9] has listed three major reasons for our interest in forecasting: (1) forecasting provides a means for determining if, when, and where a given management practice should be applied (2) forecasting helps us to identify the gaps that remain in our knowledge. Forecasting provides a systematic way to assess our knowledge of environmental factors and their influence on the progress of disease, and (3) forecasting is also important because it is intellectually stimulating.

Forecasting and warning systems involve considerable expense and, therefore, must be reliable. The following points must be taken into account:[10] (1) For a disease whose severity is most strongly influenced by variation in the amount of initial inoculum, the initial inoculum should be considered in the forecast; (2) for a disease whose severity is most strongly affected by the efficacy of initial inoculum, factors that influence the efficacy must be included in a forecast; (3) forecasts for diseases whose severities are most strongly affected by the number of secondary cycles need to be based on factors that affect the speed of secondary cycles.

In addition to reliability, disease forecasts must satisfy four additional criteria:[4] (1) a forecast is necessary only if the disease is important and sporadic; (2) development of a forecast is most easily justified for an important crop, because the forecast is likely to have a large economic impact; (3) forecasts can only be useful if appropriate disease management technology exists; (4) appropriate communications are necessary for successful implementation of disease forecasts.

III. METHODS USED IN FORECASTING

A. WEATHER CONDITIONS DURING THE INTERCROP PERIOD

The survival of plant pathogen or its vector depends on the weather conditions during the intercrop period. The intercrop weather which favors overwintering or over-summering of the inoculum can be utilized for disease prediction. The severity of blue mold of tobacco (*Peronospora tabacina*) in southern U.S. is related to temperatures in the winter season. When January temperatures are above normal, blue mold can be expected to appear early in seedbeds in the following season and to cause severe losses. In recent years, a supplementary blue mold warning system has been operated in North America by the Tobacco Disease Council and the Cooperative Extension Service. The warning system keeps the industry aware of locations and times of appearance and spread of blue mold and helps growers, to some extent, with the timing and intensity of controls.[11]

Another example is that of Stewart's wilt of corn (*Erwinia stewartii*). Disease was most severe after mild winters and was least severe after cold ones. From these observations, Stevens[12] predicted that disease would be severe if the average monthly mean temperature from December to February was greater than or equal to 33.3°F (+0.7°C). The disease would be mild if this average was less than 30°F (−1.1°C). This prediction scheme was developed even without understanding that the bacterium overwintered in the corn flea beetle.[13]

B. BASED ON AMOUNT OF INITIAL INOCULUM OR INITIAL DISEASE

Three types of diseases can be predicted from the knowledge of initial inoculum or initial disease: monocyclic diseases; polycyclic diseases for which the pathogen has few generations; polycyclic diseases for which the initial inoculum is potentially large.[10]

Scientists in Wisconsin used a simple technique to demonstrate whether root rot of peas (*Aphanomyces euteiches*) might be a problem. In this test, susceptible plants are planted in the greenhouse in soil obtained from pea fields. If the greenhouse tests show that severe root rot developed in a particular soil, the field from which the soil was taken is not planted with susceptible crop. This technique is equally useful for many other soil-borne pathogens like *Pythium*, *Phytophthora*, *Fusarium*, *Sclerotium*, *Rhizoctonia*, *Heterodera*, *Globodera*, etc.

Indirect assessment of pathogen propulations have proved useful in forecasting fire blight of apple and pear (*Erwinia amylovora*) in California.[15] The pathogen multiplies much more slowly at temperatures below 15°C than at temperatures above 17°C.

Some diseases caused by aerially dispersed polycyclic pathogens are forecast on the basis of pathogen monitoring. Spores in the air are trapped, and population size is determined. Spore trapping has been used to forecast early blight of potato caused by *Alternaria solani*. Early in the season most spores are produced from mycelium on plant debris; later spores from mycelium in infected tissue contribute to the aerial population. As plants mature they become more susceptible and fungicidal spray is necessary to suppress disease if the pathogen population is large. A sudden large increase in the number of spores trapped is associated with a need for fungicide.[16] Another notable example is that of *Venturia inaequalis* (apple scab) which is a polycyclic pathogen, but the initial inoculum is usually present in large amount and there are only a few secondary cycles of pathogenesis. Early in the season the inoculum consists of ascospores produced in perithecia in overwintered leaves on the orchard floor. The forecast assumes that initial inoculum is abundant and then predicts, on the basis of leaf wetness and temperature, whether slight, moderate, or severe disease will result.[10]

C. FORECASTS BASED ON SECONDARY INOCULUM
1. Forecast Based on Weather Favorable for Secondary Cycles

In late blight of potato and tomato (*Phytophthora infestans*), the initial inoculum is usually low and generally too small to detect and measure directly. Even with low initial inoculum, the initiation and development of a late blight epidemic can be forecast with

TABLE 1
Relationship of Temperature and Periods of High Moisture (⩾90% RH) to Late Blight Development Indicated by Severity Values[17,20]

Temperature (°C)	Intervals of time (h) at RH ⩾90%				
7.2—11.6	0—15	16—18	19—21	22—24	24+
11.7—15.0	0—12	13—15	16—18	19—21	22+
15.1—26.6	0—9	10—12	13—15	16—18	19+
Severity values	0	1	2	3	4

TABLE 2
Relationship of Spray Recommendation in BLITECAST to Favorable Days and Severity Values[17]

Recommendation	Severity values required to generate each recommendation	
	FD ⩾ 5	FD < 5
No spray	0—2	0—3
Warning	3	4
7-d spray schedule	4	5—6
5-d spray schedule	⩾5	⩾7

FD = favorable day.

reasonable accuracy if the moisture and temperature conditions in the field remain within certain ranges favorable to the pathogen. When constant cool temperatures between 10 and 24°C prevail, and the relative humidity remains over 75% for at least 48 h, infection will take place and a late blight outbreak can be expected from 2 to 3 weeks later. If within that period and afterward, several hours of rainfall, dew, or relative humidity close to saturation point occur, they will serve to increase the disease and will foretell the likelihood of a major late blight epidemic.[11]

In recent years, computerized predictive systems for late blight epidemics have been developed. One such system is called BLITECAST. It was developed by plant pathologists at Pennsylvania State University.[8,17] BLITECAST combined two older late blight forecasting techniques. The first of these uses the accumulation of favorable days to indicate when the first spray should be applied. A favorable day is one for which mean temperature of the previous 5 d is less than 25.6°C and total rainfall for the 10 preceeding days was less than 3 cm or more.[18,19] Late blight was predicted to occur within 1 to 2 weeks after the occurrence of 10 consecutive favorable days. The second technique incorporated in BLITECAST is one based on relative humidity and temperature.[20] In this system "severity" values are assigned to different combinations of high relative humidities (⩾90%) and temperature for different periods of time (Table 1). With this technique late blight is expected to occur 1 to 2 weeks after 18 severity values have accrued, and, therefore, a fungicide spray is recommended.

BLITECAST combines these methods as given in Table 2. Relative humidity and temperature data are required in beginning when about half of the plants have emerged. Sensors that measure these parameters are placed within the plant canopy to measure the effect of microenvironment on the fungus. Initially, BLITECAST was implemented with computer technology. A computer was programmed to process weather data, determine the hazard of infection, and provide growers with spray recommendations.[18]

MacHardy[21] developed a simplified, non-computerized program known as the "New Hampshire Late Blight Forecasting Program". The criteria established for forecasting and spray recommendations are derived from BLITECAST. The two programs differ mainly in the means by which the final decision for fungicide application is made. BLITECAST requires a computer to issue recommendations whereas the "New Hampshire Program" relies on the grower.

Several leaf spots (such as those caused by *Cercospora* on peanuts and celery and *H. turcicum* on maize) can be predicted by taking into account the number of spores trapped daily, the temperature, and the duration of periods with relative humidity near 100%. For example, in case of *Cercospora* leaf spot of peanuts rapid epidemics are associated with periods of high relative humidity ($\geq$95%) for more than 10 h.[22] The relationship of temperature and duration of high relative humidity to the rate of disease increase is estimated.[23] Parvin et al.[24] incorporated this information into a computer program which is used to generate a daily peanut leaf spot spray advisory service and this also enhances the efficacy of fungicide application.[25]

2. Forecasts Based on Trapping Secondary Inoculum

The work of Berger[26,27] on celery early blight (*Cercospora apii*) is a notable one. In the Belle Glade area of Florida where much celery is grown, the growers make 20 to 35 fungicide applications during 6 months growth period of the crop if weather is favorable to disease development.[26,27] The forecast relates the number of spores detected in a spore trap to probable disease intensity and a spray frequency is recommended. It was found that if spore count per day was 100 to 300, 300 to 500, or <500, 2, 3, or 3 to 7 fungicide sprays per week, respectively, are needed to suppress the disease adequately.[27]

D. FORECASTS BASED ON AMOUNTS OF INITIAL AND SECONDARY INOCULUM

In wheat leaf rusts (*Puccinia recondita*) and stem rust (*P. graminis* f. sp. *tritici*) short forecasts have been developed by taking into account disease incidence, stage of plant growth, and spore concentration in the air. When the regression model was tested it gave a reasonably accurate estimate of disease intensity 1 week into the future, but it was less accurate in forecasting disease intensity 3 weeks ahead.[28]

In many insect-transmitted virus diseases (for example, barley yellow dwarf, cucumber mosaic, sugarbeet yellows), the likelihood, and sometimes the severity, of epidemics can be predicted. This is accomplished by determining the number of aphids, especially viruliferous ones, coming into the field at certain stages of the plant growth. A number of aphids trapped are tested for virus by inoculating them on healthy plants, or by testing them for virus serologically through the ELISA technique. The more numerous the viruliferous aphids and the earlier they are detected, the more rapid and more severe will be the virus infection.[11]

Such predictions can be further improved by taking into account late winter and early spring temperatures, which influence the population size of the overwintering aphid vectors. Above-average temperatures during late winter and early spring were associated with a high incidence of yellows,[29] and low temperatures during this time of year were associated with relatively little disease. Presumably, low temperatures reduced the overwintering populations of vectors, whereas higher temperatures permitted a greater proportion to survive and increase to a large early population.[30]

E. BASED ON METEOROLOGICAL FACTORS

Attempts have been made to correlate the effect of temperature, relative humidity, total rainfall, number of rainy days, and duration of sunshine on disease severity. Based on these

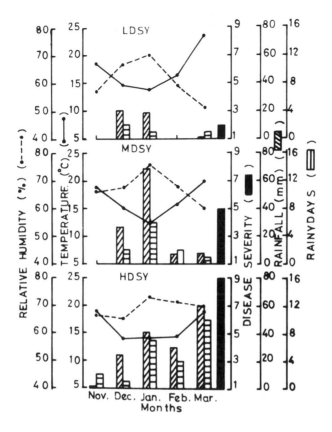

FIGURE 1. Influence of meteorological factors on Ascochyta blight of chickpea and its prediction, LDSY—Low disease severity year, MDSY—Moderate disease severity year, and HDSY—high disease severity year.

parameters linear equations have been derived to predict disease severity for maize northern leaf blight,[31] Erwinia stalk rot of maize,[32] and ascochytosis of chick pea.[33]

In ascochytosis of chickpea,[33] based on severity of disease the crop seasons were categorized as high disease severity year (HDSY), moderate disease severity year (MDSY), and low disease severity year (LDSY) (Figure 1). Number of rainy days and total rain fall during the crop season had significant effect on disease severity, while ambient temperatre, relative humidity, and sunshine duration had a less pronounced effect. Severe disease years had a maximum of 24 rainy days with 166 mm rainfall. Based on the partial regression coefficients, a linear equation was derived to predict the severity of disease depending on weather conditions prevailing during the crop season. The equation reads as

$$Y = a + (0.39)X_1 + (0.20)X_2 + (-0.0035)X_3 + (0.95)X_4 + (0.64)X_5$$

where Y = disease severity, $a = -22.17$, X_1 = average temperature, X_2 = average relative humidity, X_3 = total rainfall, X_4 = total rainy days, and X_5 = sunshine hours.

REFERENCES

1. **Miller, P. R. and O'Brien, M. J.,** Plant disease forecasting, *Bot. Rev.,* 18, 547, 1952.
2. **Bourke, P. M. A.,** Potato blight and the weather in Ireland (Tech. Note No. 15) in 1953, *Irish Meteorol. Serv., Dublin,* 1937, 15.
3. **Cox, A. E. and Large, E. C.,** Potato blight epidemics throughout the world, *Agric. Handb.* No. 174, USDA Washington, 1960, 230.
4. **Bourke, P. M. A.,** Use of weather information in the prediction of plant disease epiphytotics, *Annu. Rev. Phytopathol.,* 8, 345, 1970.
5. **Miller, P. R.,** Plant disease epidemics—their analysis and forecasting, in *Paper presented at the F.A.O. Symp. Crop Losses, 1967,* F.A.O. Rome, 1967, 330.
6. **Miller, P. R. and O'Brien, M. J.,** Prediction of plant disease epidemics, *Annu. Rev. Microbiol.,* 11, 77, 1957.
7. **Waggoner, P. E.,** Forecasting Epidemics, in *Plant Pathology,* Vol. 3, Horsfall, J. G. and Dimond, A. E., Eds., Academic Press, New York, 1960, 291.
8. **Krause, R. A. and Massie, L. B.,** Predictive systems: modern approaches to disease control, *Annu. Rev. Phytopathol.,* 13, 31, 1975.
9. **Shrum, R. D.,** Forecasting of Epidemics, in *Plant Diseases: An Advanced Treatise,* Vol. 2, Horsfall, J. G. and Cowling, E. B., Eds., Academic Press, New York, 1978, 436.
10. **Fry, W. E.,** *Principles of Plant Disease Management,* Academic Press, New York, 1987, 378.
11. **Agrios, G. N.,** *Plant Pathology,* 3rd ed., Academic Press, New York, 1988, 803.
12. **Stevens, N. E.,** Stewart's disease in relation to winter temperatures, *Plant Dis. Rep.,* 18, 141, 1934.
13. **Elliott, C. and Poos, F. W.,** Over-wintering of *Aplanobacter stewartii, Science,* 80, 289, 1934.
14. **Sherwood, R. T. and Hagedorn, D. J.,** Determining the common root rot potential of pea field, *Wis. Agric. Exp. Bull.,* 1958, 531.
15. **Thomson, S. V., Schroth, M. N., Reil, W. O., Beutel, J. A. and Davis, C. S.,** Pesticide applications can be reduced by forecasting the occurrence of fireblight bacteria, *Calif. Agric.,* 31, 12, 1977.
16. **Harrison, M. D., Livingston, C. H., and Oshima, N.,** Control of potato early blight in Colorado, I. Fungicidal spray schedules in relation to the epidemiology of the disease, *Am. Potato J.,* 42, 319, 1965.
17. **Krause, R. A., Massie, L. B., and Hyre, R. A.,** Blitecast: A computerized forecast of potato late blight, *Plant Dis. Rep.,* 59, 95, 1975.
18. **Hyre, R. A., Bonde, R., and Johnson, B.,** The relation of rainfall, relative humidity, and temperature to late blight in Maine, *Plant Dis. Rep.,* 43, 51, 1959.
19. **Hyre, R. A., Bonde, R., and Manzer, F. E.,** Re-evaluation in Maine of three methods for forecasting late blight of potato, *Plant Dis. Rep.,* 44, 235, 1960.
20. **Wallen, J. R.,** Summary of recent progress in predicting late blight epidemics in United States and Canada, *Am. Potato. J.,* 39, 306, 1962.
21. **Mac Hardy, W. E.,** A simplified, non-computerized program for forecasting potato late blight, *Plant Dis. Rep.,* 63, 21, 1979.
22. **Jensen, R. E. and Boyle, L. W.,** The effect of temperature, relative humidity and precipitation on peanut leaf spot, *Plant Dis. Rep.,* 49, 975, 1965.
23. **Jensen, R. E. and Boyle, L. W.,** A technique for forecasting leaf spot on peanuts, *Plant Dis. Rep.,* 50, 810, 1966.
24. **Parvin, D. W. Jr., Smith, D. H., and Crosby, F. L.,** Development and evaluation of a computerized forecasting method for Cercospora leaf spot of peanuts, *Phytopathology,* 64, 385, 1974.
25. **Smith, D. H., Crosby, F. L., and Ethredge, W. J.,** Disease forecasting facilitates chemical control of Cercorpora leaf spot of peanuts, *Plant Dis. Rep.,* 58, 666, 1974.
26. **Berger, R. D.,** Forecasting of Cercospora blight of celery in Florida, *Phytopathology,* 59, 1018 (Abstr.), 1969.
27. **Berger, R. D.,** A celery early blight spray program based on disease forecasting, *Proc. Fla. State Hortic. Soc.,* 82, 107, 1969.
28. **Burleigh, J. R., Eversmeyer, M. G., and Roelfs, A. P.,** Development of linear equations for predicting wheat leaf rust, *Phytopathology,* 62, 947, 1972.
29. **Hurst, G. W.,** Forecasting the severity of sugarbeet yellows, *Plant Pathol.,* 14, 47, 1965.
30. **Hull, R.,** Research on the sugarbeet crop, *Ann. Appl. Biol.,* 82, 1, 1976.
31. **Levi, Y. and Noy Meir, I.,** A simulation model of the northern leaf blight of corn, *Phytoparasitica,* 9, 242, 1981.
32. **Saxena, S. C. and Lal, S.,** Use of meteorological factors in prediction of Erwinia stalk rot of maize, *Trop. Pest Manage.,* 30, 82, 1984.
33. **Chaube, H. S.,** Ascochyta blight of chickpea, *Tech. Report,* Expt. St., G.B. Pant University, Pantnagar, India, 110, 1987.

Chapter 12

REGULATORY METHODS

I. INTRODUCTION

Unrestricted movement of plants and plant products within and between countries has resulted in worldwide distribution of many plant pathogenic organisms. While it is not always possible for alien organisms to survive under new ecological and climatic conditions, many of them can adapt themselves to new environments to become eventually potentially dangerous to the crop plants of those areas. The international exchange of plants and their parts is now practiced widely to improve the crops of a country and their genetic base.[1] In addition, shiploads of grains for consumption or large quantity of seeds for sowing is practiced in many countries.[2] Even minute quantities of soil and plant debris contaminating true seeds can disseminate pathogens.[3] Kahn[4] has very rightly pointed out that a segment of scientific community has not fully accepted the concept that risks may be associated with imported germplasms particularly in small lots. Therefore, in order to prevent introduction and spread of plant pathogens and their races/biotypes into the country or states where they are not known to exist, certain federal and state laws must regulate the exchange of such materials.

II. PRINCIPLES

One of the six principles of management of plant diseases is the exclusion of the inoculum by preventing the inoculum from entering or establishing into the field or area where it does not exist. In this commercialized world today, through easy and quick means of transportation, there is exchange of not only grains and other food stuffs between nations but also of germplasms or seeds for the improvement of crops especially in developing nations. Past experiences have revealed that no country or state can afford unrestricted movement of plant and planting materials. Thus, the knowledge and methodology of exclusion of the pests and pathogens are enforced by a legally constituted authority to prevent the entry and thus spread of injurious crop pests and pathogens in the public interest.

III. PRACTICES EMPLOYED FOR DISEASE MANAGEMENT

A. PLANT QUARANTINE
1. Concept and Importance
Quarantine is derived from the Latin word *quarantum* meaning 40. It refers to a 40-d period of detection of ships arriving from countries with bubonic plague and cholera in the middle ages.[3] Plant quarantine may be defined as ''the restriction imposed by duly constituted authorities on the production, movement and existence of plants or plant materials, or animals or animal products or any other article or material or normal activity of persons and is brought under regulation in order that the introduction or spread of a pest may be prevented or limited or in order that the pest already introduced may be controlled or to avoid losses that would otherwise occur through the damage done by the pest or through the continuing cost of their control''.[5] Thus, plant quarantines promulgated by a government or group of governments, restrict the entry of plant or plant products, soil, culture of living organisms, packing materials, and commodities as well as their containers and means of conveyance.

The adoption of Quarantine Regulations and Acts by different countries of the world has arisen out of the fact that extensive damages, often sudden in nature, have been caused not by indigenous organisms but by exotic ones which have been introduced along with plants, plant parts or seeds in the normal channel of trade or individual transit. Instances may be cited of the introduction of grape *Phylloxera (Phylloxera vitifolii)* from the U.S. to France which caused destruction of French vineyards; Mexican boll weevil (*Anthonomus grandis*) whose original home was in Mexico or Central America, about 1892 entered the U.S. and later to various countries of the world, causing extensive damage to cotton; European corn borer (*Ostrinta nubilalis*) which reached North America probably through broom corn from Italy or Hungary and has since become a major pest there. Pink boll worm (*Pectinophora gossypiella*) considered to be one of the six most destructive insects of the world probably a native of India, is now established as a highly destructive pest in almost all cotton growing areas of the world.

Some of the examples cited above thus, clearly reveal that through unrestricted movement of plants and planting materials, diseases and pests have been introduced in regions, where they were not known to exist. Some of the important plant pathogens introduced into different countries of the world and where they caused severe losses, are listed in Table 1.

2. Principles, Basis, and Justification

All the nations of the world are passing through a period of intensive agricultural development in an effort to accelerate food production. All available resources are being mobilized to step up food production which include introduction of high yielding exotic germplasm. However, associated with these introductions is the danger of introducing some serious pathogens or their biotypes or races unknown in the importing country. Sometimes grains imported for milling purposes find their way into the farmers field thereby introducing organisms. Thus, international trade with regards to plant and plant products has eventually put a great burden on quarantines all over the world.

Plant quarantines promulgated by a government or group of governments, restrict the entry of plants, plant products, soil, culture of living organisms, packing materials, and commodities as well as their containers and means of conveyance and thus, help to protect agriculture and environment from avoidable damages by hazardous organisms. However, quarantines are justified only if the organism has little or no chance of spreading naturally. Thus, before its enforcement, the nature of the pathogen, its mode of spread/transmission, host range, and natural barriers should be properly understood. In addition, socioeconomic and geopolitical factors likely to work against the interests of exporting and importing countries too should be taken into account.

Plant quarantine regulations in order to be effective have to be based on sound scientific principles.[1,7-9] They are (1) the biology and ecology of the organism against which quarantine measure is proposed to be enforced should be known; only those organisms which are supposed to pose threat to major crops and forests should be taken into consideration, (2) in the event of its introduction whether the organism is likely to be established and cause damage of any consequence, (3) formulated to prevent or control the entry of the organisms and not to hinder trade or attainment of other objectives, as quarantine measures are for the crop and not for trade protection, (4) derived from adequate legislation and operated solely under the law, (5) amended as conditions change or further facts become available, and (6) attended by trained and experienced workers and the public must cooperate on an international scale for effective operation of the quarantine regulations.

3. Plant Quarantine as a Management Practice

Management of plant diseases refers to prevention of infection or to a reduction in incidence and severity. Plant quarantine as a practice covers two basic principles of disease management: exclusion and eradication. When germplasm is regulated by quarantine, the entry of organism is prevented by inspection and treatment, or the host is banned or otherwise restricted. According to Kahn[4] "exclusion" is a positive-image term that relates to keeping

TABLE 1
Some Examples of Plant Pathogens Introduced into Some Other Countries[3,6]

Disease and Pathogen	Introduced from	Introduced into	Year of introduction
American goose berry mildew (*Sphaerotheca morsuvae*)	N. America	England	1899
Bacterial canker of tomato (*C. michiganense* pv. *michiganense*)	U.S.	U.K.	1942
Bacterial leaf blight of paddy (*X. campestris* pv. *oryzae*)	Phillipines	India	1959
Black rot of crucifers (*X. campestris* pv. *campestris*)	Java	India	1929
Black shank of tobacco (*Phytophthora nicotianae*)	Holland	India	1938
Blister rust of pines (*Cronartium ribicola*)	Europe	U.S.	1910
Bunchy top of banana (Viral)	Sri Lanka	India	1940
Chestnut blight (*Endothia parasitica*)	Asia	U.S.	1904
Citrus canker (*X. campestris* pv. *citri*)	Asia	U.S.	1907
Powdery mildew of cucurbits (*Erysiphe cichoracearum*)	Sri Lanka	India	1910
Downy mildew of grapes (*Plasmopara viticola*)	U.S. Europe	France India	1878 1910
Downy mildew of maize (*Sclorospora phillipinensis*)	Java	India	1912
Dutch elm (*C. ulmi*)	Holland	U.S.	1928—30
Flag smut of wheat (*Urocystis tritici*)	Australia	India	1906
Fire blight of apple (*E. amylovora*)	N. America	New Zealand	1919
Golden nematode of potato (*G. rostochiensis*)	Europe	U.S., Mexico, India	1881 1961
Hairy root of apple (Viral)	England	India	1940
Late blight of potato (*Phytophthora infestans*)	S. America U.K.	Europe India	1830 1883
Leaf rust of coffee (*Hemileia vastatrix*)	Sri Lanka Asia, Africa	India Brazil	1879 1970
Onion Smut (*Urocystis cepulae*)	Europe	India	1958
Paddy blast (*Pyricularia oryzae*)	S.E. Asia	India	1918
Peanut rust (*Puccinia arachidis*)	Brunei U.S.	Brazil	—
Powdery mildew of grape (*Uncinula necator*)	North America	England	1845
Powdery mildew of rubber (*Oidium heavea*)	Malaya	India	1938
Rye grass seed infection (*Gleotinia temulenta*)	New Zealand	Oregon	1940
Wart of potato (*Synchytrium endobioticum*)	Netherlands	India	1953
Wheat bunt (*Tilletia caries*)	Australia	California, U.S.	1854
Witches broom of cocoa (*Marasmius perni*)	Trinidad	South America	1974—75

organisms out whereas "plant quarantine" is a negative image that relates to keeping plants out. This concept has been recognized by the California Department of Agriculture which employs a "detection and exclusion officer" rather than a "plant protection officer".

4. Some Important Quarantines

Strict and timely application of quarantine laws along with suitable eradication practice have resulted in control of several plant diseases. For example, rubber plantation in Southeast Asia. In Southeast Asia, rubber trees are susceptible to *Mycrocyclus ulei* (South American leaf blight). The environment is also congenial for the development of the pathogen and disease. The quarantine regulations, publicity, surveillance, and eradication of diseased trees prevented the pathogen from being established in Southeast Asia. Similar is the case with citrus canker (*X. campestris* pv. *citri*) in U.S. The disease was first observed in 1912. Early in 1914 the disease was confirmed in all Gulf Coast states and Florida and prohibited importation of citrus stock. Subsequently the U.S. government supported eradicatory efforts. Burning the infected trees in nurseries and groves was undertaken. Subsequently, continued vigilance and eradication practices have prevented establishment of *X. campestris* pv. *citri*.[10]

Wart of potato caused by fungal pathogen *Synchytrium endobioticum* was observed in 1919 in West Virginia.[11] Eventually, 70 garden sites were found infested with the pathogen. In 1921, the area was quarantined. Only varieties of potato immune to *S. endobioticum* were allowed to be grown and movement of soil and root crop vegetables required a special permit. Persistent quarantine and eradication efforts resulted in extinction of the disease. By 1973, all infested areas were declared free of the pathogen.[11] Similar is the story of this disease in India. *S. endobioticum* was introduced in the Darjeeling hills in 1913 from Netherlands. Quarantine and other eradicatory and preventive efforts have succeeded in restricting the disease only in Darjeeling hills.

Golden nematode of potato (*Globodera rostochiensis*) was first discovered in New York state (Long Island) in 1941, and subsequently in Western New York, Delaware, and New Jersey. Infestations in Delaware and New Jersey appear to have been contained.[12] With active and coordinated efforts of state and federal agencies, growers and universities, the population of the nematode has been brought below levels at which dispersal is unlikely to occur.[13] The effort has prevented crop loss as well. This pathogen was introduced into Nilgiri hills, India in 1961 from Europe. Quarantine regulations and continuous monitoring of the disease have resulted in restriction of the disease in Nilgiri areas so far.

5. Organisms of Quarantine Significance and Risk Analysis

Organisms of quarantine significance may include any pest or pathogen that a government (or intergovernment organization) considers to pose a threat to the country's (or region's) agriculture and environment. Such organisms are usually exotic to that country or region but may also include exotic strains or races of domestic organism. For larger countries the definition may also be extended to include foreign isolates of domestic organisms.

Analysis of pest and pathogen risks is the decision making process that brings the biological approach into play. The analysis is based on two general precepts:[4] (1) the benefits must exceed the risk, and (2) the benefits must exceed the cost. The benefits consist of the opportunity to introduce new crops or new varieties of old crops or to introduce new genes to improve existing varieties. Such improvement may consist of increased yields through increased resistance, changes in maturity periods, better quality, increased protein or oil levels, etc. When costs are entered into the pest risk analysis, the costs of adequate safeguards are taken into account. But infact, costs in a risk analysis involve more than the costs of safeguards. Costs enter the picture in a more general sense in that the predicted benefits resulting from the improvement in agriculture must greatly exceed the potential losses should a pest or pathogen of quarnatine significance be introduced and require eradication or suppression.

In the U.S., estimation of the expected economic impact has been used to help establish priority of quarantine activity. The expected economic impact is a function of the estimated probability of successful establishment and the expected economic impact of the pest once established.[14] In a report published in 1973, the expected economic impacts of 551 exotic plant pathogens were established and the pests considered to be most serious economically were identified.[14] This can be used by the U.S. Animal and Plant Health Inspection Service (APHIS) to focus its activities on the most important pests. Descriptions of tropical diseases of global importance were developed by Thurston[15] and these could aid countries in the tropics to establish priorities for their quarantine efforts. A list of threatening diseases of global importance is given in Table 2.

6. Entrance and Establishment of Pathogens

An exotic species must first gain entry and then it must become established.

It is an extremely difficult propostion to predict accurately whether an exotic organism will become established and, once established, become economically important. In a relatively few cases, the pathogeographical approach has led to prediction of the occurrence of pathogens based on knowledge of the life cycles, distribution of the pathogen, and ecological characteristics of the host and pathogen. Their assessment revealed that two organisms, the potato wart fungus (*Synchytrium endobioticum*) and the pathogen of citrus scab (*Elsinoe fawcetti*), could not survive unfavorable seasons. Thus, the potato wart fungus does not survive in soil where the temperature rises to 30°C in any season. Consequently, it was recommended that quarantine regulations against this pathogen be relaxed in tropical countries.

Some factors which affect entry and establishment include:[4] (1) hitchhiking potential compared with natural dispersal; (2) ecological range of the pest as compared with ecological range of its host, i.e., climate vs. life-cycle vs. natural enemies vs. population of susceptible hosts, etc; (3) weather; (4) ease of colonization, including reproductive potential; and (5) agricultural practices including pest management.

7. Biogeographical Regions

One of the basic tenets of plant quarantine is that the larger the landmass covered by uniform quarantine regulations, the greater the protection to any area therein. The regionalization concept was discussed by Mathys,[16,17] who proposed eight such regions (Figure 1) to serve as basis for coordinated action. Within each region harmful organisms of quarantine significance should be defined. Similarly, those noxious organisms already present in some parts of the region and for which prevention of spread, control or eradication by central or joint quarantine actions are likely to succeed have to be determined.

8. Plant Quarantine Regulations

a. Brief History

In 1660, France promulgated first quarantine law against barberry. In 1873, an embargo was passed in Germany to prevent importation of plant and plant products from the U.S. to prevent the introduction of the Colorado potato beetle, and in 1877, the United Kingdom Destructive Pests Act prevented the introduction and spread of this beetle. In 1891, the first plant quarantine measure was initiated in the U.S. by setting up a seaport inspection station at San Pedro, California, and the first U.S. quarantine law was passed in 1912. The Federal Plant Quarantine Service was established in Australia in 1909.[3,7] In India, a Destructive Insect and Pests Act was passed in 1914. Since then most of the countries have formulated quarantine regulations.

On a global basis, the first International Plant Protection Convention (the *Phylloxera* Convention) was signed in 1881, with the objective of preventing the spread of severe pests.

TABLE 2
Threatening Plant Diseases of Global Importance[15]

Disease	Pathogen

Limited Threat Potential

Disease	Pathogen
American leafspot of coffee	*Mycena citricolor*
Coconut cadang cadang	Viroid
Enanisimo of barley, oats, wheat	Etiology unknown
Potato rust	*Puccinia pittieriana*

Intermediate Threat Potential

Disease	Pathogen
Bunchy top of banana	Virus
Cocoa swollen shoot	Virus (CSSV)
Gummosis of imperial grass	*Xanthomonas axonoperis*
Hoja blanca of rice	Virus (HBV)
Lethal yellowing of coconut	M L O
Monilia pod rot of cocoa	*Monilia roreri*
Red ring of coconut	*Rhadinaphelenchus cocophilus*
Streak disease of maize	Virus
Stunting virus of pangola grass	Virus

High Threat Potential

Disease	Pathogen
African cassava mosaic	Virus
Bacterial leaf blight of rice	*X. campestris* pv. *oryzae*
Downy mildew of maize	*Sclerospora* spp.
	Sclerophthora spp.
Moko disease of bananas and plantains	*Pseudomonas solanacearum*
South American leaf blight of rubber	*Microcyclus ulei*

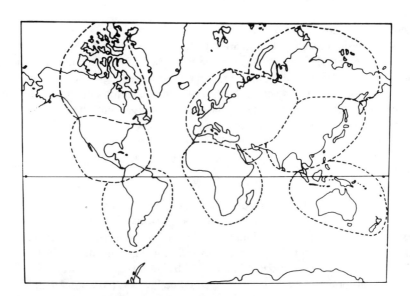

FIGURE 1.   Proposed repartition in biogeographical world regions for quarantine purpose (Mathys, G., *EPPO Bull.*, 5, 55, 1975. With permission).

This convention was amended in 1889, 1929, and 1951. The International Plant Protection Convention (IPPC or Rome Convention) under the FAO was established to prevent the introduction and spread of diseases and pests through legislation and organizations across international boundaries.[8] This convention provided a model phytosanitary certificate (Rome certificate) to be adopted by member countries. Within this convention, ten regional plant protection organizations have been established on the basis of biogeographical areas.[2,7,8,18] These are European and Mediterranean Plant Protection organization (EPPO), Inter-African Phytosanitary Council (IAPSC), Organismo International Regional de-Sanidad Agropecuaria (OIRSA), Plant Protection Committee for the South-East Asia and Pacific Region (SEAPPC), Near East Plant Protection Commission (NEPPC), Comité Interamericano de Protection Agricola (CIPA), Caribbean Plant Protection Commission (CPPC), North American Plant Protection Organization (NAPPO), Organismo Bolivariano de Sanidad Agropecuaria (OBSA), and ASEAN region grouping Indonesia, Malasia, Phillipines, Thailand and Singapore. The regional organizations are concerned with the coordination of legislation and regulations within their area, agreement on the quarantine objects, inspection procedures etc.[2,3,18]

b. Plant Quarantine in U.S.

The control of plant introduction into the U.S. began in 1829, when the U.S. Congress allotted $1000 for the import of rare plants and seeds.[3,19] The office of the U.S. Patent Commissioner was authorized for the introduction of germplasm between 1836 to 1862. With the establishment of U.S. Department of Agriculture (USDA) in 1862, a Commissioner of Agriculture was made responsible for collection, testing, and distribution of potentially valuable plant germplasm. A section, Seed and Plant Introduction, was established in 1898. This system was continued with minor changes.[1,3] Over 465,000 plants or seeds have been introduced since 1898. At present, about 7500 new introductions are made each year. The Plant Protection and Quarantine Program (PPQ) is planned and executed by the Animal and Plant Health Inspection Service (APHIS) of the USDA.[2,3] At present three quarantine acts are in operation in the U.S.[1,3]

i. Plant Quarantine Act of 1912

The first U.S. federal plant quarantine law known as the Plant Quarantine Act of 1912 was passed after the establishment of white pine blister rust and chestnut blight fungi, and the citrus canker bacterium. The act controls the introduction of exotic pests and the spread of plants new to U.S.and within the U.S. as a domestic quarantine.[20]

ii. Organic Act of 1944

This act is mainly for pest management strategies, but gives an authority for issuance of phytosanitary certificates in accordance with the requirements of the importing states and foreign countries.

iii. Federal Plant Pest Act of 1957

This act authorizes emergency actions to prevent the introduction or interstate movement of plant pests not covered under the act of 1912.

Nearly all imported germplasms fall into one of the three categories: restricted, postentry, or prohibited. Restricted germplasm is inspected and chemically treated, and can be imported easily. In the postentry category, seeds or other materials, after inspection and treatment, are grown under close observation. If no pest is found, the germplasm is released. Prohibited materials must meet certain specific requirement before being imported since they may pose serious threat to agriculture.[1] Import of some major crops seeds prohibited in the U.S. are given in Table 3.

Postentry surveillance for the detection and interception of seed-borne pathogens on introduced plants and the production of disease-free seeds are accomplished at the regional

TABLE 3
Import of Some Major Crops Seeds Prohibited in the U.S.[1]

Crop	Country	Disease
Gossypium spp. (cotton)	All countries	Various diseases
Lens spp. (lentil)	South America	Rust
Oryza sativa (paddy, rice)	All countries	Smuts, viruses and other diseases
Sorghum vulgare (milo, sorghum)	Africa, Asia, Brazil	Smuts
Triticum aestivum (wheat)	Asia, Australia, Eastern Europe	Flag smut
Zea mays (corn, maize)	Africa, Asia	Downy mildews

plant introduction stations. These stations are operated by co-operative agreement between the USDA and land-grant colleges and universities. Here, the plants are subjected to inspection, detection, postentry surveillance, and release of seeds.[21]

c. Plant Quarantine in U.K.

The Plant health legislation in the U.K. was passed as the Destructive Insects Act of 1877 to prevent the entry and establishment of Colorado beetle. It was extended by the Destructive Insects and Pests Act of 1907 to check the entry of American gooseberry mildew (*Sphaerotheca morsuvae*) and all insects, fungi, or other destructive pests of plants. To cover bacteria and viruses as well as invertebrate pests, the act was extended as the Destructive Insects and Pests Act of 1927. These three acts were consolidated and formulated in a Plant Health Act of 1967.[3,22] This act was amended by the European Committees Act of 1972. The most familiar activity is the inspection of plants or produce either before export or after import.[23]

d. Plant Quarantine in India

Wadhi[24] has reviewed plant quarantine activities in India. The earlier activities concerned with the introduction of plant pests and diseases with plant material was in early 1900s. Fumigation of all imported cotton bales was required to prevent introduction of the Mexican boll weevil (*Anthonomus grandis*). In 1914, the Government of India passed the Destructive Insects and Pests Act prohibiting or restricting the import of plant and plant materials, insects, fungi, etc., to India from foreign countries. This comes under foreign quarantine. Rules and regulations have been made prohibiting or restricting the movement of certain diseased and pest-infested materials from one state to another in India. This comes under domestic quarantine.

The enforcement of plant quarantine regulations is carried out by the technical officers of the Directorate of Plant Protection and Quarantine, Ministry of Agriculture, Government of India, under the overall supervision of Plant Protection Advisor. There are eight quarantine stations at seaports, seven at airports and seven at land frontiers. Research material is examined by three agencies, the National Bureau of Plant Genetic Resources, New Delhi, for agricultural and horticultural crops; the Forest Research Institute, Dehradun, for forest plants, and Botanical Survey of India, Calcutta, for all other plants of general economic significance.

e. International Organizations

The first successful attempt towards solving a major pest problem on a collaborative basis resulted in the establishment in Europe of the *Phylloxera* Convention of 1881. Each contracting government had to guarantee that only American rootstocks were used in vine growing so as to secure containment of *Phylloxera vitifolia,* the root aphid which had been devastating the European vine yards. In 1929, a first world Plant Protection Convention was signed in Rome. In 1945, the United Nations and its specialized agency, the Food and Agriculture Organization (FAO), were established. This provided the best basis for the development of plant protection on a global level. The real breakthrough in this respect occurred with the establishment of the 1951 FAO International Plant Protection Convention. For implementation of various tasks the signatory governments have to make provision for: (1) the establishment of an official plant protection service, mainly in charge of quarantine matters, including plant health certification, and surveillance of the phytosanitary situation within the country and controlling harmful organisms of major importance; (2) a system securing technology transfer; and (3) a research organization.

The FAO Plant Protection Convention stipulates that governments have to cooperate with one another through regional plant protection organizations of the appropriate areas. At present there are eight regional plant protection organizations covering different areas of the world[17] (Table 4).

The regional organizations, listed in Table 4, differ from each other in scope and functions, some of them dealing with all protection technologies and others specifically with quarantine. However, they all serve as advisory and coordinating bodies to participating governments. They operate as a network to promote concerted action for combating pests of international importance and to propose quarantine measures for preventing the spread of these pests.

9. Plant Quarantine Measures

Plant quarantine measures operative in different countries are summarized in flow chart given below:

Factors Affecting Disease Development

Host factors	Pathogen factors	Environmental factors	Biotic factors
–Host reaction	–Aggressiveness	–Temperature	–Rhizosphere
–Plant nature	–Virulence	–Moisture	–Phyllosphere
–Plant type	–Parasite fitness	–RH	
–Plant vigor	–Variability	–Shade	
–Plant density	–Latent infection	–Light	–Vectors
–Plant age	–Inoculum potential	–Soil fertility	
–Host exudates		–Soil reaction	

Enforcement of embargoes is the most effective measure to exclude diseased plant materials. The seed lots/samples must be examined by using prescribed, sensitive and reliable methods for detection of pathogens listed under plant quarantine regulations. To ensure that consignment is free from pathogens, the materials must be subjected to postentry quarantine. For example, seeds are subjected to a period of growth at a quarantine station under strict supervision of the officials of the importing country.[25,26] The plants are kept under close

TABLE 4
Regional Government Organizations[17]

Region	Organization[a]	Member governments	Establishment
Western hemisphere	Central America (OIRSA)	7	1955
	FAO Caribbean commission	12	1967
	South America: northern part (OBSA)	3	1965
	Southern part (CIPA)	6	1965
West and east palaearctic	EPPO	35	1951
	FAO Near-East commission	16	1963
Africa	North and Central Africa (IAPSC)	41	1967
	South Africa (SARCCUS)	8	1950
Asia	FAO Committee (SEAPPC)	18	1956

[a]   See text for abbreviations.

observation in isolation so that any disease which appears can be detected immediately. Plants are grown under optimum conditions so that symptoms are not masked.[3] Pathogen free seed is then produced from the imported seed material for distribution. Thus, valuable germplasm can be saved for breeding and crop improvement programs without any danger of introducing prohibited plant pathogens.[3]

Under export control the crop growing in the field is inspected regularly. Infected plants are rogued out. In case of seed lots, the seed samples are thoroughly examined before export. The sample should meet the standards of the importing country. The seeds should be treated with effective chemical or as per the requirement of the importing country. Phytosanitary certificates are issued by the exporting country along with the seeds as per International Plant Protection Convention of 1951. Intermediate quarantine is an international cooperative effort to lower the risk of introducing a pathogen to one country with the germplasm from another by passing this germplasm through isolation or quarantine in a third country. The selection of the third country is important. The pathogen in question should not pose a threat to the third country because either the crop is not grown there or the pathogen, even if it escapes, will not become established because of environment.[3,4] Third country plant quarantine locations are Plant Quarantine Facility, Glenn Dale Md; the U.S. Subtropical Horticulture Research Unit, Miami; Kew Botanical Gardens, U.K.; Royal Imperial Institute, Wageningen, The Netherlands; and IRAT at Nogentsur Marne, France. The U.S. serves as a third country for the international exchange of coffee, tea, rubber, and cacao.[3,4]

10. Guidelines for Import of Germplasm[3]

1.   Import from a country where the pathogen(s) is absent.
2.   Import from a country with an efficient plant quarantine service, so that inspection and treatment is done.
3.   Obtain plant material from the safest known source within the selected country.
4.   Obtain untreated seeds so that detection of seed borne pathogens is facilitated.
5.   Obtain clean healthy looking seeds free from impurities.
6.   Obtain an official certificate of freedom from pests and diseases from the exporting country.
7.   Import the smallest possible amount of planting material; the smaller the amount the less the chance of its carrying infection. It will also simplify postentry inspection.
8.   Inspect material carefully on arrival and treat.

9. If other precautions are not adequate, subject the material to intermediate or postentry quarantine.

11. Problems in Plant Quarantines

Quarantines serve as a filter against the introduction of hazardous pathogens but still pathogens are introduced. Possible reasons are that:[3]

1. It is difficult to detect all types of infectious pathogens by conventional methods
2. The methods may not be sensitive enough to reveal traces of infection
3. The latent infections may pass undetected under postentry quarantine
4. Destruction of all infected or suspected material, and
5. Lack of sensitive methods for testing fungicide treated seeds.

Plant pathogens may be introduced on inert material such as packing material, dried root bits, plant debris, soil clods etc. Cysts of *Heterodera schachtii* and *Heterodera goettingiava* nematodes have been intercepted on such materials.[3,27]

REFERENCES

1. **Waterwarth, H. E. and White, G. A.,** Plant introduction and quarantine: the need for both, *Plant Dis.,* 66, 87, 1982.
2. **Neergaard, P. A.,** A review on quarantine for seed, in *Golden Jubilee Commemoration Volume,* National Academy of Sciences, New Delhi,1980.
3. **Agrawal, V. K. and Sinclair, J. B.,** *Principles of Seed Pathology,* Vol. II, CRC Press, Inc., Boca Raton, FL, U.S.A., 1987, 168.
4. **Kahn, R. P.,** Plant Quarantine: principles, methodology and suggested approaches, in *Plant Health and Quarantine in International Transfer of Genetic Resources,* Hewitt, W. B., and Chiarappa, Eds., CRC Press, Cleveland, 1977, 289.
5. **Reddy, D. B.,** The international plant protection convention for Southeast Asia and Pacific region, *Plant Prot. Bull.,* 25, 157, 1977.
6. **Mehrotra, R. S.,** *Plant Pathology,* Tata McGraw-Hill, New Delhi, 1980, 771.
7. **Mathys, G. and Baker, E. A.,** An appraisal of the effectiveness of quarnatines, *Annu. Rev. Phytopathol.,* 18, 85, 1980.
8. **Chock, A. K.,** The international plant protection convention, in *Plant Health,* Ebbels, D. L. and King, J. E., Eds., Blackwell Scientific, Oxford, 1979, 1.
9. **Morrison, L. G.,** Quarantine Principles and Policy, *SPC Workshop and Training Courses in Plant Quarantine,* Suve, Fiji, SPC, Suva, 1977, 2.
10. **Fry, W. E.,** *Principles and Practices of Plant Disease Management,* Academic Press, New York, 1987, 378.
11. **Brooks, J. L., Given, J. B., Baniecki, J. F., and Young, R. J.,** Eradication of potato wart in West Virginia, *Plant Dis. Rep.,* 58, 291, 1974.
12. **Evans, K. and Brodie, B. B.,** The origin and distribution of the golden nematode and its potential in the U.S.A., *Am. Potato J.,* 57, 79, 1980.
13. **Mai, F. W.,** World-wide distribution of potato cyst nematodes and their importance in crop protection, *J. Nematol.,* 9, 30, 1977.
14. **McGregor, R. C.,** People placed pathogens: The emigrant pests, in *Plant Diseases: An Advanced Treatise,* Vol. II, Horsfall, J. G. and Cowling, E. B., Eds., Academic Press, New York, 1978, 383.
15. **Thurston, H. D.,** Threatening plant diseases, *Annu. Rev. Phytopathol.,* 11, 27, 1973.
16. **Mathys, G.,** Thoughts on quarantine problems, *EPPO Bull.,* 5(2), 55, 1975.
17. **Mathys, G.,** Society sponsored disease management activities, in *Plant Diseases: An Advanced Treatise,* Vol. I., Horsfall, J. G. and Cowling, E. B., Eds., Academic Press, New York, 1977, 363.
18. **Singh, G. K.,** Regional ASEAN collaborations in plant quarantine, *Seed Sci. Technol.,* 11, 1189, 1983.
19. **Hodge, W. H. and Erlanson, C. O.,** Federal plant introduction: a review, *Econ. Bot.,* 10, 229, 1956.

20. **Rohwer, G. C.,** Plant quarantine philosophy of the United States, in *Plant Health—The Scientific Basis for Administrative Control of Plant Diseases and Pests,* Ebbels, D. L. and KIng, J. E., Eds., Blackwell Scientific, Oxford, 1980, 23.

21. **Leppik, E. E.,** Introduced seed-borne pathogens endanger crop breeding and plant introduction, *FAO Plant Prot. Bull.,* 16, 57, 1968.

22. **Ebbels, D. L. and King, J. E., Eds.,** *Plant Health—The Scientific Basis for Administrative Control of Plant Diseases and Pests,* Blackwell Scientific, Oxford, 1979, 322.

23. **Southey, J. F.,** Preventing the entry of alien diseases and pests into Great Britain, in *Plant Health—The Scientific Basis of Administrative Control of Plant Diseases and Pests,* Ebbels, D. L. and King, J. E., Eds., Blackwell Scientific, Oxford, 1979, 63.

24. **Wadhi, S. R.,** *Plant Quarantine Activity at the National Bureau of Plant Genetic Resources,* NBPGR Sc. Mongr. No. 2., National Bureau of Plant Genetic Resources, New Delhi, 1980, 99.

25. **Sheffield, F. M. L.,** Requirements of a post entry quarantine station, *FAO Plant Prot. Bull.,* 6, 149, 1958.

26. **Sheffield, F. M. L.,** Closed quarantine procedures, *Rev. Appl. Mycol.,* 47, 1, 1968.

27. **Sethi, C. L., Nath, R. P., Mathur, V. K., and Ahuja, S.,** Interception of plant parasitic nematodes from imported seed material, *Indian J. Nematol.,* 2, 89, 1972.

Chapter 13

PHYSICAL METHODS

I. INTRODUCTION

As early as 1832, Sinclair suggested that hot air treatment in an oven may control smuts of oats and barley.[1] On the other hand, hot water therapy was first employed by gardeners in Scotland for treating the bulbs of different ornamental plants.[1] However, the credit for conclusively demonstrating the therapeutic nature of heat must go to Jensen,[2] who successfully employed hot water for controlling loose smut of cereal grains and suggested that moisture played some role other than heat transfer. This method was followed in Denmark and later adopted in the U.S. on the recommendation of Swingle in 1892.[3]

Wilbrink[4] advocated the use of hot water treated sugarcane cuttings for planting nurseries and new fields. Encouraged by the results of Wilbrink,[4] hot water (50°C for 2 h) was tried with success against ratoon stunting disease of sugarcane in Australia. In Louisiana (U.S.), hot air was preferred over hot water for controlling the ratoon stunting disease because of immature nature of the cane.[6]

II. PRINCIPLES

The scientific principle involved in thermotherapy is that the pathogens present in seed material are inactivated or eliminated at temperatures nonlethal for the host tissues.[7] The exact mechanism by which heat inactivates the pathogen is not fully understood.[8-10] However, it is universally accepted that heat causes inactivation and not immobilization of the pathogen.

There are two schools of thought regarding inactivation of pathogen (viruses) by heat. One school of thought holds the opinion that the heat treatment stimulates enzymes that cause the degradation of virus, though according to Benda[10] this has not been established so far. The other school pursues the idea that heat causes what is known as "loosening" of bonds both in nucleic acid and the protein components of the pathogen. In the nucleic acid, when the bonds are disrupted the linear arrangement of nucleotides is disturbed and thus the virus looses infectivity. In proteins, the bonds holding the chains of amino acids together may be broken, or more likely, the architecture of the folding of the chain may be destroyed.[10] Disruption of bonds causes denaturing of protein molecules, which become less soluble in water, and finally leads to coagulation. It is still a matter of conjecture whether breakage in the nucleic acid chain or protein denaturation is responsible for the inactivation of the pathogen in host tissues. The rate at which the pathogen is inactivated is determined by temperature; the higher the temperature, the faster is the inactivation. At constant temperature, the drop in the density of pathogenic inoculum is exponential.

III. METHODS

Different physical methods employed for reduction and/or elimination of primary inoculum are as follows

A. HOT-WATER TREATMENT

Hot water is widely used for the control of seed-borne pathogens, especially bacteria and viruses. A list of some important seed-borne diseases claimed to have been controlled by hot-water treatment is given in Table 1, 2, and 3.

TABLE 1
Control of Seed-Borne Pathogens Through Hot-Water Treatment of Seed

Crop	Disease	Causal organism	Treatment	Ref.
Brassica spp.	Black rot	*X. campestris* pv.*campestris*	50°C for 20 or 30 min	11 3
Clusterbean, Guar (*Cyamopsis tetragono-loba*)	Blight	*X. campestris* pv. *cy-amopsidis*	56°C for 10 min	12
Cucumber (*Cucumis sativus*)	Seedling blight	*Ps. syringae* pv. *lachry-mans*	50°C and 75% RH for 3 days	13
Lettuce (*Lactuca sativa*)	Leaf spot	*X. campestris* pv. *vitians*	70°C for 1 to 4 d	14
Peanut (*Arachis hypogea*)	Testa nematode	*Aphlenchoides arachidis*	60°C for 5 min after soaking for 15 min in cool water	15
Pearl millet (*Pennisetum typhoides*)	Downy mildew	*Sclerospora graminicola*	55°C for 10 min	16
Potato (*Solanum tuberosum*)	Potato phyllody	MLO	50°C for 10 min	17
Rice (*Oryza sativa*)	Udbatta	*Ephelis oryzae*	54°C for 10 min	18
	White tip	*Aphlenchoides besseyi*	51—53°C for 15 min after dipping for 1 d in cool water	19
Safflower	Leaf spots	*Alternaria* spp.	50°C for 30 min	20
Teasel (*Dipsacus* spp.)	Stem nematode	*Ditylenchus dipsaci*	1 h at 50°C or 48.8°C for 2 h	21
Tobacco (*Nicotiana tabacum*)	Hollow stalk	*E. carotovora* pv. *caro-tovora*	50°C for 12 min	22
Tomato (*Lycopersicon esculen-tum*)	Black speck	*Ps. syringae* pv. *tomato*	52°C for 1 h	23

TABLE 2
Control of Seed-Borne Diseases Through Hot-Water Treatment of Sugarcane Cuttings

Disease	Pathogen	Treatment	Ref.
Downy mildew	*Perenosclerospora sacchari*	54°C for 1 h, dried at room temperature for 1 d and again treated in 52°C for 1 h	24
Grassy shoot	MLO	54°C for 2 h	25
Leaf scald	*Xanthomonas albilineans*	Soaking in cold water for 1 d and then treating the cuttings at 50°C for 2—3 h	26
Mosaic	Virus (potato virus Y group)	20 min treatment each day on 3 successive days at 52, 57.3 and 57.3°C, respectively	10
Ratoon stunting	*Clavibacter xyli* ssp. *xyli*	50°C for 3 h	5
Red rot	*Colletotrichum falcatum* (*Physalospora tucumanensis*)	54°C for 8 h	27
Smut	*Ustilago scitaminea*	55 to 60°C for 10 min	28
Spike	Virus	52°C for 1 h	29
White leaf	MLO	54°C for 40 min	30
Wilt	*Acremonium* sp., *Fusarium moniliforme*	50°C for 2 h	31

TABLE 3

Time and Temperature Recommendations for Hot-Water Treatment for Denematizing Planting Stocks[32]

Nematode	Planting stock	Time (min)	Temp. (°C)
Aphelenchoides ritzemabosi	Chrysanthemum stools	15	47.8
		30	43.0
A. fragariae	Easter lily bulbs	60	44.0
Ditylenchus dipsaci	Narcissus bulbs	240	43.0
D. destructor	Irish bulbs	180	43.0
Meloidogyne spp.	Cherry root stocks	5—10	50—51.0
	Sweet potatoes	65	45.7
	Peach root stocks	5—10	50—51.1
	Tuberose tubers	60	49.0
	Grapes rooted	10	50.0
	Cuttings	30	47.8
	Begonia	30	48.0
	Tubers	60	45.0
	Caladium tubers	30	50.0
	Yam tubers	30	51.0
	Ginger rhizomes	10	55.0
	Strawberry roots	5	52.8
	Rose roots	60	45.5

B. HOT-AIR TREATMENT

Hot-air treatment is less injurious to seed and easy to operate but also less effective than hot water. It has been used against several diseases of sugarcane. Singh[27] claimed complete control of red rot in varieties Co 527, CoS 510, Bo 3, and Bo 32 by hot-air treatment at 54°C for 8 h. It is used for treating sugarcane stalks on a commercial scale in Louisiana[33] to control ratoon stunting disease (RSD). It is employed for treating canes which are soft and succulent. Lauden[34] working in Mainland U.S., reported that hot-air treatment at 54°C for 8 h, effectively eliminates RSD pathogen without impairing the germination of buds. After 3 years, Steib and Chilton[35] confirmed Lauden's finding using thermocouple. Antoine[36] reported that in hot-air units the temperature of canes comes to 54°C after 4 h of introducing hot air at 58°C at inlet point. Similarly, grassy shoot disease of sugarcane has been controlled by hot air at ≈54°C for 8 h.[37]

C. STEAM AND AERATED STEAM

The use of aerated steam is safer than hot water and more effective than hot air in controlling seed-borne infections.

The heat capacity of water vapor is about half that of water and 2.5 that of air; hence the temperature and time required may be higher than that of hot water and lower than that of hot air.[7] The advantages of this method include easier drying of seeds, low loss in germination, easy temperature control and no damage to seed coat of legumes. Besides its use in controlling sugarcane diseases, it has been used against citrus greening.[38]

Most frequent application of steam and aerated steam has been in greenhouses where steam also provides heat during cold seasons. As a gas, it moves readily through soil, in contrast to the slow, inefficient movement of water. Steam raises temperatures efficiently. As gaseous water molecules (steam) condense into a liquid they give up much more heat than that given up as liquid water cools (540 cal/g relative to 1 cal/g°C). Thus much more heat is available in 1 g of steam at 100°C than in 1 g of liquid water at 100°C.[39] Steam leaves no toxic residue. Aerated steam provides an opportunity to treat soil at temperatures lower than those possible with pure steam. When air is mixed with steam at 100°C, the temperature of steam is reduced. The final temperature is determined by the amount and temperature of air mixed with steam.[40] The steam retains the heat of vaporization, however, so that it retains most of its efficiency in heating soil.

D. SOLAR HEAT TREATMENT

Solar heat treatment is effective in controlling both, seed-borne, and soil-borne diseases.

1. Seed-Borne Diseases

Solar heat treatment controls effectively the loose smut of wheat (*Ustilago nuda tritici*). In this method the seed is soaked in water for 4 h (8 a.m. to 12 noon) on a bright summer day. After this presoak, that seed is dried in the sun for 4 h from 12 noon to 4 p.m.[41-43] Mitra and Taslim[44] found the solar energy and sun-heated water methods are suitable for controlling the disease in north Bihar (India) conditions. Loose smut of barley (*Ustilago nuda*) has also been effectively controlled by solar heat. Bedi[45] recommended use of cloth sheet instead of brick floor.

Ascochyta rabiei, the causal organism of Ascochytosis of chickpea survives in seed.[46,47] Chaube[48] studied the effect of sun drying on the survival of the pathogen in chickpea seeds. The seeds were exposed to bright sunlight during the last week of May and the first week of June. The seeds were spread on cemented floor from 8 a.m. to 4 p.m. daily for 15 d. Direct exposure of seeds on cemented floor reduced the recovery of the fungus from 31.5 to 16%. In seeds covered with polyethylene sheets on cemented floor the reduction in survival of *A. rabiei* was of higher magnitude. In both the exposure methods, there was no effect on germination of the seed.

2. Soil Solarization
a. Introduction

A solar heating or soil solarization[49] or plastic/polyethylene tarping or polyethylene or plastic mulching[50] of soil is a soil disinfestation method which aims to reduce or eradicate the inoculum existing in soil. It was to the credit of Israel extension workers and growers who suggested that intensive heating that occurs in mulched soil might be used for the control of soil-borne diseases. Since then, this approach to control soil-borne pathogens and weeds has been widely used in Israel and other countries.[50]

b. The Principles

Solar heating method for disease control is similar, in principle, to that of artificial soil heating by steam or other means, which are usually carried out at 60 to 100°C.[50-52]

There are, however, biological and technological differences;[50] with soil solarization there is no need to transport heat from its source to the field. Solar heating is carried at relatively low temperatures, as compared to artificial heating; thus, its effects on living and nonliving components of soil is likely to be less drastic. Of the four components of disease severity,[53] inoculum density is the most affected component by solarization, either through the direct effect of the heat or by microbial processes induced in the soil. The other components except the host susceptibility which is genetically controlled, might also be affected.[50]

c. Solar Radiation and Soil Temperature

Absorption of solar radiation varies according to the colour, moisture, and texture of the soil.[50] In general, the soil has a relatively high thermal capacity and is a poor heat conductor. This results in a very slow heat penetration. On an average, a one square centimeter area outside the earth's atmosphere and parallel to its surface receives 2 cal/cm^2/ min (solar constant) of energy in the form of solar radiation, but only about half of it finally

TABLE 4
Effect of Soil Tarping on Temperature at Different Soil Depths[54]

Treatment	Maximum Soil Temperature (°C)			
	5 cm	15 cm	30 cm	45 cm
Shafter[a]				
Untarped	46	38	32	30
Tarped 1 mil	60	50	42	39
Tarped 4 mil	57	48	40	38
Davis[b]				
Untarped	42	32	27	NM[c]
Pre-irrigation one week before tarping + 4 mil	53	44	38	NM
Pre-irrigation one week before tarping + 4 mil + a single irrigation under the tarp	55	45	39	NM

[a] Temperatures, 29 June 1977—26 July 1977, maximum air temperature 41°C.
[b] Temperatures, 19 July 1977—23 September 1977, maximum air temperature 41°C.
[c] NM = not measured.

reaches the ground.[50] The heat that does penetrate the soil surface is stored in the soil and at night, when the thermal gradient is reversed, it is lost again, resulting in a cyclic reversal in the direction of heat flow.

An example,[54] showing the course and effect of heating of a nonmulched and polyethylene mulched wet soil is given in Table 4.

d. Soil Solarization and Plant Disease Control

Soil solarization as a practice for management of soil-borne plant pathogens has been demonstrated in several cases. A list of pathogens controlled by solarization is given in Table 5.

e. Beneficial Side Effects
i. Weeds and Other Pests

Researches conducted on soil solarization have revealed that polyethylene mulching of soil could be an effective method for weed control, lasting in cases for the whole year or even longer.[50] Grinstein et al.[74] observed that weed population of both perennial grasses and broad leaves were highly reduced by polyethylene mulching (Table 6). This treatment continued to be effective until the end of the season, 140 d after treatment.

In general, most of the annual and many perennial weeds such as *Amaranthus* spp., *Amsinckia douglasiana, Anagallis* sp., *Avena fatua, Chaenopodium* spp., *Convolvulus* spp., *Cynodon dactylon, Digitaria sanguinalis, Echinochloa crus-galli, Eleusine* sp., *Fumaria* sp., *Lactuca* sp., *Lamium amplexicaule, Mercuriales, Molucella, Montia, Notabaris* sp., *Phalaris* spp., *Poa* sp., *Portulaca oleraceae, Sisymbrium* sp., *Senecio vulgaris, Solanum halpense, Sorghum, Stellaria,* and *Xanthium pensylvanicum* have been controlled. Many graminae are especially sensitive, while others like *Melilotus,* remain unaffected. *Cyperus rotundus* was only partially controlled and *Orobanche* was effectively controlled in a number of studies.[50,51]

The possible mechanisms of weed control by solarization are: (1) thermal killing of seeds, (2) thermal killing of seeds induced to germinate, (3) breaking seed dormancy and consequent killing of the germinating seed, and (4) biological control through weakening or other mechanisms. Volatiles that accumulate under the mulch[51] may play a role in direct killing of weeds or may affect seed dormancy or germination.

TABLE 5
Effect of Solarization on Soil-Borne Plant Pathogens

Pathogen	Disease and crop	Ref.
Fungal Pathogens		
Bipolaris sorokiniana	Crater disease of wheat	55
Didymella lycopersici	A tomato disease	51
F. oxysporum f. sp. *melonis*	Wilt of watermelon	56
F. oxysporum f. sp. *vasinfectum*	Wilt of cotton	57
F. oxysporum f. sp. *ciceri*	Wilt of chickpea	58, 59
F. oxysporum f. sp. *lentis*	Wilt of lentil	60
Phytophthora cinnamomi	Root rot of several plants	61
Pyrenochaeta lycopersici	Brown root rot or corky root disease of tomato	51
Pyrenochaeta terrestris	Pink root rot of onion	62
Pythium ultimum	Damping off and root rot of several crops	63, 64
Rhizoctonia solani	Root rot of several crops	59
Sclerotinia minor	Root diseases of row crops	65, 66
Sclerotinia sclerotiorum		66
Sclerotium cepivorum	White rot of onion	65
Sclerotium oryzae	Stem rot of paddy	67
Sclerotium rolfsii	Root rot of several crops	68, 69
Thielaviopsis basicola	Root rot of several crops	52
Verticillium dahliae	Wilt of cotton, tomato, potato and egg plant	63, 64
Nematodes		
Ditylenchus dipsaci	Garlic bulb nematode	70
Globodera rostochiensis	Golden nematode of potato	71
Helicotylenchus digonicus	The spiral nematode	72
Heterodera trifolii	Clover cyst nematode on carnation	51, 73
Meloidogyne hapla	Root knot nematode	72
Meloidogyne javanica	Root knot nematode	65
Pratylenchus thornei	Pin nematode of potato and other crops	68

TABLE 6
The Effect of Soil Mulching and Nematicide Treatment on Weed Infestation 140 d After Potato Planting[74]

Treatment	Weed infestation (% coverage)
Polyethylene mulching	9.1
Nematicide	30.0
Control	49.1

Gerson et al.[75] have reported that populations of bulb mite *Rhizoglyphus robini,* which causes heavy damage to certain crops, are drastically reduced by soil solarization.

ii. Increase in Yield

The amount of yield increase is a function of disease reduction as well as of level of soil infestation, and the damage caused to the crop by the concerned disease. For example, Grinstein et al.[74] observed that soil solarization and nematicide treatment significantly in-

TABLE 7
Selected Examples of Yield Increase by Soil Solarization

Crop	Pathogen	Increase over control (%)	Ref.
Carrot	*Orobanche aegyptiaca*	ca. 70 ton/ha as compared to 0 in untreated control	76
Cotton	*F. oxysporum* f. sp. *vasinfectum*	40—70	50
	Verticillium dahliae	60	64
Egg plant	*V. dahliae*	215	51
Onion	*Pyrenochaeta terrestris*	60—125	77
Peanut	*Sclerotium rolfsii*	42—64	68
Potato	*V. dahliae*	35	74
Tomato	*Pyrenochaeta lycopersici*	100—300	50
Safflower	*V. dahliae*	113	50

creased potato yield in soil naturally infested with *Verticillum dahliae* and *Pratylenchus thornei* by 34.7 and 29.8%, respectively.

Some important examples of yield increase in solarized soil are given in Table 7.

iii. Solarization and Integrated Control

Being a nonchemical method, integrating soil solarization with chemical, biological and cultural practices of control, is both possible and promising. Findings of Chet et al.[78] and Elad et al.,[79] who combined the antagonist *Trichoderma harzianum* with heating or solarization in *Rhizoctonia solani* infested soil, to achieve improved disease and pathogen control and delayed inoculum build up, is an invitation to scientists to try to integrate solarization with biocontrol agents for long lasting disease control. Similarly, fungicide vapam and solarization when combined together had synergistic effect in controlling delimited shell spots of groundnut pods.[51] A proven cultural practice, crop sequence when combined with solarization resulted in an improved and longer-lasting control of cotton wilt (*F. oxysporum* f. sp. *vasinfectum*). Pal[60] used PCNB alongwith solarization for the control of Fusarium wilt of lentil and recorded significantly higher control than solarization alone.

f. Mechanisms of Disease Control

Solarized soils undergo changes in their temperature and moisture regimes, the inorganic and organic composition of their solid, liquid, and gaseous phases, and their physical structure, all of which in turn affect the biotic components. Thus, reduction in disease incidence occurring in such soils, results from effects exerted on each of the three living components, i.e., the host, pathogen and microorganisms alongwith physical and chemical environment of the soil (Figure 1). Although these processes occur primarily during solarization, they may continue to various extents and in different ways, after the removal of the polyethylene sheets and planting.

Of the four components of disease severity,[53] inoculum density is the one most affected by solarization, either through the direct physical effect of heat or by microbial processes induced in the soil. The other components, however, (except for susceptibility which is genetically controlled), might also be affected.

E. RADIATION

Electromagnetic radiations have been studied for the control of seed-borne pathogens as well as post-harvest diseases. Ionizing radiation, including electromagnetic energy with a wave length of less than 100 nm (e.g., X-ray and gamma ray) and charged or uncharged particles with energy generally above 10 eV, is sufficient to kill pathogen or host cells.[80]

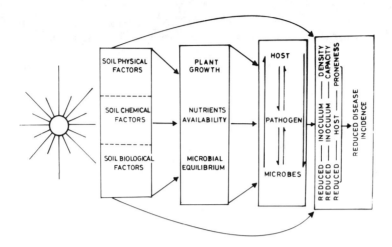

FIGURE 1.   Schematic diagram of the mechanisms of disease management through soil solarization.

These radiations release a large amount of energy in a small volume and thus cause much more biological and chemical effect than UV light, which generally produces only molecular excitation.[80] Ultraviolet rays (nonionizing radiations) have less penetrating power, and therefore, can be used only for surface sterilization. Infrared radiation is used for killing microorganisms in food stuffs, but is very selective.[80]

Various theories have been proposed to account for the radiation effects on plant pathogens. Lea[81] proposed the "one hit" hypothesis, while Kleezkowski[82] explained the "disequilibrium" hypothesis. Raychaudhary and Verma[80] favor the view that the primary action of radiation is to transform cell metabolites into poisonous substances.

Soybean seeds were subjected to treatments by radiation, high voltage of electric currents, ultrasonic radiations, and very high frequency (VHF) radio waves. On passing electricity at 4 KW/g for 30 s. through seeds, bacterial infections decreased from 5 to 9 to 2%, the degree of disinfection depending on voltage and exposure. Bacterial infection in plants grown from seeds when exposed to ultrasonic radiation decreased with exposure of 21.3 KC/s for 15 min. Cotyledon bacteriosis (*Pseudomonas syringae* pv. *glycinea* and *Ps. solanacearum*) and angular leaf spot (*Ps. syringae* pv. *glycinea*) were 16.5 and 40.4% respectively, compared with 30.8 and 73.2% in untreated plants. No specific results were obtained with VHF waves, though leaf spot bacteria were inhibited. Germination of seeds was not affected.[14,83] Treatment of tobacco seeds with 625-W microwave radiation for 20 min eliminated *Erwinia carotovora* var. *carotovora* without affecting germination. The number of infected seeds decreased by 68 and 99% by a 10- and 15-min treatment, respectively.[84] According to Halliwell and Longston,[85] gamma irradiation of barley seeds increased the ratio of healthy plants over plants infected with barley stripe mosaic. Radiation of *Prunus* seeds caused reduction in seed transmission of necrotic ring spot and prune dwarf viruses.[86]

Gamma rays have been used for the control of post harvest fungal diseases of peaches, strawberries, and tomatoes. *Phytophthora infestans,* which is highly sensitive to gamma rays, could be controlled during storage at 40 krad.[87] Aspergilli and Fusaria, although radioresistant, can be inhibited by the combined actions of fungicides (1000 ppm aureofungin or captan) and gamma radiation (200 krad); captan combined with radiation could prevent the progress of black-rot infection in apples.[80]

F. SONIC WAVES

Sonic energy has also been tried for the control of viruses. Sonic waves apparently break the viruses into shorter particles with reduced infectivity.[80] Tomlinson and Walkey[88] observed that susceptibility to sonic rupture may depend on the shape of the virus. The process has not yet found its practical use in therapy but may be useful for viruses which show exceptional resistance to heat.[89]

G. OSMOTIC PRESSURE

Feder[90] claimed that up to 100% of the nematodes were killed when sucrose or dextrose was added to nematode infested soil at the rate of 1 to 5% by weight. Feder and his collegues[91] showed that the osmotic effect was enhanced by the addition of 1 ton/acre of monosodium-lauryl-sulphate detergent to sugar soil mixture. The application of 20 to 30 ton/acre of sugar to soil is impracticable and uneconomical.

H. BURNING

Controlled burning may alter the environment and affect plant disease response, providing both a temperature effect and a means of destroying the pathogen.[92] Parmeter and Uhrenholt[93] introduced a new aspect of burning by demonstrating that smoke may kill a number of plant pathogens in tissues.

According to Hardison[94,95] burning is a single most important practice in grass seed production in the Pacific Northwest. It was initiated there to control the blind seed disease of perennial rye grass caused by *Gloeotinia temulenta*. It also effectively controlled *Claviceps purpurea* (ergot of rye), *Anguina agrostis* (seed nematode) and silver top.[94] Most of the crops to which fire can be applied are cereals in which inoculum can be destroyed after harvest, or pasture grasses which can periodically be freed from inoculum before they make new growth. In addition, where residues are dry and inflammable enough to be burn without fuel, this can be a cheap and fairly cost effective. It has also been realized that the increasing success of nontillage in some crops, and the resultant problems of debris management, make burning an attractive and effective proposition for reduction of inoculum, provided the cost involved is reasonably low.

Diseases that have been successfully managed by burning or flaming crop residue are given in Table 8.

I. FLOODING

Flooding fields and orchards to reduce or eliminate soil-borne inoculum of plant pathogens is an ancient practice. According to Kelman and Cook[102] flooding has been recognized to be one of the key factors for the low incidence of soil-borne diseases in present day chinese agriculture. In Lower Yangtze and South China, the plots where vegetables are grown, are covered to one or two crops of rice or planted to spinach, water chestnut, water bamboo, or lotus.[103]

Several explanations of the harmful effects of prolonged flooding on soil-borne pathogens have been suggested. Lack of oxygen may be involved in some cases or, more often perhaps, accumulation of CO_2 in the soil. The survival of *F. oxysporum* f. sp. *cubense* in soil after 2 weeks depends on formation of chlamydospores, since the conidia are not apparently long lived in soil, and Newcombe[104] found that CO_2 and flooded soil both largely inhibited chlamydospore formation, whereas they at first stimulated the production of conidia. Consequently the fungus, although able to survive in banana plantation soil containing organic matter, is likely to die out in a fallow, flooded field where organic matter is in short supply. Newcombe[104] concluded that the main factor in the elimination of the fungus by flooding is a high CO_2 content in the flooded soil combined with a decreased availability of colonizable substrate. In flooded soil, CO_2 stimulates germination of conidia, presumably by overcoming the fungistatic factor present in soil, but prevents the formation of chlamydospores so that

TABLE 8
Diseases Controlled by Fire and Flame[94-101]

Pathogen	Disease	Pathogen	Disease
Anguina sp.	Seed nematode of *Lolium rigidum*	*Anguina agrostis*	Seed nematode of *Festuca rubra*
Anguillulina tumefaciens	Leaf gall nematode of *Cynodon transvalensis*	*Corticium sasaki*	Sheath blight of rice
Claviceps paspali	Ergot of *Paspalum dilatatum*	*Claviceps purpurea*	Ergot of *Lolium perenne*
Cuscuta sp.	Parasite of lucern	*Diaporthe vaccinii*	Die back of low bush blueberry
Drechslera poae	Leaf mold of *Poa pratensis*	*Gaeumannomyces graminis*	Take-all of wheat
Gerlachia nivalis	Snow mold of wheat and barley	*Glueotinia temulenta*	Blind seed of *Lolium perenne*
Godronia cassandrae	Canker of *Vaccinium* sp.	*Leptosphaeria* sp.	Leaf blight of sugarcane
Phleospora idahoensis	Stress eye spot of *Festuca rubra*	*Pleiochaeta setosa*	Brown spot of Lupin
Pseudocercosporella herpotrichoides	Eye spot of wheat	*Puccinia menthae*	Rust of peppermint
Puccinia asparagi	Asparagus rust	*Puccinia poaenemoralis*	Leaf rust of *Poa pratensis*
Puccinia graminis	Stem rust of *Poa pratensis*	*Puccinia striiformis*	Stripe rust of *Poa pratensis*
Rhynchosporium secalis	Eye blotch of barley	*Sclerotium oryzae*	Stem rot of rice
Septoria avenae	Leaf blotch of oat	*Septoria nodorum, Septoria tritici*	Leaf blotch of wheat
Selenophoma bromigera	Leaf spot of *Bromus inertis*	*Urocystis agropyri*	Flag smut of wheat
Verticillium dahliae	Wilt of potato and peppermint		

TABLE 9
Plant Diseases Controlled by Flooding[96,105-111]

Pathogen	Disease	Pathogen	Disease
Alternaria porri f. sp. *solani*	Alternaria blight of tomato and potato	*Alternaria dauci*	Blight of carrot
Aphelenchoides besseyi	White tip of rice	*F. oxysporum* f. sp. *cubense*	Wilt of banana
Meloidogyne sp.	Root knot of celery	*Orobanche* spp.	Phanerogamic plant parasite of several crops
Phytophthora parasitica var. *nicotianae*	Black shank of tobacco	*Pyrenophora teres*	Canker and blight of barley
Radopholus similis	Burrowing nematode of banana	*Sclerotinia sclerotiorum*	White mold of vegetables
Trichodorus sp.	Stubby root nematode of celery	*Tylenchorhynchus* sp.	Stunt nematode of celery
Verticillium dahliae	Wilt of cotton		

the fungus dies out when the organic matter is exhausted. A similar situation perhaps holds for other soil borne fungi, but few cases have been investigated.[105]

A list of plant diseases that have been successfully controlled by flooding fields and orchards is given in Table 9.

REFERENCES

1. **Zendebergen, M.,** Hot water treatment for bulbs, *R. Hort. Soc. Daggodit Tulip Year Book,* 30, 187, 1964.
2. **Hermansen, J. E. and Jorgensen, J.,** Historical aspects of the control of seed-borne cereal diseases in Denmark, *Seed Sci. Technol.,* 11, 1005, 1983.
3. **Sharvelle, E. G.,** *Plant Disease Control,* AVI Publishing, West Port, Conn. 1979, 331.
4. **Wilbrink, G.,** Warm water behandeling Van stekken als genees, midel tegende serehziekte Van hot suikerriet, *Arch. Suikerind, Ned. Ind. Surabaya,* 3, 1, 1923.
5. **Steindl, D. R. L.,** Ratoon stunting disease, *Proc. Intern. Soc. Sugarcane Technology,* 7th Congress, 457, 1950.
6. **Schexnayder, C. A.,** The ratoon stunting disease of sugarcane in Louisiana with notes on its control, *Proc. Intern. Soc. Sugarcane Technology,* 9th Congress, 1058, 1956.
7. **Baker, K. F.,** Thermotherapy of planting materials, *Phytopathology,* 52, 1244, 1962.
8, **Bawden, F. C.,** *Plant Viruses and Virus Diseases,* Ronald Press, New York, 1964.
9. **Broadbent, L.,** Control of plant virus diseases, in *Plant Virology,* Corbett, M. K. and Sisler, H. D., Eds., University of Florida Press, Gainsville, FL, 1964, 527.
10. **Benda, G. T. A.,** Hot water treatment for mosaic and RSD control, *Sugar J.* 34, 32, 1972.
11. **Shekhawat, P. S., Jain, M. L., and Chakravarti, B. P.,** Detection and seed transmission of *X. campestris* pv. *campestris* causing black rot of cabbage and cauliflower and its control by seed treatment, *Indian Phytopathol.,* 35, 442, 1982.
12. **Srivastava, D. N., and Rao, Y. P.,** Epidemiology and control of bacterial blight of guar *(Cyamopsis tetragonoloba), Bull. Indian Phytopathol. Soc.,* 6, 1, 1970.
13. **Laben, C.,** Control of *Pseudomonas lachrymans* in cucumber seed by a temperature - RH method, *Phytopathology,* 71, 235, 1981.
14. **Agrawal, V. K. and Sinclair, J. B.,** *Principles of Seed Pathology,* Vol. I, CRC Press, Boca Raton, FL, 1987.
15. **Bridge, J., Bos, W. S., Page, L. T., and Mc Donald, D.,** The biology and possible importance of *A. arachidis,* a seed-borne ectoparasite nematode of groundnut from northern Nigeria, *Namatologica,* 23, 253, 1977.
16. **Thakur, D. P. and Kanwar, J. S.,** Internal seed-borne infection and heat therapy in relation to downy mildew of *Pennisetum typhoides* Stapf. and Hubb., *Sci. Cult.,* 43, 433, 1977.
17. **Paul Khurana, S. M., Singh, V., and Nagaich, B. B.,** Hot water treatment of tubers for elimination of the potato phyllody pathogen, *Indian Phytopathol.,* 32, 646, 1979.
18. **Mohanty, N. N.,** Control of Udbatta disease of rice, *Proc. Indian Nat. Sci. Acad. B,* 37, 432, 1971.
19. **Todol, E. H., and Atkins, J. G.,** White tip disease of rice. II. Seed treatment studies, *Phytopathology,* 49, 184, 1959.
20. **Zazzerini, A., Campeli, C., and Panattoni, L.,** Use of hot water treatment as a means of controlling *Alternaria* spp. on safflower seeds, *Plant Dis.,* 69, 350, 1985.
21. **Thorne, G.,** *Principles of Nematollgy,* McGraw Hill, New York, 1961, 553.
22. **McIntyre, J. L., Sands, D. C., and Taylor, G. S.,** Overwintering, seed disinfestation, and pathogenicity studies of the tobacco hollow stalk pathogen, *E. carotovora* var. *carotovora, Phytopathology,* 435, 1978.
23. **Devash, Y., Okon, Y., and Henis, Y.,** Survival of *Pseudomonas tomato* in soil and seed, *Phytopathol. Z.,* 99, 175, 1980.
24. **Wang, C. S.,** Internal hot water treatment as a method to control downy mildew disease (*Sclerospora sacchari*) of sugarcane, *Taiwan Sugar,* 4, 21, 1957.
25. **Gumaste, J. A., Soman, B. A., and Patil, B. L.,** Curative effect of different treatments on sugarcane setts from canes possibly infected by the presence of virus, *Proc. DSTA,* 12, 13, 1955.
26. **Steindl, D. R. L.,** The elimination of leaf scald from infected planting material, *Proc. Intern. Sugarcane Technol.,* 14th Congress, 925, 1971.
27. **Singh, K.,** Hot air therapy against red rot of sugarcane, *Plant Dis. Rep.,* 57, 220, 1973.
28. **Chona, B. L.,** Sugarcane smut and its control, *India Fmg.,* 4, 401, 1943.
29. **Jha, A., Prasad, H. C., and Misra, B.,** Hot water treatment for control of spike and grassy shoot disease of sugarcane, *Indian Sugar,* 23, 677, 1973.
30. **Ling, K. C. and Chung, Y., C.,** Studies on the white leaf disease of sugarcane. II. Efficacy of hot water treatment in the disease control, *Taiwan Sugar, Expt. St. Report,* 30, 75, 1963.
31. **Srinivasan, K. V. and Rao, J. T.,** Hot water treatment of sugarcane seed material, *Indian Fmg.,* 18, 25, 1968.
32. **Reddy, P. P.,** *Plant Nematology,* Agricole Publishing Academy, 1983, 287.
33. **Steib, R. J. and Forbes, I. L.,** Hot air for control of the ratoon stunting disease of sugarcane in Louisiana, *Phytopathology,* 48, 398, 1958.
34. **Lauden, L.,** Decision on the control of the stunting disease, *Sugar Bull,* 31, 382, 1953.

35. **Steib, R. J. and Chilton, C. J. P.,** Recent studies conducted on the ratoon stunting disease of sugarcane in Louisiana, *Sugar Bull.,* 34, 238, 1956.

36. **Antoine, R.,** Cane Diseases, *Rept. Sug. Ind. Res.,* Mauritius, 53, 1957.

37. **Singh, K.,** Grassy shoot disease of sugarcane. II. Hot air therapy, *Curr. Sci.,* 37, 592, 1968.

38. **Cheema, S. S., Chohan, J. S., and Kapur, S. P.,** Effect of moist hot air treatment on citrus greening infected bud wood, *J. Res. Punjab Agric. Univ.,* 19, 97, 1982.

39. **Fry, W. E.,** *Principles and Practices of Plant Disease Management,* Academic Press, New York, 1987, 378.

40. **Baker, K. F.,** Selective killing of soil-microorganisms by aerated steam, in *Root Diseases and Soil Borne Pathogens,* Toussoun, T. A., Bega, R. V., and Nelson, P. E., Eds., Univ. of California Press, Berkeley, 1970, 234.

41. **Luthra, J. C. and Sattar, A.,** Some experiments on the control of loose smut of wheat, *Indian J. Agric. Sci.,* 4, 117, 1934.

42. **Luthra, J. C.,** Solar treatment for loose smut of wheat, *Indian Fmg.,* 2, 416, 1941.

43. **Luthra, J. C.,** Solar energy treatment of wheat loose smut, *Ustilago tritici, Indian Phytopath.,* 6, 40, 1953.

44. **Mitra, M. and Taslim, M.,** The control of loose smut of wheat in north Bihar by solar energy and sun heated water method, *Agric. Livestock India,* 6, 43, 1936.

45. **Bedi, K. S.,** Further studies on the control of loose smut of wheat in Punjab, *Indian Phytopath.,* 10, 133, 1957.

46. **Maden, S., Singh, D., Mathur, S. B., and Neergaard, P.,** Detection and location of seed borne inoculum of *Ascochyta rabiei* and its transmission in chickpea *(Cicer arietinum), Seed Sci. Tech.,* 3, 667, 1975.

47. **Nene, Y. L.,** A review of Ascochyta blight of chickpea, *Tropical Pest Management,* 28, 61, 1982.

48. **Chaube, H. S.,** *Ascochyta Blight of Chickpea,* Technical Report, Expt. St., G. B. Pant Univ. of Agric. Technol., Pantnagar, India, 1987, 110.

49. **Adams, P. B.,** Effect of soil temperature and soil amendments on *Thielaviopsis* root rot of sesame, *Phytopathology,* 61, 93, 1971.

50. **Katan, J.,** Solar heating (solarization) of soil for control of soil borne pests, *Annu. Rev. Phytopathol.,* 19, 211, 1981.

51. **Katan, J.,** Soil solarization, in *Innovative Approaches to Plant Disease Control,* Chet, I., Ed., John Wiley & Sons, New York, 1987, 372.

52. **Baker, K. F. and Cook, R. J.,** *Biological Control of Plant Pathogens,* W. H. Freeman & Co., San Francisco, 1974, 433.

53. **Baker, R.,** Mechanisms of biological control of soil-borne pathogens, *Annu. Rev. Phytopathol.,* 6, 263, 1968.

54. **Pullman, G. S., DeVay, J. E., Garber, R. H., and Weinhold, A. R.,** Control of soil-borne pathogens byplastic tarping of soil, in *Soil-Borne Plant Pathogens,* Schippers, B. and Gams, W., Eds., Academic Press, New York, 1979, 686.

55. **Smith, E. W., Wehner, F. C., and Kotze, J. M.,** Effect of soil solarization and fungicide soil drenches on crater disease of wheat, *Plant Dis.,* 68, 582, 1984.

56. **Katan, J., Grinstein, A., Fishler, G., Frank, A. Z., Rabinowitch, H. D., Greenberger, A., Alon, H., and Zig, U.,** Long term effects of solar heating of soil, *Phytoparasitica,* 9, 236, 1981.

57. **Katan, J., Fishler, G., and Grinstein, A.,** Short and longterm effects of soil solarization and crop sequence on Fusarium wilt and yield of cotton in Israel, *Phytopathology,* 73, 1215, 1983.

58. **ICRISAT,** *Pulse Pathology (Chickpea) Progress Report 47,* ICRISAT, Patancheru, A. P., India, 1986.

59. **Tripathi, H. S.,** personal communication.

60. **Pal, B. P.,** personal communication.

61. **Pinkas, Y., Kariv, A., and Katan, J.,** Soil solarization for the control of *Phytophthora cinnamomi:* thermal and biological effect, *Phytopathology,* 74, 796, 1984.

62. **Katan, J.,** Solar pasteurization of soils for disease control: Status and prospects, *Plant Dis.,* 64, 450, 1980.

63. **Pullman, G. S., DeVay, J. E., and Garber, R. H.,** Soil solarization and thermal death: A logarithmic relationship between time and temperature for four soil borne pathogens, *Phytopathology,* 71, 959, 1981.

64. **Pullman, G. S., DeVay, J. E., Garber, R. H., and Weinhold, A. R.,** soil solarization: Effects on Verticillium wilt of cotton and soil-borne populations of *Verticillium dahliae, Phythium* spp., *Rhizoctonia solani* and *Thielaviopsis basicola, Phytopathology,* 71, 954, 1981.

65. **Porter, I. J. and Merriman, P. R.,** Effect of solarization of soil on nematode and fungal pathogens at two sites in Victoria, *Soil Biol. Biochem.,* 15, 39, 1983.

66. **Porter, I. J. and Merriman, P. R.,** Evaluation of soil solarization for control of root diseases of row crops in Victoria, *Plant Pathol.,* 34, 108, 1985.

67. **Usmani, S. M. H. and Ghaffar, A.,** Polyethylene mulching of soil to reduce viability of sclerotia of *Sclerotinia oryzae, Soil Biol. Biochem.,* 14, 203, 1982.

68. **Grinstein, A., Katan, J., Abdul-Razik, A., Zeidan, O., and Elad, Y.,** Control of *Sclerotium rolfsii* and weeds in peanuts by solar heating of soil, *Plant Dis. Rep.,* 63, 1056, 1979.

69. **Mihail, J. D. and Alcorn, S. M.,** Effect of soil solarization on *Macrophomina phaseolina* and *Sclerotium rolfsii, Plant Dis.,* 68, 156, 1984.

70. **Siti, E., Cohen, E., Katan, J., and Mordechai, M.,** Control of *Ditylenchus dipsaci* in garlic by bulb and soil treatment, *Phytoparasitica,* 10, 93, 1982.

71. **La Mondia, J. A., and Brodie, B. B.,** Control of *Globodera rostochiensis* by solar heat, *Plant Dis.,* 68, 474, 1984.

72. **Stapleton, J. J., and DeVay, J. E.,** Response of phytoparasitic and free-living nematodes to soil solarization and 1,3 dichloropropene in California, *Phytopathology,* 73, 1429, 1983.

73. **Hadar, E., Sofer, S., Brosh, S., Mordechai, M., Cohn, E., and Katan, J.,** Control of clover cyst nematode on carnation, *Hadesseh,* 63, 1698, 1983.

74. **Grinstein, A., Orion, D., Greenberger, A., and Katan, J.,** Solar heating of the soil for the control of *Verticillium dahliae* and *Pratylenchus thornei* in potatoes, in *Soil-Borne Plant Pathogens,* Schipper, B. and Gams, W., Eds., Academic Press, New York, 1979, 686.

75. **Gerson, U., Yathom, S., and Katan, J.,** A demonstration of bulb mite control by solar heating of the soil, *Phytoparasitica,* 9, 153, 1981.

76. **Jacobson, R., Greenberger, A., Katan, J., Levi, M., and Alon, H.,** Control of Egyptian broomrape (*Orobanche aegyptiaca*) and other weeds by means of solar heating of the soil by polyethylene mulching, *Weed Sci.,* 28, 313, 1980.

77. **Katan, J., Rotem, I., Finkek. Y., and Daniel, J.,** Solar heating of the soil for the control of pink root and other soil-borne diseases in onions, *Phytoparasitica,* 8, 39, 1980.

78. **Chet, I., Elad, Y., Kalfon, A., Hadar, Y., and Katan, J.,** Integrated control of soil-borne and bulb borne pathogens in Iris, *Phytoparasitica,* 10, 229, 1982.

79. **Elad, Y., Katan, J., and Chet, I.,** Physical, biological and chemical control integrated for soil-borne diseases in potatoes, *Phytopathology,* 70, 418, 1980.

80. **Raychaudhary, S. P. and Verma, J. P.,** Therapy by heat, radiation, and meristem culture, in *Plant Diseases: An Advanced Treatise* Vol. I., Horsfall, J. G. and Cowling, E. B., Eds., Academic Press, New York, 1977, 177.

81. **Lea, D. E.,** *Actions of Radiations on Living Cells,* MacMillan, New York, 1947, 402.

82. **Kleezkowski, A.,** Effect of non-ionizing radiation on plant viruses, *Ann. N.Y. Acad. Sci.,* 83, 661, 1960.

83. **Krasnova, M. V.,** The effect of some physical factors on the causal agents of bacteriosis in soybean seeds, *J. Microbiol. (Kiev),* 25, 50, 1963.

84. **Hankin, I. and Shands, D. C.,** Microwave treatment of tobacco seeds to eliminate bacteria on seed surface, *Phytopathology,* 67, 794, 1977.

85. **Halliwell, R. S. and Longston, R.,** Effect of gamma radiation on symptom expression of barley stripe mosaic virus disease and on two viruses *in vivo, Phytopathology,* 55, 1039, 1965.

86. **Megahed, E. S. and Moore, A.,** Inactivation of necrotic ring spot and prune dwarf viruses in seeds of some *Prunus* spp., *Phytopathology,* 59, 1758, 1969.

87. **Beraha, L., Ramsey, G. B., Smith, M. A., and Wright, W. R.,** Effect of gamma radiation on some important potato tuber decays, *Am. Potato. J.,* 36, 333, 1959.

88. **Tomlinson, J. A. and Walkey, D. G. A.,** Effect of ultrasonic treatment on turnip virus and potato virus. X, *Virology,* 32, 267, 1967.

89. **Semanick, J. S. and Weathers, L. G.,** Exoccrtis virus of citrus: association of infectivity with nucleic acid preparations, *Virology,* 36, 326, 1968.

90. **Feder, W. A.,** Osmotic destruction of Plant parasitic and saprophytic nematodes by the addition of sugars to soil, *Plant Dis. Rep.,* 44, 883, 1960.

91. **Feder, W. A., Eichhorn, J. L., and Hutchins, P. C.,** Sugar induced osmotic dehydration of nematodes enhanced by the addition of detergents, *Phytopathology,* 52, 9, 1962.

92. **Zentmyer, G. A. and Bald, J. G.,** Management of the environment, in *Plant Diseases: An Advanced Treatise,* Vol. 1, Horsfall, J. G. and Cowling, E. B., Eds., Academic Press, New York, 1977, 122.

93. **Parmeter, J. R., Jr. and Uhrenholt, B.,** Some effects of pine needle or grass smoke of fungi, *Phytopathology,* 65, 28, 1975.

94. **Hardison, J. R.,** Fire and flame for plant disease control, *Annu. Rev. Phytopathol.,* 14, 355, 1976.

95. **Hardison, J. R.,** Role of fire for disease control in grass seed production, *Plant Dis.,* 64, 641, 1980.

96. **Palti, J.,** *Cultural Practices and Infectious Crop Disease,* Springer Verlag, Berlin, 1981, 343.

97. **Gray, P. M. and Guthrie, J. W.,** The influence of sprinkler irrigation on post-harvest residue removal practices on the seed-borne population of *Drechslera poae* on *Poa pratensis* ''Merion'', *Plant Dis. Rep.,* 61, 90, 1977.

98. **Huber, D. M. and Watson, R. D.,** Effect of organic amendment on soil-borne plant pathogens, *Phytopathology,* 60, 22, 1970.

99. **Horner, C. E.,** Control of mint rust by propane gas flaming and contact herbicides, *Plant Dis. Rep.,* 49, 393, 1965.

100. **Yarham, D. J.,** The effect on soil-borne diseases of changes in crop and soil management, in *Soil-Borne Plant Pathogens,* Schippers, B. and Gams, W., Eds., Academic Press, London, New York, San Francisco, 1979, 686.

101. **Webster, R. K., Bolstad, J., Wick, C. M., and Hall, D. H.,** Varietal distribution and survival of *Sclerotium oryzae* under various tillage methods, *Phytopathology,* 66, 97, 1976.

102. **Kelman, A. and Cook, R. J.,** Plant Pathology in the People's Republic of China, *Annu. Rev. Phytopathol.,* 15, 409, 1977.

103. **William, P. M.,** Vegetable crop protection in the People's Republic of China, *Annu. Rev. Phytopathol.,* 17, 311, 1979.

104. **Newcombe, M.,** Some effects of water and anaerobic condition on *Fusarium oxysporum* f. sp.*cubense* in soil. *Trans. Brit. Mycol. Soc.,* 43, 51, 1960.

105. **Tarr, S. A. J.,** *Principles of Plant Pathology,* Mac Millan Publishers, 1972, 632.

106. **Rotem, J., and Palti, J.,** Irrigation and plant diseases, *Annu. Rev. Phytopathol.,* 7, 267, 1969.

107. **Stover, R. H.,** Flood-fallowing for eradication of *Fusarium oxysporum* f. sp. *cubense.* II. Some factors involved in fungus survival, *Soil Sci.,* 77, 401, 1954.

108. **Stover, R. H.,** Flood-fallowing for eradication of *Fusarium oxysporum* f. sp. *cubense* III. Effect of oxygen on fungus survival, *Soil Sci.,* 80, 397, 1955.

109. **Johnson, S. R. and Berger, R. D.,** Nematode and soil fungi control in celery seed beds on muck soil, *Plant Dis. Rep.,* 56, 661, 1972.

110. **Kasasian, L.,** *Orobanche* spp., *PANS,* 17, 35, 1971.

111. **Moore, W. D.,** Flooding as a means of destroying the sclerotia of *Sclerotinia sclerotiorum, Phytopathology,* 39, 920, 1949.

Chapter 14

BIOLOGICAL CONTROL

I. INTRODUCTION

Biological control of plant pathogens accomplished through host resistance and cultural practices has been working for decades and continues to be a predominant disease control strategy. In contrast, biological control accomplished through introduction or encouragement of microorganisms antagonistic to plant pathogens has been slow to develop.[1] Extensive efforts were made in the 1920s and 1930s to introduce antibiotic producing microorganisms into soil for the control of root diseases, but the attempts were so unsuccessful that research on this method of biological control virtually ceased for about three decades. The failures in these earlier years also fostered negative attitudes among the plant pathologists toward biological control.[2] A turning point for research on biological control of plant pathogens occurred in 1963 when, at Berkeley, California, an international Symposium was held on *Ecology of Soil-Borne Plant Pathogens—Prelude to Biological Control*.[3] Since then more than 15 international symposia have been held on this topic and the first book entitled *"Biological Control of Plant Pathogens"*[1] devoted wholly to the subject was published. This was followed by some excellent books entitled *"Biological Control in Crop Production"*,[4] *"Principles and Practices of Biological Control of Plant Pathogens"*,[5] *"Microbial Control of Plant Pests and Diseases,"*[6] and *"Innovative Approaches to Plant Disease Control"*.[7]

The art and science of plant disease control must continue to move in the direction of biological control of plant pathogens, including use of introduced antagonists.[2]

II. DEFINITION

There is much disagreement on what constitutes biological control. DeBach,[8] defined it as "the action of predators, parasites, or pathogens in maintaining another organism's population density at a lower average than would occur in their absence". This was meant specifically in relation to insect pests and weeds. In plant disease control, Garrett[9] defined biological control as "any condition under which or practice whereby survival or activity of a pathogen is reduced through the agency of any other living organism (except man himself) with the result that there is a reduction in the incidence of the disease caused by that pathogen". This definition excluded the role of the host in reducing disease and also the role of man. Antagonism was the major basis for this definition and all those practices by which antagonism (antibiosis, competition and exploitation) could be achieved were included under biological control methods.

Baker and Cook[1] have given a broader definition of biological control. According to them "biological control is the reduction of inoculum density or disease producing activities of a pathogen or parasite in its active or dormant state, by one or more organisms, accomplished naturally or through manipulation of the environment, host, or antagonist, or by mass introduction of one or more antagonists".

Deacon[6] proposed a definition of biological control based on Garrett's definition but covering both pests and pathogens of plants as "biological control is the practice in which, or process whereby, the undesirable effects of an organism are reduced through the agency of another organism that is not the host plant, the pest or pathogen, or man." In other words biological control is mediated by a "third party", as in case of microbial control it is the microorganism.[6] He finally concludes that microbial/biological control can be achieved in several different ways but all of the cases included in the definition have one thing in common: "an integrated part of the control process is the activity of a microorganism". So, in effect, microbial control is one of the practical applications of microbial ecology.

We feel that the definition given by Baker and Cook[1] which is based on equation[10] disease severity = inoculum potential (inoculum density × inoculum capacity) × disease potential (host proneness × susceptibility), for relating the various factors involved in biological control is most suited from a plant pathologist's point of view.

III. THEORIES AND MECHANISMS

Management of the associated microbiota is a major form of biological control.[1] Cook[11] has discussed five elements in the theoretical base. The theory behind this type of disease management is to encourage the soil microbiota performs the same job as the man does for suppression of plant pathogens and for helping the plant to resist attack of pathogens.

A. REDUCTION OF INOCULUM DENSITY

Inoculum density can be reduced by destroying propagules or by preventing their formation. Crop rotation adds chemically different plant residues to soil and, therefore, helps in complexities of soil microbiota. It starves the pathogen due to absence of host and weakens it to the extent that it is more rapidly destroyed by the microflora.

Dormant sclerotia are killed off by *Trichoderma, Fusarium roseum, Coniothyrium minitans,* and other fungi and bacteria. Microbial activities helping in decay of organic residues and release of acetaldehyde and methanol, etc., stimulate germination of sclerotia and reduce their resistance so that they are easily colonized by above-mentioned saprophytes and destroyed. The germ tube that comes out is lysed without forming new sclerotia. Other treatments that predispose sclerotia to microbial decay are wetting and drying, flooding, and sublethal doses of fumigants.

Destruction of resting structures (conidia, chlamydospores and sclerotia) occurs in soil by direct activities of bacteria, and actinomycetes. There is evidence that with increased organic matter content in soil perforations appear in the walls of these structures due to activity of some organisms. Microbes enter the structure through these perforations and destroy them.

Lysis of germ tubes or germlings before they can produce resting structures is another method of reducing inoculum through increased microbial activity in soil. To achieve good results it is essential that resting structures should be first induced to germinate. Decomposing organic matter releases substances which stimulate germination of sclerotia, chlamydospores or conidia as well as the activity of saprophytic microflora. In such situations germination is followed by lysis.

B. DISPLACING THE PATHOGEN FROM HOST RESIDUES

This approach applies to those pathogens that depend for survival on occupancy of the host remains during the host free period. The pathogens use the residues both as shelter and as a food base. This gives them the advantage of pioneer colonization. The system of residue possession by root pathogens is either *passive, active,* or both. *Pythium* spp., that attack succulent roots exemplify *passive possession.* They invade thoroughly, digest extensively, store the surplus food in their resting bodies (oospores), and then abandon the fragile, exhausted host remains to other sarprophytes.[12] In *active possession* the organisms invade the residue, usually as a parasite while the tissues are still alive and active, become established in some tissues in which they persist and are metabolically active within the dead host remains. Utilization of the substrate is slow and the pathogen persistently defends the substrate against saprophytes. Normally the active possessor does not retreat into a dormant structure.

Cephalosporium gramineum is a typical example. *Gaeumannomyces graminis* and *Fusarium graminearum* have some characteristics of active possessors. It is for these pathogens that efforts are to be made for their displacement or for nullifying their pioneer colonization. *F. oxysporum, F. solani,* and *F. culmorum* come under the category of combination possessors, that is, they are both active as well as passive possessors. They also need displacement for control through cultural practices.

The hold of *C. gramineum* on wheat straw is weakened in alkaline soil or by reducing moisture in the soil. These conditions permit entry of *Penicillium* spp. without much resistance from *C. gramineum*. Nitrogen deficiency reduces the hold of the take all fungus which is displaced by other saprophytes. *Armillaria mellea* on citrus wood is weakened by application of carbon disulfide and loses its capacity to produce antibiotics. This enables its antagonist *T. viride* to displace it from the wood.

C. SUPPRESSION OF GERMINATION AND GROWTH OF PATHOGEN

This form of biological control has two aspects: (1) reduction or prevention of germination (soil fungistasis), and (2) slowing down of growth of germlings due to starvation, antibiotics, bacteriocin, mycoviruses, etc. The phenomenon of fungistasis has been discussed separately.

D. PROTECTION OF AN INFECTION COURT

This approach aims at encouraging the soil microlfora in or on the infection court which slow or prevent infection by the particular pathogen. Such protection mainly includes conditions where a weak pathogen or nonpathogenic organism takes possession of the sites of infection on the host. The mechanisms by which these precolonizers may protect the infection court include (1) prior use of essential nutrients or oxygen needed by the pathogen, (2) modification of the rhizosphere pH, redox potential, and other environmental factors that places the pathogen at a competitive disadvantage, (3) production of antibiotics, (4) hyperparasitism or exploitation of the pathogen, and (5) modification of the host resistance.

E. STIMULATION OF RESISTANCE RESPONSE OF THE HOST

This includes cross protection provided by a weak strain of the same pathogen or by another pathogen. A tomato variety resistant to *F. oxysporum* f. sp. *lycopersici,* if inoculated with that pathogen, becomes resistant to *Verticillium dahliae* mint is resistant to *V. dahliae* if inoculated first with *V. nigrescens*. Take all of wheat (*Gaeumannomyces graminis* var. *tritici*) is reduced if the roots are precolonized by *Phialophora radicicolia* or *Gaeumannomyces graminis* var. *graminis*.

IV. MECHANISM AND PROCESS OF PATHOGEN DECLINE

A. ANTAGONISM

Antagonism is one main subdivision of microbial associations in soil. It implies that in any association of two or more species at least one of the interacting species is harmed due to activity of the one or more of the rest. The mechanisms of antagonism could be as given below.

Antibiosis is defined as the condition in which one or more metabolites excreted by an organism have a harmful effect on one or more other organisms.[13] In this type of antagonistic relationship species A produces a chemical substance that is harmful or inimical to species B without species A deriving any direct benefit.[14] However, the species A may have an indirect benefit in having a better competitive ability thereby getting an advantage over species B for substrate colonization.

Exploitation in which species A inflicts harm by the direct use of species B for its own benefit (parasitism and predation). Exploitation is a condition wherein an organism directly harms another organism to get benefit out of the harm done to the organism. This type of antagonism is operated through parasitism and predation. These two terms have basically same effect, i.e., destruction of the host or the prey. From a plant pathologist's view point predation is a form of parasitism (living on another organism). However, the mode of operation makes the two terms somewhat distinct. In parasitism some sort of etiological relationship between the parasite and the host is established and the host is not rapidly eliminated as in predation. A successful parasite does not eliminate its host. A predator physically eliminates its prey by direct feeding on it without establishing any etiological relationship.

Mycoparasitism, hyperparasitism, direct parasitism, or interfungus parasitism are terms interchangeably used to refer to a phenomenon in which one fungus is parasitic on another through a nutritional relationship established during life of the host. Almost all taxonomic groups of fungi are found to be involved in this phenomenon and often species within the same genus (e.g., *Pythium*) interact as host and parasite. Details of mycoparasitism is duscussed elsewhere in this chapter.

Competition which is found in the indirect rivalry of two species for some feature of the environment that is in short supply.[14] Broadly speaking competition could involve all kinds of interplay between organisms in which one is favored at the expense of other. But in strict sense, if we keep antibiosis or even exploitation restricted to their specific mode of action and result, competition has been defined as ''a more or less active demand in excess of the immediate supply of material or condition on the part of two or more organisms.[15]

B. FUNGISTASIS

Soil fungistasis, a phenomenon of inhibition of spore germination in soils, is an eco-logically important mechanism in maintaining the biological balance of the soil. Since Dobbs and Hinson,[16] it has received considerable attention of many workers who endeavored to unravel the mystery of fungistatic nature of soil, but it still remains a controversy. The three decade investigations on the nature of fungistasis have attributed it to the involvement of chemical substances of nonvolatile and volatile nature of either biotic or abiotic origin, nutrient status of the soil and the kind of fungal propagules. Nevertheless, it is generally accepted that more than one factor is responsible for the widespread occurrence of fungistasis in soil.

The term fungistasis, originally proposed by Dobbs and Hinson,[16] describes the phe-nomenon whereby viable propagules not under the influence of endogenous or constitutive dormany do not germinate in the soil in conditions of temperature and moisture favorable for germination.

1. Theory of Stimulators and Inhibitors

Watson and Ford[17] proposed a theoretical explanation for the dynamic phenomenon of soil fungistasis. According to them, there are three stages in fungistasis, namely, induction, maintenance and release of fungistasis. These stages are controlled by ''a complex balance of stimulators and inhibitors'' present in soil microenvironments. Both ''exogenous and endogenous factors, acting concurrently, consecutively, or both, may be involved in causing fungistasis''. The stimulators are of biotic origin, and may act as nutrients, while inhibitors are of both biotic and abiotic origin.

The organic substances present in root exudates and various sugars and nitrogen sources neutralizing fungistasis are not considered as stimulators of spore germination. The advocates of this theory believe that there should be some stimulators analogous to inhibitors. They discussed this point with the evidence that the sclerotia of *Sclerotium cepivorum* were freed from fungistasis by volatiles associated with the root of *Allium* species. These substances have been identified as allyl sulfides from allyl cystein sulphoxides originating from *Allium* roots.[18] Similarly, the volatile compounds like alcohols and aldehydes with low molecular weights released from plant residues were also found to be stimulatory to soil fungi.[18]

The germination of nutritionally independent spores of *Thielaviopsis basicola* was not stimulated by the addition of certain sugars including glucose to natural soils while the complex materials such as carrot juice or alfalfa meal stimulated germination.[18]

2. Annulment of Fungistasis

Freeing the fungal propagules from soil fungistasis and allowing a spore germination and lysis are important aspects for those using soil fungistasis as a tool in controlling soil-borne pathogens. Fungistasis is generally annulled by the addition of organic substances or energy rich compounds to the soil. The fungal spores readily germinate in the region of rhizosphere because of the presence of various organic substances in the root exudates.

Watson and Ford[17] were of the opinion that besides the energy rich sources and organic substances there might be some stimulatory substances analogous to the volatile inhibitors. The S-containing volatile compounds associated with *Allium* roots stimulated germination of the sclerotia of *Sclerotium rolfsii*.

3. Soil Fungistasis and Biological Control

The reviewers have suggested the possibility of using soil fungistasis in controlling plant diseases caused by soil-borne plant pathogens. They have also suggested that researches should be diverted to pinpoint how to overcome fungistasis and to allow the germination lysis, and to elevate the fungistasis to a point at which germination cannot occur even in the presence of a susceptible host.[19] It is also possible to manipulate the soil environment to elevate the levels of toxicity to such an extent that propagules remain inactivated even in the presence of root exudates.[19]

Fungistasis in natural soil is usually nullified by enriching the soil with energy sources or organic matter. It was observed that fungistasis was established within 7 d in soil amended with either alfalfa or oilseed meals.[20] Such enhanced fungistasis[18] occurred in soil amended with organic substrates like alfalfa, corn stover, oat straw, soybean, cotton and linseed meal, cellulose, and chitin.

The manipulation of the soil environment by organic amendment to enhance the liberation of volatile fungistatic substances is useful for the biological control. The enhanced production of ammonia was reported from soil amended with chitin, organic matters of crucifers and soybean, linseed and cotton seed meals, and it was found that ammonia was suppressive to some of the soil-borne fungal pathogens like *Fusarium oxysporum* and *F. solani*, *F. solani* f. sp. *cucurbitae*, *Macrophomina phaseolina*, and *Phytophthora cinnamomi*.[18]

In the studies of soil amended with alfalfa or oil seed meals, it was found that the fungistatic property of the soil was nullified temporarily, but restored within 7 d after the incorporation of amendments.[20] *F. oxysporum* and *F. solani* were completely suppressed in soil amended with oilseed meals in closed containers in the laboratory, but attempts to reduce populations in the field condition failed. This has been attributed to the unfavorable soil moisture, inadequate mixing with the soil or escape of volatile products from microbial degration.[21]

Lewis and Papavizas[22] reported the production of sulphur containing volatile compounds like methanethiol, dimethyl sulfide in soils amended with organic matters of crucifers, and were found strongly suppressive to pea root rot pathogen, *Aphanomyces euteiches*. Since the use of crucifer amendments is uneconomical, it has been suggested to use S-containing commercial fumigants like Vapam, Vorlex, and dazomet.[18]

V. NATURAL BIOLOGICAL CONTROL: SUPPRESSIVE SOILS

The inhospitality of certain soils to some pathogens is such that either the pathogen can

TABLE 1
Examples of Disease-Suppressive Soils[24-26]

Pathogen	Disease(s) caused	Pathogen	Disease(s) caused
Armillaria mellea	Root rot of conifers	*Cephalosporium graminearum*	Stripe of wheat
Didymella lycopersici	Stem rot of tomato	*Fusarium avenaceum*	Root rot of many crops
F. oxysporum f. sp. *batatos*	Wilt of sweet potatoes	*F. oxyspurum* f. sp. *cubense*	Wilt of banana
F. oxysporum f. sp. *cucumerinum*	Wilt of cucurbits	*F. oxysporum* f. sp. *cyclaminis*	Wilt of cyclamen
F. oxysporum f. sp. *dianthi*	Wilt of carnation	*F. oxysporum* f. sp. *lini*	Wilt of flax
F. oxysporum f. sp. *lycopersici*	Wilt of tomato	*F. oxysporum* f. sp. *melonis*	Wilt of melon
F. oxysporum f. sp. *pisi*	Wilt of pea	*F. oxysporum* f. sp. *raphani*	Wilt of radish
Fusarium udum	Wilt of pigeonpea	*Fusarium solani*	Root rot of bean
Gaeumannomyces graminis	Take all of wheat	*Heterodera avenae*	Cereal cyst nematode
Olpidium brassicae	Lettuce big vein	*Phomopsis sclerotoides*	Root rot of cucurbits
Phytophthora cinnamomi	Root rots of various crops	*Poria weirii*	Root rot of conifers
Pseudocercosporella herpotrichoides	Root rot of cereals	*Pseudomonas solanacearum*	Soft rots
Pythium aphanidermatum	Root rot of radish	*Phythium ultimum*	Root rots
Pythium spp.	Root rots	*Rhizoctonia solani*	Root rots of many crops
Sclerotium rolfsii	Root rot of tomato	*Sclerotium cepivorum*	Wilt of onion
Streptomyces scabies	Scab of potatoes	*Verticillium alboatrum*	Wilt of potatoes

not establish themselves, or they become established but fail to cause disease, or they become established and initiate disease, but diminish in severity with continued culture of the crop.[1]

Thus, "suppressive soil" is an umbrella term encompassing fungistasis, competitive saprophytic ability and other disease and pathogen interactions where the defined relationship of reduced disease in the presence of the pathogen and susceptible host exist.[23] The term which is used in this chapter are "suppressive soils" and their opposite, "conducive soils".

A. EXAMPLES OF DISEASE SUPPRESSIVE SOILS

The list of pathogens reported to have been suppressed in certain soils appears in Table 1.

B. PHYSICAL AND CHEMICAL CHARACTERISTICS OF SUPPRESSIVE SOILS

Burke[27] was not able "to transfer the rot suppressive properties of the resistant soil to others by transfer of microorganisms" even with the addition of nutritive substances or by autoclaving the receptor soil. This "suggests that rot suppression depends upon the predominance of both physical and microbiological components of the resistant soil and that neither component alone is sufficient to produce the effect." Examples of suppressive and/or conducive soils influenced by physical and chemical properties of soil are summarized in the Table 2.

VI. BIOLOGICAL CONTROL BY INTRODUCED ANTAGONISTS

The severity of plant diseases can be reudced by several different means, for example:

TABLE 2
**Suppressive and Conducive Soils Influenced by Soil Physical and Chemical
Properties[24-27]**

Disease(s)	Soil properties
Where the pathogen does not establish	

Disease(s)	Soil properties
Fusarium wilt of banana	Sandy soil (C)
	Clay soil (S)
Fusarium wilt of cotton	Acidic soil (C)
Fusarium wilt of peas	Heavy clay soil (S)
Fusarium root rot of wheat	Sandy soil, low organic matter and low rainfall (C)
	Fine textured, high organic matter and high rainfall (S)
Fusarium root rot of bean	Loessial (C)
	Lacustrine (S)
Aphanomyces root rot of peas	Aluminum ions (S)
Phymatotrichum root rot of cotton	Alkaline soils (C)
Pine root rot	Light soil (C)
	Heavy soil (S)
Verticillium wilt of sunflower	Aluminum ions (S)

Pathogen establishes but fail to cause disease

Disease(s)	Soil properties
Aphanomyces root rot of peas	Compact soil (C)
Fusarium root rot of beans	Compact soil (C)
Phytophthora root rot of avocado	High organic matter, high exchangeable calcium (S)
Phytophthora root rot of avocado	Abiotic fungistatic factors (S)
Verticillium wilt of cotton	Copper induced fungistasis (S)

Pathogen establishes, produces disease for a while but then declines

Disease(s)	Soil properties
Fusarium root rot of bean	Continuous crop (S)
Phymatotrichum root rot of cotton	High exchangeable sodium (S)
Scab of potato	Continuous potatoes (S) In rotation with sugarbeet, oats and maize (C)
Rhizoctonia damping off of radish	Continuous crop (S)

Note: C = conducive, S = suppressive.

organic amendment, use of resistant varieties, foliar sprays, etc. From numerous experiments
that have been conducted all over the world, a common fact emerges: reduction in the acivity
of a pathogen is often correlated with an increase in the populations of antagonists as assessed
in the soil. The use of selected antagonists is a direct method based on the theory that these
antagonists when introduced into the soil, can act directly on the behavior of the pathogen.

A. EXAMPLES OF BIOLOGICAL CONTROL WITH SOIL AUGMENTATION

With few exceptions, the biological control of plant pathogens by augmenting soils with
antagonists has remained restricted to research studies. However, continued scientific efforts
have resulted in identification of several antagonists which have given successful biological
control in the experimental plots and field when augmented in soil (Table 3).

B. MODES OF ACTION INVOLVED

The entire basis of biological control is exploitation of an antagonist. The mechanism
by which an antagonist adversely effects the target pathogen can be termed as "antagonism".

TABLE 3
Examples of Biological Control by Introduced Antagonists[1-7,28,29]

Host plant and disease	Pathogen	Antagonist(s)
Carnation disease	*F. oxysporum* f. sp. *dianthi*	*Bacillus subtilis*
	F. roseum/culmorum	*Pseudomonas* sp.
	R. solani	*Trichoderma harzianum*
Chickpea wilt	*F. oxysporum* f. sp. *ciceri*	*Trichoderma harzianum*
Cotton root rot	*R. solani*	*T. harzianum*
Cottonwilt	*F. oxysporum* f. sp. *vasinfectum*	Actinomycetes
Crown gall of woody trees	*Agrobacteriam tumifaciens*	*Agrobacterium radiobacter* var.*termifaciens* (Non Pathogenic strain)
Cucumber black root rot	*Phomopsis sclerotioides*	*Gliocladium roseum*
Cucumber fruit rot	*R. solani*	*T. harzianum*
Damping off of pepper	*R. solani*	*B. subtilis*
		Streptomyces sp.
Damping off of radish	*R. solani*	*B. subtilis*
Damping off of bean, cotton, cruciferous and solanaceous vegetables, sugarbeet, tobacco	*Phythium* spp. *R. solani* *S. rolfsii*	*T. harzianum*
Lentil wilt	*F.oxysporum* f. sp. *lentis*	*Trichoderma* sp.
Mushroom dry bubble	*Verticillium malthousei*	*Trichoderma* spp.
Peanut stem rot disease	*S. rolfsii*	*T. harzianum*
Phytophthora blight of pigeon pea	*P. drechsleri* f. sp. *cajani*	*T. harzianum*
Potato scab	*Streptomyces scabies*	*B. subtilis*
Seedling blight of corn	*F. roseum* f. sp. *cerealis*	*B. subtilis, Chaetomium globosum*
Take all of grasses	*Gaeumannomyces graminis*	*Phialophora radicicola*
Take all of cereals	*G. graminis* var. *tritici* and var. *avenae*	*G. graminis* var. *graminis* *Bacillus mycoides* *Phialophora radicicola*
Tomato stem rot	*S. rolfsii*	*T. harzianum*
Tomato wilt	*F. oxysporum* f. sp. *lycopersici*	*Cephalosporium* sp.
Wilt of *Lens culinaris*	*S. rolfsii*	*Trichoderma viride*
	F. oxysporum	*Streptomyces* sp.
Wheat root rot	*R. solani*	*Streptomyces griseus* *Bacillus* sp.
White rot of onion	*S. cepivorum*	*Coniothyrium minitans*

Antagonism includes antibiosis, competition, parasitism, predation, and lysis. Here the most important mechanism "mycoparasitism" is discussed.

1. Mycoparasitism

Mycoparasitism is an act where one fungus parasitizes on another. Barnett and Binder[30] classified the mycoparasitism into two main groups, i.e., necrotrophic and biotrophic, on the basis of nutritional relationship of parasite with the host. The necrotrophic (destructive) parasite makes contact with its host, excretes toxic substance which kills the host cells and utilizes the nutrients that are released. The destructive mycoparasites vary in host range from those capable of attacking only one species. The biotrophic (balanced) mycoparasites are able to obtain their nutrients from the living host cells, a relationship that normally exists in nature. It causes little or no harm to the host at least in the early stages of development. They are generally incapable of survival in the absence of their hosts. However, the necrotrophic parasites are capable of indefinite saprophytic survival in absence of hosts and are also termed as facultative or opportunistic having enzymes that enable them to compete strongly with other microorganisms for space and nutrients.

TABLE 4
Some Examples of Mycoparasitism[1-7,30-32]

Mycoparasite	Host	Type of parasitism
Cephalosporium sp.	*Phytophthora megasperma* var. *sojae*	Produce chlamydospores within oospores and abundant hyphae extended out from the oospores
Coniothyrium minitans	*Sclerotinia sclerotiorum*	Invasion of sclerotia
Corticium sp.	*Rhizoctonia solani*	Hyphal invasion *in vitro*, reduced saprophytic colonization
Dactylella spermatophaga	*P. megasperma* var. *sojae*	Parasitism of oospores
Didymella exitialis	*Gaeumannomyces graminis*	Cell wall penetration by infection pegs, grows into thallus, break down of cell wall by chitinase
Fusarium oxysporum	*R. solani*	Coiling and penetration
Fusarium roseum 'Sambucinum'	*Claviceps purpurea*	Mycoparasitism of sclerotia
F. semitectum	*R. solani*	Coiling and penetration
Humicola fuscoatra	*P. megasperma* var. *sojae*	Destruction of oospores
Hypochytrium catenoides	*P. megasperma* var. *sojae*	Destruction of oospores
Penicillium vermiculatum	*R. solani*	Penetration and branching of the parasitic hyphae inside the host
Pythium sp.	*P. megasperma* var. *sojae*	Destruction of oospores
Rhizoctonia solani	*Rhizopus stolonifer*	Coiling and penetration of the Sporangiophores and Hyphae
Septoria nodorum	*Botyris cinerea*	Hyphal bursting

a. Examples of Mycoparasitism

The mycoparasitism is of common occurrence and examples can be found among all groups of fungi from chytrids to the higher basidiomycetes. Some examples of mycoparasitism, including the recent works, have been enumerated in Table 4.

b. Mechanism of Mycoparasitism

The parasitism includes different kinds of interactions, viz., coiling of hyphae, penetration, production of haustoria and lysis of the hyphae (Figure 1). Ikediugwu[33] studied the ultrastructure of hyphal interference between *Coprinus heptamerus* and *Ascobolus crenulatus* and observed that due to hyphal contact vacuolation and swelling of organelles occur in the host. He, using electron microscopy, made a detailed study of hypahl interference by *Peniophora gigantea* against *Heterobasidium annosum* and observed the presence of an extra plasmalemmal zone. An electron dense material in *H. annosum* was confined to the region around the zone of contact. It was concluded that this electron dense material offered a measure of protection to the cell against the hyphal interference factor. Huang and Hoes[34] found that due to the infection, the host cytoplasm disintegrated and the cell collapsed as a result of infection. Generally, destructive parasites do not initiate their parasitic activity at a distance and, therefore, it appears that an intimate association of the host and parasite is a prerequisite for the production of chemical substance initiating parasitism.

Investigations have shown that necrotrophic parasites kill the susceptible hosts by the action of toxins, antibiotics, or enzymes or by the complementary action of both and utilize the host nutrients.[30] Informations on the mechanism of coiling is insufficient. However, Butler[35] and Dennis and Webster[36] demonstrated that coiling is partly a thigmotropic and partly a chemotropic response.

c. Trichoderma spp. and Antagonism (Mycoparasitism)

Hyphae of majority of *Trichoderma* coil around hyphae of different host fungi. Metabolite

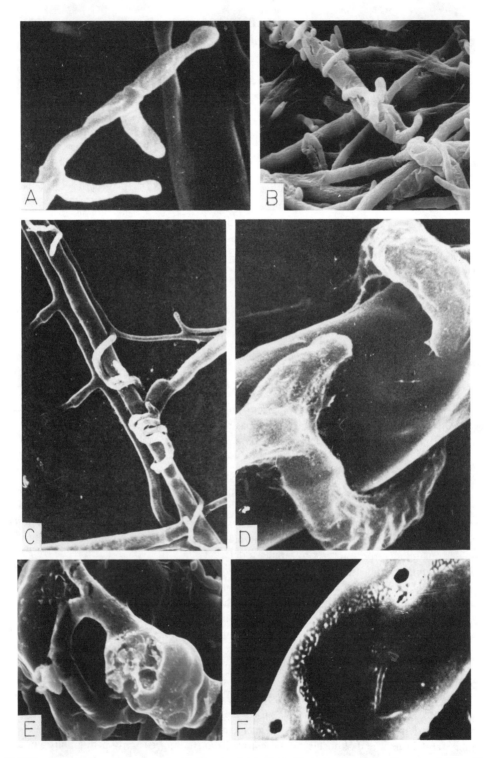

FIGURE 1. Scanning electron micrographs showing interaction of hyperparasite *Trichoderma harzianum* with *Rhizoctonia solani* (A-E) or *Sclerotium rolfsii* (F); chemotropic growth of *T. harzianum* toward hypha of *R. solani* (A), coiling around (B & C) and penetration (D); damaged sclerotium o *R. solani* (E); hypha of *S. rolfsii* from which a coiling hypha of *T. harzianum* was removed, showing digested zone with penetration sites caused by the antagonist (F). (Source: (A & C) - Chet, I., ed., *Innovative Approaches to Plant Disease Control,* John Wiley & Sons, New York, 1987, 372. (With permission); (B & E)-courtesy Dr. M. A. Bari; (D & F) Elad et al., *Phytopatholoy,* 73, 85, 1983. (With permission).

TABLE 5
Mechanism of Parasitism of *Trichoderma* spp. on Different Plant Pathogenic Fungi[1-6,37-46]

Plant pathogen	Mode of antagonism
Botrytis cinerea	Hyphal interaction
Fusarium spp.	Parasitism
F. oxysporum	Hyphal interaction
Helminthosporium teres	Inhibition by contact action
Many test fungi	Hyphal interaction
Mucor mucedo	Hyphal interaction
Pythium spp.	Parasitism
P. aphanidermatum	Hyphal interaction, coiling, penetration and lysis of hyphal cells, antibiosis
P. nunn	Hyphal interaction
Pyricularia oryzae	Inhibition of mycelial growth of pathogen *in vitro*
Rhizoctonia solani	Hyphal interaction, penetration, cell wall lytic enzymes, etc.
Sclerotium rolfsii	Hyphal interaction, coiling and penetration of sclerotia, lysis of mycelial cells, agglutination, etc.
Sclerotinia sclerotiorum	Plasmolysis and parasitism of sclerotia
Seiridium cardinale	Production of toxic metabolites, strong inhibition of conidial germination, germ-tube development and mycelial growth

produced from *T. harzianum* inhibit growth and sporulation of *Aspergillus niger* and *Pestalotia rhododendri*.[37] Fajola and Alasoadura[38] reported the antagonistic effects of *T. harzianum* on *Pythium aphanidermatum*. *T. harzianum* directly attacked the mycelium of *R. solani* when the two fungi were grown together.[39] Elad et al.[40] reported that *T. harzianum* grew better than *S. rolfsii* in culture and invaded mycelium under growth conditions adverse to the pathogen. Elad et al.[41,42] made ultra structural study on the interaction between *Trichoderma* spp. and plant pathogenic fungi. They also detected high β-(1-3) glucanase and chitinase activities in dual culture. Henis et al.[43] reported penetration of sclerotia of *S. rolfsii* by *Trichoderma* spp. and also detected factors affecting germinability and susceptibility to attack sclerotia in field soil. Barak et al.[44] correlated the ability of mycoparasitism by *Trichoderma* to agglutination of conidia of *Trichoderma* by *S. rolfsii*. They also reported that lectins could be involved in the interaction of *Trichoderma* and *S. rolfsii*. Table 5 lists the mechanism of parasitism of *Trichoderma* spp. on different plant pathogenic fungi.

d. Mycoparasites and Biological Control

The plant pathologists are very optimistic regarding the use of mycoparasites for biological control of plant pathogens. Pathogenicity of *G. graminis* causing take-all disease, was reduced by *Didymella oxitialis*. This mycoparasite penetrated and killed the hyphae of *G. graminis* in the rhizosphere as a result of which the pathogen died and subsequently broken down. *Trichoderma harzianum* has been successfully used on a field scale as biocontrol agent.[1-6] It was recorded that mycoparasite inoculum applied to peanut fields after 70 to 100 d of planting reduced Southern blight caused by *S. rolfsii*.[5] Hadar et al.[39] found that an isolate of *T. harzianum* directly attacked the mycelium of *R. solani* when grown together. This strain, when applied in the form of wheat-bran culture into *R. solani* infested soil, effectively controlled damping off of bean, tomato, and egg plant seedlings. In several cases, integrated use of mycoparasites with soil solarization, fumigants and fungicides have given excellent disease control.

Mycoparasitism of *C. minitans* have also been recorded to show promise for biological control of white rot of onion caused by *S. cepivorum*. It was found that the treatment of the pathogen infested soil with dust of *C. minitans* applied either to the seed furrow or as seed dressing at the planting time resulted in 57 to 61% disease control. *Fusarium roseum*

TABLE 6
List of Growth Media Used for Production of *Trichoderma*[1-5,7,29,39,41,43,47-52]

Antagonist	Growth media	Country
Trichoderma harzianum	Grain bran (rye grass) soil medium	U.S.
	Diatomaceous earth granule impregnated with 10% molasses solution	U.S.
	Attapulgus clay granules	U.S.
	Wheat bran	Israel
	Wheat bran-saw dust	Israel
	Sand-cornmeal	U.S.
	Barley grain	U.S.
	Chopped straw	France
	Sorghum seed	India
	Low quality oat seeds	U.S.
	Finger millet seeds	India
	Oat seeds	Bulgaria
	Lignite and stillage	U.S.
	Wheat bran-saw dust	Israel
	Wheat bran	India
	Wheat bran-saw dust	India
	Bran peat	Israel
T. harzianum and *T. viride*	Liquid fermentation technology, molasses and brewers yeast	U.S.
T. viride	Sand-sorghum	India
T. hamatum	Wheat bran	U.S.
T. koningii	Wheat bran-saw dust	India
Trichoderma spp.	Cereal meal-sand	France

"Sambucinum" a hyperparasite of *Claviceps purpurea* infected sclerotia of the pathogen due to which a reduction in the infection was noted. The disease due to *Phomopsis sclerotiodes* was found to be reduced by addition of inoculum of *Gliocladium roseum*. *Trichoderma harzianum* is a potential biocontrol agent for several diseases. It has been successfully used for biological control of diseases caused by *R. solani* on bean, tomato, peanut, carnation, strawberry, and cucumber, and by *S. rolfsii* on the lupine, tomato, peanut, carnation, tomato and strawberry.

2. Systems of Growth and Delivery of the Biocontrol Agents

One of the most critical obstacles to biological control by direct massive soil augmentation has been the lack of methods for mass culturing and delivering antagonists to soil. The production and commercialization of biological products is a dynamic process. Successful introduction is dependent on a number of factors such as production, formulation, toxicity evaluation, quality assurance, marketing, government regulation and efficacy. Despite the limited progress, scientists around the world are attempting to develop effective experimental system for growth and delivery of biocontrol agent. An exhaustive list of growth media used for the mass culture of *Trichoderma* is presented in Table 6. Backman and Rodriguez-Kabana[47] developed a method for mass production and delivery of *Trichoderma* using diatomaceous earth impregnated with molasses as the food base. Commercial products of *Trichoderma* prepared by this and other methods are now available.[4] This material has low bulk, no residue, and can be applied with standard agricultural granule machinery. Recently, an unique achievement in the field of biological control by mycoparasite is the invention of BINAB-T PELLETS,[48] a *Trichoderma* formulation based on *T. harzianum* and *T. polysporum* as active ingredients, manufactured by BINAB, Sigtune, Sweden. This is being used as wood protectants in Sweden and against Dutch elm disease in Belgium in U.K. Another product containing spores of *Trichoderma* and sold under the trade name BINAB-T SEPPIC is applied to casing soil in commercial mushroom houses in France to control *Verticillium malthousei*. In this case *Trichoderma* is applied as a spray at 1 l/m² with 1 × 10⁸ *Trichoderma* spores/l and significant control is achieved.[5]

VII. MYCORRHIZAL FUNGI AND THE CONTROL OF ROOT DISEASES

Probably the most universal and important mutualistic associations between microorganisms and plants are those which develops between fungi and roots, the mycorrhizas (myco meaning fungus, and rhiza meaning root). Most, if not all, plants establish and benefit from mycorrhizal relationships, and a large but uncertain number of fungal species are involved. There are four main types of mycorrhizas: *ectotrophic, vesicular-arbuscular* (V-A), *orchidaceous,* and *ericaceous* (Figure 2). The last three used to be lumped together and described as *endotrophic,* but this is misleading as there are enormous differences between them.

A. ECTOMYCORRHIZAS

Marx[53] reviewed the mechanisms of resistance due to ectomycorrhizas to pathogenic root infections. Ectomycorrhizal fungi can: (1) utilize various chemicals in the root and at the root surface, thereby reducing the amount of nutrients available to pathogens, (2) provide mechanical barrier, the fungal mantle, to penetration of primary cortical cells by pathogens, (3) support, along with the root, an antagonistic rhizosphere population of microorganisms, and (4) induce production of inhibitors in the cortical cells and inhibit infection and spread of pathogens.

The evidence for a role of mycorrhizal fungi in protecting plants against disease concerns the sheathing mycorrhizas of pine trees and the pathogen *Phytophthora cinnamomi*.[53,54]

B. ENDOMYCORRHIZAS

Reports concerning the interaction of V-A mycorrhizal fungi with other microorganisms present in soil are increasing. These fungi interact with soil inhabiting microorganisms such as other fungi, bacteria, actinomycetes, insects, and nematodes. They also interact with various root-borne pathogens.[55]

There is a distinct possibility that colonization of roots by V-A mycorrhizal fungi confers resistance to invasion by other root pathogens. However, a few studies reveal that V-A mycorrhizal fungi also make the roots susceptible to root rot pathogens, thereby increasing severity of disease rather than reducing it. Some workers could not find any definite effect of V-A mycorrhizal fungi on the fungal pathogens.

There are clear indications that mycorrhiza formation induces alteration in host cells which make them more resistant to soil and root-borne pathogens. It has been reported that the disease caused by *Olpidium brassicae* in lettuce and tobacco decreased when inoculated with *Glomus mosseae* while diseases caused by *Helminthosporium sativum* and *Erysiphe graminis* in barley, *Colletotrichum lindemuthianum* and *Uromyces phaseoli* in bean, *Botrytis cinerea* in lettuce, *Erysiphe cichoracearum* in cucumber and TMV in tobacco increased on inoculation with *Glomus mosseae*.[55] It was concluded that mycorrhiza reduced the infection by root invading fungi but increased the incidence of infection on aerial parts by fungi or viruses. It is also reported that *Glomus mosseae* induces higher chitinase and arginine accumulation which caused development of resistance in mycorrhizal plants.[55]

VIII. ORGANIC AMENDMENTS AND BIOLOGICAL CONTROL

Modification of soil environment is one of the methods of biological control. Amendment

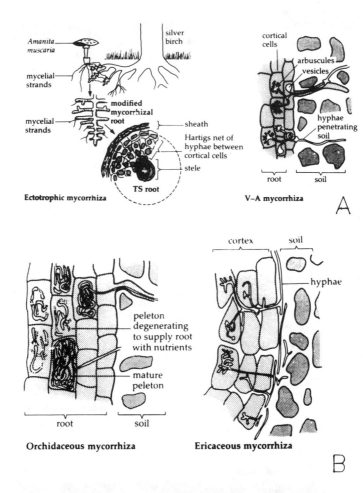

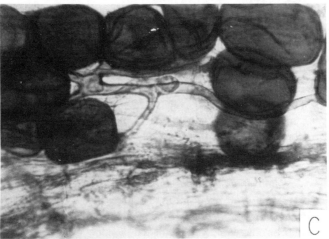

FIGURE 2.   (A and B) Main categories of mycorrhizal associations (Ingle, M. R., *Microbes and Biotechnology,* Blackwell, Basil, U. K., 1986, 78. (With permission). (C)-VA-mycorrhizae (Courtesy Dr. A. K. Sharma).

of soil with decomposable organic matter is recognised as an effective method of changing the soil and rhizosphere environment, thereby adversely affecting the life cycle of pathogens and enabling the plant to resist their attack through better vigor and/or altered root physiology. There are many advantages from this type of disease control such as possibility of multiple pathogen suppression, lasting effects, less cost, absence of hazards usually associated with chemicals, and improvement of soil fertility and nutrient uptake by plants.

Crop residue decomposition in soil is known to encourage microbial activity both quantitatively and qualitatively. The enhanced microbial activity has two effects: it increases the variety of complex organic compounds in soil and promotes the population of antagonists in the soil. Together, these two factors increase the biological buffering capacity of the soil. This capacity of the soil helps in disease prevention in areas where the pathogen is not yet established.

Amendments act against a disease through (1) direct effect on the active pathogen on the root or in the rhizosphere, (2) direct effect on the pathogen during its survival in the absence of the host, and (3) indirect effect on the pathogenesis through the host. In the latter are included all those consequences of amendment which result from uptake of organic compounds by the plant root, changed host physiology and possible development of resistance. Therefore, the amendments can reduce inoculum density, inoculum capacity, host proneness and can also increase host resistance; the net result being reduction in disease severity.

A. DISEASES CONTROLLED

The list of soil-borne diseases that have been controlled in glasshouse, microplots or field plots by organic amendments of soil, is quite exhaustive. Some of the important ones are given in Table 7.

B. MECHANISMS OF DISEASE MANAGEMENT IN AMENDED SOIL

It is the decomposition of the organic matter at the site where pathogen is present that causes suppression of disease in amended soil. The material used for amendment may or may not contain the inhibitory factor. Even if it does contain such a substance, its status in the soil is difficult to define because of microbial action on the substance. The wide variety of substances used for soil amendment, such as green or dry crop residues, oil-cakes, sawdust, different types of meals (both of plant or animal origin), chitinous materials, etc., widely differ in their composition and, therefore, in speed of their decomposition, preference for microorganisms, and in the nature of decomposition products. Stover[57] proposed the mechanisms (Figure 3) which might operate in biological control through organic amendments in soil. Singh and Sitaramaiah[58,59] have also proposed a broad concept of mechanisms in biological control of plant diseases (Figures 4 and 5).

1. Host Mediated Disease Management in the Amended Soil

Organic matter influences soil physical characters such as pore size, aeration, temperature, water retention capacity, etc. The modifications in the structural characters of the soil help in better solubilization of mineral plant nutrients, which together with the nutrients released by microbial action on the organic matter help in rapid extension of the root system, quicker replacement of damaged roots, better uptake of nutrients, retention of added nitrogen for a longer period, and finally better plant vigor. In many diseases this offsets the damage by pathogens. This is one part of the host management. It is known that decomposition of organic matter releases many organic acids, aromatic compounds, phenols, etc., both volatile and nonvolatile. Although the volatile compounds are of transitional nature, they have profound effect on microorganisms in the bulk soil as well as on the root-soil interface.

TABLE 7

Suppression of Soil-Borne Diseases by Organic Amendments of Soil[1-6,10,11,56-58]

Pathogen	Crop and Disease
Fungal Diseases	
Aphanomyces euteiches	Root rot of peas
Fusarium oxysporum f. sp. *udum*	Wilt of pigeonpea
F. oxysporum f. sp. *cubense*	Wilt of banana
F. oxysporum f. sp. *lini*	Wilt of linseed
F. oxysporum f. sp. *pisi*	Wilt of pea
F. oxysporum f. sp. *corianderi*	Wilt of coriander
F. oxysporum f. sp. *ciceri*	Wilt of chickpea
F. solani f. sp. *phaseoli*	Wilt and root rot of bean
Fusarium coeruleum	Wilt and root rot of guar
Helminthosporium sativum	Root rot of wheat
Macrophomina phaseolina	Root rot of cotton
Ophiobolus graminis	Take-all disease of wheat
Phymatotrichum omnivorum	Root rot of cotton
Phytophthora sp.	Root rots of ornamentals
Phytophthora cinnamomi	Root rot of avocado
Pythium spp.	Browning root rot of wheat
Pythium aphanidermatum	Soft rot of ginger
Phythium ultimum	Seedling blight of alfalfa
Rhizoctonia solani	Black scurf of potato
Sclerotium rolfsii	Wilt of *piper beetle*
Sclerotium graminis	Foot rot of wheat
Streptomyces scabies	Common scab of potato
Thielaviopsis spp.	Root rots of ornamentals
Thielaviopsis basicola	Root rot of bean and sesamum
Verticillium albo-atrum	Wilt of potato, tomato and cotton
Diseases Caused by Nematodes	
Belanolaimus longicaudatus	Sting nematode
Heterodera avenae	Cereal cyst nematode
Heterodera major	Cereal cyst nematode
Heterodera rostochidensis	Potato cyst nematode
Heterodera schachtii	Sugarbeet cyst nematode
H. tabacum	Tobacco cyst nematode
Hoplolaimus indicus	Lance nematode
Hoplolaimus tylenchiformis	Lance nematode
Meloidogyne javanica	Root knot nematode of vegetables
M. incognita	Root knot nematode of vegetables
Pratylenchus penetrans	Lesion nematode
Tylenchulus semipenetrans	Citrus nematode

These effects can be mediated through the host. The host may absorb some of the compounds or in some cases the superficial cells of the roots may be damaged, increasing the CO_2 release. The root physiology is changed and may affect the pathogens in many ways:

1.   Orientation of infective propagules (namatode larvae or fungal germlings) toward the root may be disturbed. In the case of root knot nematode, the roots did not appreciably attract the larvae.[58]

PARASITISM		ANTAGONISM					HOST NUTRITION	CHANGES IN PHYSICAL ENVIRONMENT
		ANTIBIOSIS			PATHOGEN NUTRITION			
		PLANT FUNGISTASIS	MICROBIAL FUNGISTASIS		INDIRECT STARVATION	DIRECT STARVATION		
DIRECT	INDIRECT LYSIS AND DECOMPOSITION	TOXIC COMPONENTS OF ORIGIN	SPECIFIC FUNGISTASIS (ANTIBIOTIC)	GENERAL NON-SPECIFIC FUNGISTASIS	NUTRIENTS ADEQUATE BUT POOR COMPETITIVE OR ANTIBIOSIS PREVENT UTILIZATION	LOW LEVEL OR ABSENCE OF ESSENTIAL NUTRIENTS CAUSES REDUCTION OR CESSATION OF ACTIVITY	RESISTANCE OR TOLERANCE OF HOST INCREASED BY IMPROVED NUTRITION THROUGH (A) MORE VIGOROUS HEALTHIER ROOTS (B) NUTRIENTS IN AMENDMENTS	pH CO_2

FIGURE 3. Mechanisms of disease control through organic amendments (Stover, R. H., *Recent Prog. Microbiol*, 8, 267, 1962. With permission).

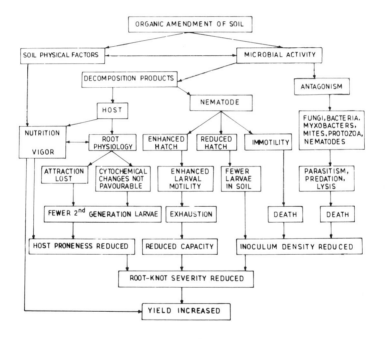

FIGURE 4. Possible pathways of action of organic amendments against a soil-borne disease (Root knot - *Meloidogyne* sp. (Courtesy - Dr. R. S. Singh).

2. Cytochemical status of roots may become unfavorable for pathogen development within the roots. Females of *Meloidogyne javanica* lay fewer eggs in plants raised on amended soil. Tomato plants grown in amended soils (margosa cake or sawdust plus nitrogen amendments) had more total phenols than the plants grown in nonamended soil. Phenols are known to impart resistance to nematodes and also to many fungal diseases. Exposure of tomato roots to phenolics imparted some resistance and fewer larvae could enter the roots, fewer females could mature and only few eggs were produced.[59]

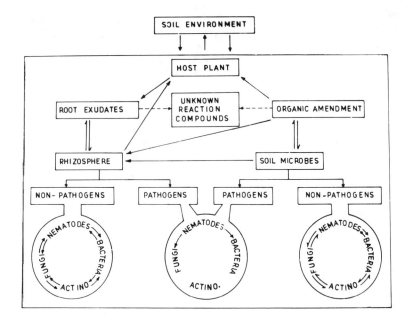

FIGURE 5.   Interaction of amendments, rhizosphere and soil microflora (Courtsey - Dr. R. S. Singh).

3.   Carbon dioxide status, pH, and microbial cover of the soil-root interface may be such that the pathogen on reaching the site of infection may fail to colonize it. In the control of soft rot of ginger by sawdust and oilcake amendment, Rajan[60] suggested that the amendments did not suppress the pathogen as such as they protected the fixed infection court. Here, it may be emphasised that environment of the root-soil interface rather than the bulk soil is important for management of root diseases through change in soil organic environment. Khanna and Singh[61,62] attempted to study the rhizosphere of plants in amended soil and noted that rhizosphere of pigeonpea and pea tended to reverse the effect of amendments in bulk soil. The stimulatory effect of oilcake amendments was suppressed and stimulation of low magnitude was reversed to inhibitory effect. Such informations related with the changes in rhizosphere, though difficult to study can definitely provide precise information leading to more effective and economical use of amendments.

The combined effect of nutrition, plant vigor and altered root physiology reduces the host proneness and also reduces the amount of secondary inoculum as a result of limited infection.

2. Effect of Decomposition Products on the Pathogen
The different organic products of decomposition also affect the pathogen directly. With normal aerated cultivated soils, most of the reports suggest an initial stimulatory effect of the decomposition products on the pathogens in the same way as other microflora. Oil cakes and green or mature crop residue and chitinous materials, increase the population of *Pythium aphanidermatum* in soil.[63] Similar stimulatory and inhibitory effects are reported for *Fusarium* spp.[62,64] The stimulation was obviously due to nutrition and due to volatiles released by decomposition or by enhanced microflora. If stimulation occurred in the absence of a host, the stimulated activity of other microorganisms could cause lysis of the mycelium in due course when nutrient supply dwindles.

Suppression of *Fusarium oxysporum* in pine forest litter has been attributed to the presence of fatty acids like shikimic, quinic, and malic or phosphoric acids which occur in pine needles.[64] These acids stimulate germination of chlamydospores and then the germlings are lysed by the microbial activity. Several alcohols and aldehydes from decomposing hay stimulate germination of secondary sclerotia of *Sclerotium rolfsii* followed by lysis or formation of secondary sclerotia by the mycelium. In studies with nematodes, Singh and Sitaramaiah[58] found that low concentrations of fatty acids enhanced larval hatch of *M. javanica*. The larvae thus released showed abnormal motility in the soil atmosphere. It caused rapid exhaustion of their lipid reserves and this either killed them or made them too weak to cause penetration of the roots.

Depending in the chemical composition of the amendment and prevailing soil conditions, inhibitory volatiles are also produced. Lewis and Papavizas[22] attributed the suppression of *Aphanomyces euteiches* by cruciferous amendment of soil to sulphur compounds released during decomposition. Similar compounds can cause suppression of nematodes also. Peethambaran[66] demonstrated that volatiles from a soil amended with sawdust had inhibitory effect on the growth of *Phythium aphanidermatum* and *P. graminicola*. Liberation of ammonia and its accumulation causes a shift in soil pH towards alkalinity which results in changes in soil microbial equilibrium may harm the pathogens. During the last two decades, the role of ethylene as a volatile inhibitor in soil has become a subject of intensive research. This volatile is produced by microbes and plant roots under conditions of low oxygen. In organic matter amended soils, when oxygen demanding bacteria as *Bacillus subtilis* become very active, low oxygen conditions are created and ethylene is produced. This causes fungistasis. An ethylene producing bacterium like *Clostridium*, if fortified by proper amendment of soil, can suppress such pathogens as *Fusarium oxysporum*. Reduced oxygen and increased CO_2 levels on root surface due to decomposition of specific types of crop residues (such as barley straw) by selected bacteria also cause destruction of pathogens by starvation.[64]

The fungitoxic role of phenolics, e.g., chlorogenic acid, caffeic acid, catechol, tannins and glucosides (that give rise to phenolics), and quinones (oxidized phenols) are well known. They form the basis of resistance in many diseases. In the root knot disease also, their role in imparting resistance is known. Singh and Sitaramaiah[58,59] demonstrated that phenol content of the amended soil increased and exposure of *M. javanica* larvae to phenolics, high concentrations of fatty acids and certain aromatic compounds reduced their egg laying capacity. At about 100 ppm concentration of pure phenol there was 96 to 100% mortality of larvae. Amount of total phenols detected in sawdust and neem cake amended soils was 400 and 40 ppm, respectively.

The above three types of effects of decomposition products (stimulation, suppression and mortality) ultimately lead to reduced inoculum density and capacity, thus disease suppression.

3. Direct Effects of Amendments on Antagonism

a. *Competition*

In amended soils there is always intense microbial activity. Although sometimes a concomitant increase in the population of pathogens occurs, the latter have generally weak saprophytic ability. In the absence of its host the pathogen may be unable to compete successfully with saprophytic microorganisms for a long period. The increased microbial activity results in demands for oxygen and nutrients more than available and hence the pathogens suffer. Clark[15] had pointed out that oxygen deficiency could occur near a root if rhizosphere organisms were exceptionally active.

Competition for nutrients or oxygen or both is probably responsible for a significant percentage of failure and endolysis of germlings of fungi. Amendments of high C: N ratio materials create deficiencies of specific nutrients. Barley straw immobilizes nitrogen and lignins create a deficiency of carbon. The control of such diseases as bean root rot (*Rhizoctonia solani, Fusarium solani* f. sp. *phaseoli* and *Thielaviopsis basicola*) and take-all of wheat by soil amendments with high C:N ratio materials, has been attributed to nitrogen starvation of the pathogen due to intense competition which causes extra accumulation of CO_2. Due to nitrogen starvation autolysis of pathogen hyphae may occur. Certain isolates of *Aspergillus* from amended soil were found to be successful competitor of *Fusarium udum* in amended soil.[64]

b. Hyperparasitism and Predation

The amendment of soil causes several fold increase in the populations of fungi, bacteria and actinomycetes and the chances of parasitization of many soil-borne plant pathogens also increase. Among the mycoparasites, ample work has been done on *Trichoderma* species.

c. Antibiosis

The production and fate of antibiotics in soil and the significance of antibiosis in the biological control of plant diseases have been elegantly elucidated and discussed by several investigators. (1) Antibiosis by fungal metabolites: saprophytic soil fungi have high competitive ability and undoubtedly play a major role in the breakdown of pathogen infested plant refuse. Plant pathogens in straw, wood, leaves, fruits, roots, or other cellulosic materials are more apt to be replaced by saprophytic fungi than other organisms. Species of *Penicillium, Aspergillus*, etc., are well known to produce antibiotic substances which inhibit plant pathogenic fungi. (2) Antibiosis or lysis by actinomycete metabolites: actinomycetes are potentially valuable as antagonists during host free periods and at that time their population can be manipulated to encourage their activities. As resident antagonists, actinomycetes are important in maintainig satisfactory biological balance in soil because of their ability to produce powerful antibiotics. About three fourth of species of *Streptomyces* produce antibiotics that might be effective against bacteria, fungi or actinomycetes. In addition, many have the ectoenzyme systems necessary to cause breakdown of protein, cellulose and chitin, which may have potential in the biological control of micro-organisms with chitinous walls. Vruggink[67] obtained some control of *Fusarium solani* f. sp. *phaseoli, F. oxysporum, F. oxysporum* f. sp. *conglutinans. F. oxysporum* f. sp. *cubense,* and *Rhizoctonia* with chitin amendment. Actinomycetes that proliferate abundantly in chitin amended soils, liberate chitinase which contributes to the destruction of *Fusarium* hyphae. Reddy and Rao[68] reported two strains of *Streptomyces ambafaciens* effective against *Fusarium oxysporum* f. sp. *vasinfectum.* Singh and Singh[64] reported that amendment of soil with margosa cake, rice husk or sawdust with or without supplemental nitrogen greatly enhanced the lytic effect of the soil on *F. udum.* The degree of lysis depended on the dosage of amendment used and the stage of its decomposition in the soil. The extent of lysis increased with the increase in the bacterial population. Amongst bacteria, *Bacillus subtilis* was ubiquitous in most of the lytic zones.

REFERENCES

1. **Baker, K. F. and Cook, R. J.**, *Biological Control of Plant Pathogens*, Freeman, San Francisco, 1974, 433.
2. **Cook, R. J.**, Biological Control of Plant Pathogens, in *Biological Control in Crop Production*, Papavizas, G. C., Ed., Allanhold, Osmun, 1981, 461.
3. **Baker, F. K. and Snyder, W. C., Eds.**, *Ecology of Soil-Borne Plant Pathogens*, Univ. of California Press, Berkeley, 1965, 571.
4. **Papavizas, G. C.**, *Biological Control in Crop Production*, Allanhold, Osmun, 1981, 461.
5. **Cook, R. and Baker, K. F.**, *The Nature and Practices of Biological Control of Plant Pathogens*, Am. Phytopathol. Soc., St. Paul, Minnesota, 1983, 539.
6. **Deacon, J. W.**, *Microbial Control of Plant Pests and Diseases*, Van Nostrand Reinhold (U.K) Co. Ltd., 1983, 88.
7. **Chet, I., Ed.**, *Innovative Approaches to Plant Diseaese Control*, John Wiley and Sons, New York, 1987, 372.
8. **DeBach, P.**, *Biological Control of Insect Pests and Weeds*, Reinhold, New York, 1964.
9. **Garrett, S. D.**, *Pathogenic Root Infecting Fungi*, University Press, Cambridge, 1970.
10. **Baker, R.**, Mechanism of biological control of soil-borne plant pathogens, *Annu. Rev. Phytopathol.*, 6, 263, 1968.
11. **Cook, R. J.**, Management of the Associated Microbiota, in *Plant Disease—An Advanced Treatise*, Vol. I, Horsfall, J. G. and Cowling, E. B., Eds., Academic Press, New York, 1977, 465.
12. **Bruehl, G. W.**, Systems and mechanisms of residue possession by pioneer fungal colonists, in, *Biology and Control of Soil-Borne Plant Pathogens*, Bruehl, G. W., Ed., Am. Phytopathol. Soc., St. Paul, Minnesota, 1975, 530.
13. **Jackson, R. M.**, Antibiosis and fungistasis of soil microorganism in *Ecology of Soil-Borne Plant Pathogens*, Snyder, W. C. and Baker, K. F., Eds., Univ. of California, Press, Berkeley, 1965, 571.
14. **Park, D.**, Antagonism—the background to soil Fungi, in *The Ecology of Soil Fungi*, Parkinson, D. and Waid, J. S., Eds., Univ. Liverpool Press, 1960, 324.
15. **Clark, F. C.**, The Concept of Competition in Microbial Ecology, in *Ecology of Soil-Borne Plant Pathogens*, Snyder, W. C. and Baker, K. F., Eds., Univ. of California Press, Berkeley, 1965, 571.
16. **Dobbs, C. G. and Hinson, H. W.**, A wide spread fungistasis in soil, *Nature*, 172, 197, 1953.
17. **Watson, A. G. and Ford, E. J.**, Soil fungistasis—a reappraisal, *Annu. Rev. Phytopathol.*, 10, 327, 1972.
18. **Dwivedi, R. S. and Sarvanakumar, R.**, Soil fungistasis and its implications in biological control of plant pathogens, in *Progress in Microbial Ecology*, Mukerji, K. G., Agnihotri, V. P., and Singh, R. P., Eds., Print House, Lucknow (India), 1984, 653.
19. **Papavizas, G. C. and Lumsden, R. D.**, Biological control of soil-borne fungal propagules, *Annu. Rev. Phytopathol.*, 18, 389, 1980.
20. **Zakaria, M. A., Lockwood, J. L., and Filonow, A. B.**, Reduction in *Fusarium* population density in soil by volatiles degradation products of oilseed meal amendments, *Phytopathology*, 70, 495, 1980.
21. **Zakaria, M. A. and Lockwood, J. L.**, Reduction in *Fusarium* populations in soil by oilseed meal amendments, *Phytopathology*, 70, 240, 1980.
22. **Lewis, J. A. and Papavizas, G. C.**, Effect of sulfur-containing volatile compounds and vapors from cabbage decomposition on *Aphanomyces euteiches*, *Phytopathology*, 61, 208, 1971.
23. **Huber, D. M. and Schneider, R. W.**, The description and occurrence of suppressive soils, in *Suppressive Soils and Plant Disease*, Schneider, R. W., Ed., Am. Phytopathol. Soc., St. Paul, Minnesota, U.S.A., 1982, 88.
24. **Schneider, R. W., Ed.**, *Suppressive Soils and Plant Diseases*, Am. Phytopathol. Soc., St. Paul, MN, 1982.
25. **Hornby, D.**, Disease suppressive soils, *Annu. Rev. Phytopathol.*, 21, 65, 1983.
26. **Chaube, H. S.**, Pathogen suppressive soils, in *Perspectives of Phytopathology*, Agnihotri, V. P., Chaube, H. S., Singh, N., Singh, U. S., and Dwivedi, T. S., Eds., Today and Tomorrow's Printers, New Delhi, 1989, in press.
27. **Burke, D. W.**, Fusarium root rot of beans and behaviour of the pathogen in different soils, *Phytopathology*, 55, 1122, 1965.
28. **Broadbent, P., Baker, K. F., and Waterworth, Y.**, Bacteria and actinomycetes antagonistic to fungal root pathogens in Australian soils, *Aust. J. Biol. Sci.*, 24, 925, 1971.
29. **Mukhopadhyay, A. N.**, Biological control of soil-borne plant pathogens by *Trichoderma* spp., *Indian J. Mycol. Plant Pathol.*, 17, 1, 1987.
30. **Barnett, H. L., and Binder, F. L.**, Fungal host-parasite relationship, *Annu. Rev. Phytopathol.*, 11, 273, 1973.

31. **Upadhyay, R. S. and Rai, B.,** Mycoparasitism with reference to biological control of plant diseases, in *Recent Advances in Plant Pathology,* Hussain, A., Singh, K., Singh, B. P., and Agnihotri, V. P., Eds., Print House, Lucknow, India, 1983, 521.

32. **Boosalis, M. G.,** Hyperparasitism, *Annu. Rev. Phytopathol.,* 2, 363, 1964.

33. **Ikediugwu, F. E. O.,** Ultrastructure of hyphal interference between *Coprinus heptemerus* and *Ascobolus crenulatus, Trans. Br. Mycol. Soc.* 66, 281, 1976.

34. **Huang, H. C. and Hoes, J. A.,** Penetration and infection of *Sclerotinia sclerotiorum* by *Coniothyrium minitans., Can. J. Bot.,* 54, 406, 1976.

35. **Butler, E. E.,** *Rhizoctonia solani* as a parasite of fungi, *Mycologia,* 49, 354, 1957.

36. **Dennis, C. and Webster, J.,** Antagonistic properties of species group of *Trichoderma.* III. Hyphal interaction, *Trans. Br. Mycol. soc.,* 57, 363, 1971.

37. **Hutchinson, S. A. and Cowan, M. E.,** Identification and biological effects of volatile metabolites from cultures of *Trichoderma harzianum, Trans. Br. Mycol. Soc.,* 59, 71, 1972.

38. **Fajola, A. O. and Alasoadura, S. O.,** Antagonistic effects of *Trichoderma harzianum* on *Pythium aphanidermatum* causing damping-off disease of tobacco in Nigeria, *Mycopathologia,* 57, 47, 1975.

39. **Hadar, Y., Chet, I., and Henis, Y.,** Biological control of *Rhizoctonia solani* damping-off with wheat bran culture of *Trichoderma harzianum, Phytopathology,* 69, 64, 1979.

40. **Elad, Y., Chet, I., and Katan, J.,** *Trichoderma harzianum:* a biocontrol agent effective against *Sclerotium rolfsii* and *Rhizoctonia solani, Phytopathology,* 70, 119, 1980.

41. **Elad, Y., Barak, R., Chet, I., and Henis, Y.,** Ultrastructural studies of the interaction between *Trichoderma* spp. and plant pathogenic fungi, *Phytopathol. Z.,* 107, 168, 1983.

42. **Elad, Y., Chet, I., Boyle, P., and Henis, Y.,** Parasitism of *Trichoderma* spp. on *Rhizoctonia solani* and *Sclerotium rolfsii:* Scanning electron microscopy and fluorescence microscopy, *Phytopathology,* 73, 85, 1983.

43. **Henis, Y., Elad, Y., Chet, I., Hadar, Y., and Hadar, E.,** Control of soil borne plant pathogenic fungi in carnation, strawberry and tomato by *Trichoderma harzianum, Proc. 9th International Cong. Pl. Prot.,* Washington, D.C., August 5-11, 1979.

44. **Barak, R., Elad, Y., Mirelman, D., and Chet, I.,** Lectins: a possible basis for specific recognition in *Trichoderma—Sclerotium rolfsii* interaction, *Phytopathology,* 75, 458, 1985.

45. **Chet, I.,** *Trichoderma—*application, mode of action, and potential as a biocontrol agent of soil borne plant pathogenic fungi, in *Innovative Approaches to Plant Disease Control,* Chet, I., Ed., John Wiley and Sons, New York, 1987, 372.

46. **Papavizas, G. C.,** *Trichoderma* and *Gliocladium:* biology, ecology and potential for biocontrol, *Annu. Rev. Phytopathol.,* 23, 23, 1985.

47. **Backman, P. A. and Rodriguez-Kabana, R.,** A system for gorwth and delivery of biological control agents to the soil, *Phytopathology,* 65, 819, 1975.

48. **Gear, A.,** *Trichoderma Newsletter,* Hendry Doubleday Research Association, U.K., No. 3, 1986.

49. **Kelley, W. D.,** Evaluation of *Trichoderma harzianum* impregnated clay granules as a biocontrol for *Phytophthora cinnamomi* causing damping off of pine seedlings, *Phytopathology,* 66, 1023, 1976.

50. **Jones, R. W., Pettit, R. E., and Taber, R. A.,** Lignite and Stillage: carrier and substrate for application of fungal bio-control agents to the soil, *Phytopathology,* 74, 1167, 1984.

51. **Sivan, A., Elad, Y., and Chet, I.,** Biological control effects of a new isolate of *Trichoderma harzianum* on *Pythium aphanidermatum Phytopathology,* 74, 498, 1984.

52. **Lewis, J. A. and Papavizas, G. C.,** Reduction of inoculum of *Rhizoctonia solani* in soil by germlings of *Trichoderma hamatum, Soil Biol. Biochem.,* 19, 195, 1987.

53. **Marx, D. H.,** The role of ectomycorrhizae in the protection of pine from root infection by *Phytophthora cinnamomi,* in *Biology and Control of Soil—Borne Plant Pathogens,* Brueh, G. W., Ed., Am. Phytopathol. Soc., St. Paul, Minnesota, 1975.

54. **Marx, D. H. and Davey, C. B.,** The influence of ectotrophic mycorrhizal fungi on the resistance of pine roots to pathogenic infections. III. Resistance of aseptically formed mycorrhizae to infection by *Phytophthora cinnamomi, Phytopathology,* 59, 549, 1979.

55. **Mosse, B.,** Advances in the study of vesicular-arbuscular mycorrhiza *Annu. Rev. Phytopathol.,* 11, 171, 1973.

56. **Huber, D. M. and Watson, R. D.,** Effect of organic amendments on soil-borne plant pathogens, *Phytopathology,* 60, 22, 1970.

57. **Stover, R. H.,** The use of organic amendment and green manures in the control of soil-borne phytopathogens, *Recent Prog. Microbiol.,* 8, 267, 1962.

58. **Singh, R. S. and Sitaramaiah, K.,** Control of Plant parasitic nematodes with organic amendments of soil, *PANS,* 16, 287, 1971.

59. **Singh, R. S., Sitaramaiah, K.,** Control of plant parasitic nematodes with organic amendments of soil, *G. B. Pant Univ. Expt. St. Res. Bull.,* 1973, 6, 289.

60. **Rajan, K. M.,** A study of Soil Factors Influencing Inoculum Potential of *Pythium aphanidermatum* with Special Reference to Organic Amendments, Ph.D. Thesis submitted to G. B. Pant Univ., Pantnagar, India, 1971, 182.

61. **Khanna, R. N. and Singh, R. S.,** Rhizosphere population of *Fusarium* spp. in amended soils, *Indian Phytopathol.,* 27, 331, 1974.

62. **Khanna, R. N. and Singh, R. S.,** Microbial populations of pigeon pea rhizosphere in amended soils, *Indian J. Mycol. Pl. Pathol.,* 5, 131, 1975.

63. **Singh, R. S. and Pandey, K. R.,** Population dynamics of *Pythium aphanidermatum* in oil-cake amended soil, *Can. J. Microbiol.,* 13, 601, 1967.

64. **Singh, N. and Singh, R. S.,** Significance of organic amendment of soil in biological control of soil-borne plant pathogens, in *Progress in Microbial Ecology,* Mukerjee, K. G., Agnihotri, V. P., and Singh, R. P., Eds., Print House (India), Lucknow, 1984, 533.

65. **Lewis, J. A. and Papavizas, G. C.,** Volatiles from decomposing amendments on *Rhizoctonia solani, Phytopathology,* 63, 803, 1973.

66. **Peethambaran, C. K.,** Studies on the Comperative Survival of *Pythium* spp. in Soil, Ph.D. Thesis submitted to G. B. Pant Univ., Pantnagar, India, 1975, 251.

67. **Vruggink, N.,** The effect of chitin amendment on actinomycetes in soil and on the infection of potato tubers by *Streptomyces scabies, Neth. J. Plant Pathol.,* 76, 293, 1970.

68. **Reddi, G. S. and Rao, A. S.,** Antagonism of soil actinomycetes to some soil borne plant pathogenic fungi, *Indian Phytopathol.,* 24, 649, 1971.

Chapter 15

CULTURAL PRACTICES

I. INTRODUCTION

Cultural practices which include manipulation and/or adjustment of crop production techniques have been as old as possibly agriculture itself. In early stages of agriculture development, the growers through their experiences and observations had known that repeated cultivation of a particular crop species or variety on a piece of land oftenly resulted in crop sickness. By proper crop rotations they had been avoiding such sickness. As a matter of fact, in the present day agriculture, cultural practices are being considered as essential backup methods for management of plant diseases.

We have already discussed in preceding chapters that for development of infectious diseases, the contact between the host surface, and pathogen propagules must occur in an environment favorable for pathogenesis. Adequate adjustment in crop production techniques can modify the environment in such a manner that it becomes unfavorable for the pathogen and pathogenesis. Based on this the disease control affected by cultural practices are preventive. These practices aim at reducing the activity and density of inoculum.

II. METHODS

Procedures for disease control through cultural practices are discussed under the following three heads:[1]

1. Production and use of pathogen-free planting material
2. Adjustment of crop culture to minimize disease
3. Sanitation

A. PRODUCTION AND USE OF PATHOGEN FREE PLANTING MATERIAL

In chapters on survival and spread of plant pathogens it has been described that many plant pathogens are transmitted by establishing themselves on or in the seed or other vegetative propagating materials or as contaminants. For successful disease control this source of primary inoculum must be destroyed. The following methods are followed to produce and use pathogen-free seed material.

1. Seed Production Areas

Seed should be produced in areas where the pathogens of major concern are unable to establish or maintain themselves at critical levels during periods of seed development. Areas with low rainfall and low relative humidity are favorable for production of high quality seeds. Some examples are anthracnose of beans and cucurbits, Asochytosis of pea and chickpea, bacterial blight of legumes, etc. Such crops can be grown in dry areas with the help of irrigation.

2. Inspection of Seed Production Plots

Periodical inspection of crops raised for seed production is an important procedure in the production of clean and healthy seeds. Destruction of diseased plants/organs at the time of inspection helps in reducing inoculum in the field and thus, the percentage of healthy seeds in the produce is increased. If disease incidence is very high, the entire crop may be rejected for seed.

3. Isolation Distance

The distance between seed production and commercial plots has been worked out for reducing seed-borne loose smut of wheat and barley. The distance between plots may vary from region to region depending upon weather conditions. In Canada, the problem of proper isolation of stock seed production areas for barley is governed by legislation.[2] Similar legislation exists in Germany for stock production of barley and wheat; seed crops are rejected if loose smut is encountered in a field within a distance of less than 50 m of the seed field.[3] In Holland, it has been demonstrated that a minimum distance of 100 m from infected fields must be secured for barley grown under seed certification programs. Barley and wheat crops should be isolated by at least 50 m from any source of loose smut infection for certified seed production in the U.K.[3]

4. Drying and Ageing of Seed

Germinability of seeds, not properly dried before storage, is reduced. Such seeds harbor several types of pathogens. Prolonged storage of seeds also reduces many pathogens. *Fusarium solani* f. sp. *cucurbitae,* infecting cucurbits, is eliminated if the seeds are stored for 2 years before sowing. Similar eradication of pathogen has been achieved in anthracnose of cotton.[3]

5. Cleaning of Seed

In many cases the pathogen is present in plant parts mixed with the seed. When such seeds are used pathogen easily gets into the field. Thus, proper cleaning of seeds before sowing is essential. Common examples of diseases disseminated in this manner are ergot and smut of pearl millet, ear cockle of wheat, white rust of crucifers, Ascochyta blight of chickpea, etc.

6. Seed Treatment

Thermal and chemical treatments of seed are also recommended before the seeds are stored or sown. These procedures are discussed in detail in chapters on physical control and chemical control of plant diseases.

7. Harvesting Time

Appropriate timing of harvest, from the point of view of crop disease, means essentially the attempt to escape disease: the intention is to harvest either before environmental conditions become very favorable to the pathogen, or before the crop becomes highly susceptible to age related pathogens, or both.[4] A number of crops are helped by early harvest to escape pathogens. For example, potato (*Globodera rostochiensis,* potato virus X, Y, and leaf roll) in U.K., France, and Germany; tomato (*Erwinia amylovora*) in U.S.; groundnut (*Pythium, Fusarium, Diplodia, Rhizoctonia*) in Libya and Israel; lucern (*Leveillula taurica, Uromyces* sp.) in Israel; red clover (*Rhizoctonia* sp., *Phoma trifolii*) in U.S.; sugarcane (*Glomerella tucumanensis*) in Australia, etc.

Disease incidence can also be minimized by reducing spread of inoculum through adjustment of harvesting time and practice. Potato harvested when tops are still green may easily get contaminated by *P. infestans* present on the leaves. One of the practices to avoid tuber infection or contamination is to first remove the tops and let them dry before digging the tubers.

B. ADJUSTMENT OF CROP CULTURE TO MINIMIZE DISEASE

The main objective of adjustment of cultural practices is to plant healthy seeds in pathogen free soil and obtain a healthy stand of crop. The necessary steps to achieve these aims are discussed here.

1. Crop Sequence and Crop Rotation

Different degrees of disease control through biological buffering of the soil against plant pathogens can be achieved through cultivation of crops in rotation, monoculture or through growing a mixture of two or more crops of different genetic composition simultaneously in the same field. Crop sequence implies cultivation of crops in monoculture or in sequence of different crops without any fixed cycle. On the other hand, crop rotation implies repeated growing of the same crop in at cycle at regular intervals.

The crops grown on a field determine its microbial makeup. Each crop species selects for a specific saprophytic or parasitic microbiota on the basis of nutrients it supplies. Monoculture of a crop on a field will favor the perpetuation of a specific and eventually a stable flora and fauna. This may ultimately give rise to suppressive soil that supports heavy incidence of a disease in the first year of cultivation but when the same crop is grown season after season without interference by any other crop, it eliminates the pathogen through development of an antagonistic flora which becomes stable in the soil unless disturbed by variation in cropping system. However, when a different crop is grown each season, the soil microflora remain in a state of unstability and result in different dominant organisms each season. Thus, it becomes a rotation of not only the crops but biota also.

a. Crop Rotation

The practice of crop rotation is known to agriculture for more than 2000 years. Since ancient time farmers have been aware of the fact that soil becomes sick and unsuitable for cultivation of a crop when the same crop is grown continuously on the same land. This can be due to several reasons such as (1) exhaustion of a particular essential nutrient due to excessive use by same type of crop, (2) accumulation of organic acids and other toxic substances released by the crop, and (3) easy survival of the pathogen due to regular presence of the susceptible host.

To overcome these problems, crop rotation is necessary in crop management. It aims at (1) better use of nutrients, (2) desirable effects on soil texture, with deep rooted crops alternating with shallow rooted crops, (3) water economy, in particular conservation of water in years of fallow, (4) weed control, and (5) suppression of soil-borne plant pathogens.

i. Basis for Disease Control

Although disease control through crop rotation is commonly viewed as a passive process (disinfestation of soil by starvation of the pathogen), some of the benefits observed are through other means (microbiological as well as chemicals released by different crops included in the rotation).

Basis for control by crop rotation could be (1) release of toxic substances by the crop effective against the pathogen, (2) starvation of the pathogen and its consequences, (3) variation in microbial activity that increases chances of antagonism, and (4) change in the physicochemical environment of the soil. Individually or in combinations these factors primarily reduce inoculum density of a pathogen. Nutritional factor (starvation of the pathogen and nutrition of saprophytic flora) remains, however directly or indirectly, the most important factor. Major root pathogens, like wilt causing fungi, are wholly or partly active possessors of host residues. After harvest of the crop the fungus remains active as saprophyte with reduced rate of metabolism. To displace such pathogens from the residue, weakening and starvation is necessary so that either they are lysed or are overcome by the saprophytic flora waiting outside.

Many plant species release toxic compounds in their root exudates. HCN is an example. These toxic compounds directly suppress the pathogen, reduce biological competition and permit activity of some microflora tolerant to them, to colonize the area and suppress the pathogen. The root exudates of non-host crops in the rotation, being nonspecific, can stimulate germination or release dormant propagules from fungistasis at a wrong time, i.e., in the absence of the host. Starvation and lysis may follow or the germlings may be destroyed by antagonists. Use of resistant cultivars of the same crop in the rotation may do the job efficiently if the pathogen is responsive to specific root exudates.

ii. Types of Pathogen Affected, Period of Rotation, and Selection of Crop

Crop rotation as a practice for disease control is most effective in case of specialized root pathogens or soil invaders having low competitive capacity, having narrow host range, and requiring continued colonization of dead root tissues. It is not effective against diseases caused by weak, unspecialized parasites or soil inhabitants and the pathogens which have a large host range among crop and weed plants.

The length of rotation for disease control will depend on how quick the pathogen can be displaced from colonized tissues and exposed to antagonistic effects. This is determined by biotic and abiotic characteristics of the soil, treatments given to the soil, crops included in the rotation, and how far chances of recontamination can be avoided. Maintenance of high organic matter content of the soil, tillage to expose the residues to weathering, inclusion of the crops which do not permit high stimulation and subsequent growth of the pathogen, crops that are likely to produce toxins, are some of the criteria to be considered at the time of preparing crop rotation. Normally, legumes should be followed by cereals and vice versa. The pigeon pea wilt fungus, *F. oxysporum* f. sp. *udum,* is specific on pigeon pea. However, many other legumes such as cowpea, soybean, etc. stimulate its germination and growth in soil although they are nonhosts. If pigeon pea is avoided and replaced with one of these legumes, the chlamydospores will germinate abundantly. If the soil is rich in organic matter and microbial activity is intense, the germlings may be destroyed by antagonism before forming secondary chlamydospores. But in a soil with low biological activity, the fungus may continue to grow and subsequently produce more chlamydospores. Cereals like sorghum, maize, etc., also stimulate germination of chlamydospores but not to the same extent as legumes. Furthermore, the root exudates have some toxic materials (HCN) that may kill the germlings before they can form spores.

Larvae of *Meloidogyne,* an obligate root parasite, are hatched with limited food reserves and cannot infect the host when this is exhausted in their search for the host. The nematode population in a field which contains no host plants will become noninfective and die sooner or later of starvation. However, the root knot nematodes have a large host range. To overcome this limitation, it has been proposed that noneconomic catch or trap crops may be included in the rotation or only resistant or immune commercial crops should be grown for some time. Tomato is highly susceptible to all species and races of *Meloidogyne.* Groundnut is immune to all known races of *M. incognita, M. javanica* and to race 2 of *M. arenaria.* It is not immune to *M. hapla.* Cotton is highly resistant to all of the common root knot nematodes except races 3 and 4 of *M. incognita.* Antagonistic plants like *Tagetes* spp. *Chrysanthemum* spp. and *Ricinus communis* can be used in rotation in heavily infested soils. Larvae enter these plants but fail to reproduce. The plants are known to contain antinemic toxins.

An extensive list of diseases in which rotation can play a significant role in disease management is given in Table 1.

b. Monoculture

Monoculture means cultivation of single or closely allied species in annual or seasonal succession, with interruption only by fallow or intermittent growing of green manure, or by application of soil amendments, not necessarily after each crop.[4] From the general biological aspect, monoculture is clearly dangerous because it does not permit diversity. Serious crop losses from diseases are much less likely to occur in crop rotations than in monoculture.

TABLE 1
Crop Diseases in Which Crop Rotation Has Had a Major Effect in Disease Control[4-6]

Crop	Disease	Pathogen	Crop interval (year)	Remarks
Onion	Bloat	*Ditylenchus dipsaci*	2	Rotate with beets, carrots, crucifers, lettuce, spinach
	Nematode	*Belonolaimus gracilis*	5—10	Use water melon or tobacco and eradicate weeds
Pea	White rot	*Sclerotium cepivorum*	8—10	
	Anthracnose	*Colletotrichum pisi*	2+	Rotate with nonhost crop
	Bacterial blight	*Pseudomonas pisi*	2+	
	Leaf spot	*Septoria pisi*	4+	
	Root rot	*Aphanomyces euteiches*	6—10	Use more than half crop sequence in corn, grains vegetables, not forage
	Seedling blight	*Rhizoctonia solani*	1+	Rotate with crops such as cereals, corn
Pepper	Anthracnose	*Colletotrichum piperatum, C. capsici*	1	Avoid solanaceous crops in rotation
	Bacterial spot.	*Xanthomonas campestris* pv. *vesicatoria*	1	Control solanaceous weeds
Potato	Corky ring spot	Tobacco-rattle virus	1	Barley in rotation
	Golden nematode	*Globodera rostochiensis*	8	Exclude potato or tomato in rotation; in short rotations use beans, corn, red clover, rye-grass
	Root knot	*Meloidogyne hapla*	1	Use corn or *Poa pratensis* in between crops
	Root rot	*Rhizoctonia solani*	3—6	Use alfalfa before potato
	Scab	*Streptomyces* sp.	1+	Soybean as green manure before potato
	Wilt	*Fusarium oxysporum*	3—6	Alfalfa before potato
Sorghum	Root knot	*Meloidogyne naasi*	1	Rotate with root crops
	Stalk rot	*Fusarium moniliforme*	2	Winter wheat-grain sorghum-fallow with no tillage
Soybean	Brown spot	*Septoria glycinea*	1	Use nonsusceptible crops
	Brown stem rot	*Cephalosporium gregatum*	5	5 years corn before soybean or corn-soybean-oats-clover
	Downy mildew	*Peronospora manshurica*	1	Use nonhost crops
	Cyst nematode	*Herterodera glycinea*	1—5	Use nonhost crops
	Leaf spot	*Cerospora sojina*	1	Use nonhost crops
		Phyllosticta sojaeicola	1	Rotate and plow residues
		Pseudomonas glycinea	1	Rotate and plow residues
Spinach	Downy mildew	*Peronospora spinaciae*	3	
Squash	Root and stem rot	*Fusarium solani*	2—3	
Sweet Potato	Black rot	*Ceratocystis fimbriata*	2+	Use any other crop
Tobacco	Black shank	*Phytophthora parasitica*	4	Rotate with cotton, peanut, soybean, oats, rye, or wheat but not legumes before tobacco
	Black root rot	*Thielaviopsis basicola*	5	Many crops in rotation but use small grains just before tobacco
	Wilt	*Pseudomonas solanacearum*	3—5	Rotate with corn, cotton, cowpea, soybean, or small grains.

TABLE 1 (continued)
Crop Diseases in Which Crop Rotation Has Had a Major Effect in Disease Control[4-6]

	Disease	Pathogen	Crop interval (year)	Remarks
	Root knot	*Heterodera marioni*	2—3	Corn-oats-tobacco; cotton-peanut-tobacco; peanuts-oats-tobacco
Tomato	Anthracnose	*Collectotrichum phomoides*	2	Rotate only if susceptible crops not nearby
	Bacterial canker	*Corynebeacterium michiganense*	1	Rotate with other crops and control solanaceous weeds
	Black speck	*Pseudomonas tomato*	1	
	Bacterial spot	*X. campestris* pv. *vesicatoria*	1	
	Blight and fruit rot	*Helminthosporium lycopersici*	1	Rotate and plow disease debris
	Septoria blight	*Septoria lycopersici*	1	Rotate other crops and control horse nettle
	Soil rot	*R. solani*	1	Rotate with pangola grass
Vegetable	Nematodes	*Pratylenchus* spp.	1	Vegetable crops after oats and peanuts then after corn or lupines
Wheat (Spring)	Common root rot	*Helminthosporium sativum* *Fusarium roseum*	4	Use noncereal crops
	Eye spot	*Cercosporella herpotrichoides*	2—3	Rotate with other crops but not wheat and barley
	Flag smut	*Urocystis tritici*	1	Any crop but susceptible wheat
	Foot rot	*H. sativum*	1	Use non-grass crops (e.g.legumes)
	Seed gall	*Anguina tritici*	1—2	Rotate with non-cereals
	Septoria leaf and glume blotch.	*Septoria avenae, S. nodorum, S. tritici*	2	Rotate with non-cereals
Wheat (Winter)	Cephalosporium stripe	*Cephalosporium gramineum*	1—2	Rotate with corn or legumes
	Snow mold	*F. nivale, Sclerotinia borealis, Typhula* spp.	1	Rotate with spring cereals or legumes
	Take-all	*Gaeumannomyces graminis*	2—3	Rotate with non-cereals

Monoculture may also exert selection pressure on pathogens, resulting in the emergence of new pathotypes. Thus, in general the plant pathologists view monoculture as a disease-perpetuating system.

Two types of disease and pathogen development are associated with monoculture: *irreversible* increase in disease incidence until a certain level is reached and maintained, and the *reversible* phenomenon of disease first rising, but subsequently declining to a more or less fixed level.[4]

The pathogens in the irreversible disease pattern group are *Verticillium albo-atrum*, V. *dahliae* (wilt of cotton), *Heterodera schachtii* on sugarbeet, *Globodera rostochiensis* on potato, *H. glycines* on soybean and *Meloidogyne* spp. on several crops. The examples of pathogens in the group showing reversible disease pattern in monoculture are those fungi which have low ability for saprophytic survival. The group includes *Phymatotrichum omnivorum*, *Streptomyces scabies*, *Gaeumannomyces graminis* and eye spot fungus *Pseudocercosporella herpotrichoides*.[1]

c. Mixed Cropping

Mixed crops such as wheat barley, wheat chickpea, pigeonpea sorghum, etc., reduce the economic loss from diseases. Since same pathogen does not attack both the crops in the mixture, at least one crop is saved if the other is badly damaged by the pathogen. Mixed crops also reduce spread of disease.

If two crops are to be sown together, different types of plant species are usually chosen for the purpose. There are only few foliage pathogens, notably some powdery mildews, that will attack crops belonging to different botanical families. Each of the two crops thus, to some extent, shields plants of the other crop from the impact of air-borne pathogens, as well as from vectors. Soil-borne diseases may also be reduced by root exudates of one crop, e.g., onions, being detrimental to root pathogens of the other crops.

The reduction in disease incidence in a mixed crop can be attributed to one or more of the following:[1]

1. Due to reduced number of host plants there is sufficient spacing between them and chances of contact between foliage or roots of diseased and healthy plants are greatly reduced.
2. The roots of nonhost plants may act as a physical barrier obstructing the movement of pathogens in soil. They may also release toxic substances in their root exudates which suppress the growth of pathogens attacking the main crop. HCN in root exudates of sorghum is toxic to *F. oxysporum* f. sp. *udum* attacking pigeon pea in the mixture.
3. Due to reduced number of host plants in a mixed crop the susceptible area for an air-borne foliar pathogen is decreased. Therefore, there is less primary infection and less production of secondary inoculum for spread of the disease. This slows down the rate of spread of the disease.
4. By proper selection of crops for the mixture, soil environment can also be changed to one that is not favorable for the pathogen. Control of root rot of cotton by growing cotton with moth is an example.
(5) The soil-borne pathogens are not uniformly distributed in the field soil. Generally they are randomly present as dormant structures. Activation of these dormant structures is often dependent on contact with the host roots, the chances of which are highly reduced in mixed crop due to spacing between plants.

2. Decoy and Trap Crops

Decoy crops are non-host crops sown with the purpose of making soil-borne pathogens waste their inoculum potential. This is achieved by stimulating the activating dormant propagules of the pathogens in the absence of the host. A list of pathogens that can be decoyed in this way is given in Table 2.

Trap crops are host crops of the pathogen, sown to attract nematodes, but destined to be harvested or destroyed before the nematodes complete their life cycle. In Germany, sowing of somewhat resistant potatoes and harvesting the crop early before the potato cyst nematode matures, is recommended. Similarly, sowing of crucifers and plowing before the beet cyst nematode can develop fully is also recommended.

3. Crop Nutrition

The macronutrients, nitrogen (N), phosphorus (P), and potassium (K), are needed in large amounts by crop plants and are frequently limiting to plant growth. The micronutrients are needed in smaller quantities. Maintenance of optimum plant health with sufficient, but not excessive, levels of fertility can be beneficial in crop resistance to stress from pathogen attack.

TABLE 2
Decoy Crops for the Reduction of Pathogen Population

Crop	Pathogen	Decoy crop	Ref.
Fungi			
Brassicae	*Plasmodiophora brassicae*	Rye grass, *Papaver rhoeas, Reseda odorata*	7
Olive	*Verticillium albo-atrum*	*Tagetes minuta*	8
Nematodes			
Egg plant	*Meloidogyne incognita* *M. javanica*	*Tagetes patula* *Sesamum orientale*	9
Oats	*Heterodera avenae*	Maize	4
Soybean	*Rotylenchus* sp. *Pratylenchus* sp.	*T. minuta* *Crotolaria spectabilis*	4
Tomato	*M. incognita Pratylenchus alleni*	*T. patula* Casterbean, groundnut, chrysanthemum.	4

a. Nitrogen

Since the effect of nitrogen on plants is most pronounced, it has been studied in great detail. Fertilization with nitrogen, especially at enhanced dosages, causes new succulent vegetative growth of the plant and delays maturity. Those pathogens which attack such plant organs are, therefore, favored by high nitrogen. It is known that excess of nitrogen predisposes the host to rusts and powdery mildew of wheat, blast of rice, etc. When the nitrogen is deficient, the plant is weak, its development is incomplete and its maturity is hastened. Pathogens favored by slow growth of the host are thus favored by low nitrogen.

Studies[10-12] on the effects of nitrate (NO_3-N) and ammonium (NH_4-N) forms of nitrogen have revealed that pathogen-crop interactions depend frequently on the form rather than the amount of N available. These interactions are, however, complex in nature. The effects of nitrogen as such and forms of N on the incidence and severity of some infectious diseases are summarized in Table 4.

i. Mechanisms

Suppression and stimulation of diseases by specific forms of nitrogen has been attributed to preference of N forms by plants, altered host resistance, modification of plant constituents and exudates, shift in soil and rhizosphere pH, effect on pathogens and modified microbial equilibrium.

In most aerated soils NO_3-N predominates and plants adapted to such soils grow well with NO_3-N as the sole source of nitrogen. Crops that utilize NH_4-N grow well following fumigation while those requiring NO_3-N may be adversely affected. Evidence that nitrogen affected resistance of wheat to pathogen causing take-all was demonstrated by Huber.[13] He recorded reduced infection and relatively smaller lesions. Increased root growth permitting wheat to escape take-all has also been reported. NH_4-N reduced bacterial canker of *Prunus* by hastening periderm formation. NH_4-N has been found to increase permeability and exudation of broad bean leaf surface resulting in higher levels of sugars and amino acids. *Botrytis cinerea* infected such leaves more frequently than leaves supplied with NO_3-N. NH_4-N forms of nitrogen generally lower the pH while nitrates raise it. Nitrate nitrogen suppresses damping off of sugarbeet (*Pythium ultimum*) but NH_4-N fails to do so. According to Smiley[14] this is due to change in rhizosphere pH toward alkalinity by NO_3-N and beets

TABLE 3

Diseases Influenced by Nitrogen Without Reference to Forms of Nitrogen[4,10-12]

Increased		Decreased	
Disease	**Pathogen**	**Disease**	**Pathogen**
Bacterial Diseases		*Bacterial Diseases*	
Angular leafspot and fire blight of tobacco	*Pseudomonas tabaci*	Bacterial wilt of cucumber	*Erwinia tracheiphila*
Citrus canker	*Xanthomonas campestris* pv. *citri*	Bacterial spot of peach	*Xanthomonas campestris* pv. *pruni*
Fire blight of apple	*Erwinia amylovora*	Bacterial blight of tomato and tobacco	*Ps. solana cearum*
Wilt of maize	*E. stewartii*	Leaf spot of peach	*Ps. pruni*
Wilt of tobacco	*Ps. solanacearum*	Wild fire of tobacco	*Ps. tabaci*
Fungal Diseases		*Fungal Diseases*	
Bean root rot	*Rhizoctonia solani*	Bunt of wheat	*Tilletia* spp.
Blast of rice	*Pyricularia oryzae*	Club root of cabbage	*Plasmodiophora brassicae*
Browning root rot of wheat	*Pythium* spp.	Root rot of cotton	*Phymatotrichum omnivorum*
Downy mildew of cabbage	*Peronospora parasitica*	Root rot of pea	*Aphanomyces euteiches*
Dutch elm disease	*Cerocystis ulmi*	Root rot wheat	*Rhizoctonia solani*
Flag smut of wheat	*Urocystis tritici*	Stem canker of soybean	*R. solani*
Graymold of grape	*Botrytis cinerea*	Scterotium rot of sugarbeet	*Sclerotium rolfsii*
Powdery mildew of cereals	*Erysiphe graminis*		
Rusts of several crops	*Puccinia* spp.	Take all of wheat	*Gaeumannomyces graminis*
Smut of maize	*Ustilago zeae*	Wilt of cotton	*Verticillium albo-atrum*
Wilts of tomato, melon, cabbage	*Fusarium oxysporum*		
Viral Diseases		*Viral Diseases*	
Bean mosaic	Common bean mosaic virus	Bean mosaic	Tobacco mosaic virus
Lettuce mosaic	Lettuce mosaic virus	Ring spot of celery	Celery ring spot virus
Tobacco mosaic	Tobacco mosaic virus		
Wheat mosaic	Barley stripe mosaic virus		

being alkali-tolerant plants develop more vigor under these conditions, thereby developing resistance quickly through tissue maturity.

Some nitrogen compounds are directly toxic to pathogens. Ammonia is known to be toxic to *S. rolfsii, P. omnivorum, P. brassical, G. graminis,* and some plant parasitic nematodes. Calcium cyanamide is toxic to *S. rolfsii, Pythium, F. oxysporum,* and many nematodes are suppressed by urea due to direct toxicity. A form of nitrogen may affect virulence of a pathogen without affecting growth. Inhibition of pectolytic and cellulolytic enzymes by a specific form of N has been implicated to reduced Rhizopus fruit rots and susceptibility of cotton to *R. solani* and *V. alboatrum*. The thallus of *F. solani* f. sp. *phaseoli* has been found larger and more extensive with NH_4-N than with NO_3-N. Reduction of saprophytic activity of *R. solani* with NH_4-N reduces survival and disease inducing potential.

Application of nutrients may modify microbial equilibrium in soil and thus, results in suppression of disease. NH_4-N and NO_3-N have differential effect on Fusarium root rot of

TABLE 4

Effect of Inorganic Forms of Nitrogen on Plant Diseases[4,10-12]

Disease	Pathogen	No₃-N	NH₄-N
		NO_3-N	NH_4-N

Disease	Pathogen	NO_3-N	NH_4-N
Bacterial Diseases			
Angular leaf spot of cotton	*Xanthomonas campestris* pv. *malvacearum*	D	
Canker of peach	*X. campestris* pv. *pruni*	D	
Canker of tomato	*Corynebacterium michiganense*	I	
Crown gall	*Agrobacterium tumefaciens*	I	
Ring spot of potato	*C. sepidonicum*	I	
Southern wilt of tobacco/tomato	*Pseudomonas solanacearum*	I	D
Stewert's wilt of maize	*Erwina stewartii*	I	
Fungal Diseases			
Blast of rice	*Pyricularia oryzae*	D	I
Black scurf of potato	*Rhizoctonia solani*	D	I
Brown spot of rice	*Drechslera oryzae*	I	D
Charcoal rot of various crops	*Macrophomina phaseolina*	D	I
Eye spot of wheat	*Pseudocercosporella herpotrichioides*	D	I
Northern leaf blight of maize	*Drechstera turcica*	D	I
Root rot of pea	*Aphanomyces euteiches*	D	I
Root rot of pea	*Pythium* sp.	I	D
Root rot of tobacco	*Thielaviopsis basicola*	I	D
Root rot of Cotton	*Phymatotrichum omnivorum*	I	D
Root rot of bean	*Fusarium solani* f.sp. *phaseoli*	D	I
Root rot of maize	*A. euteiches*	D	I
Root rot of maize	*Pythium* sp.	I	D
Scab of potato	*Streptomyces scabies*	I	D
Stem rust of wheat	*Puccinia graminis*	I	D
Stripe rust of wheat	*P. striiformis*	I	D
Stalk rot of maize	*Fusarium* sp.	D	I
Stalk rot of maize	*Diplodia zeae*	I	D
Take-all of wheat	*Gaeumannomyces graminis*	I	D
Wilt of bean	*F. oxysporum* f.sp. *phaseoli*	D	I
Wilt of potato/tomato	*Verticillium albo-atrum*	I	D
Nematode Diseases			
Root knot of lima bean	*Meloidogyne incognita*	I	D
Soybean cyst nematode	*Heterodera glycinea*	I	D
Tobacco cyst nematode	*H. tabacum*		D
Viral Diseases			
Cadang-Cadang of coconut	Viroid		D
Potato virus	Potato virus X		D
Tobacco mosaic	Tobacco mosaic virus		I

Note:: D = decrease, I = increase

bean, Aphanomyces root rot of pea, take-all of cereals and Rhizoctonia root rot of cotton.[10] Disease severity was similar with either forms of N in sterile soil which suggests that disease control results from microbial interactions. Suppression of pea root rot in natural soil by NO_3-N was correlated with enhanced activity of bacteria and actinomycetes antagonistic to *A. euteiches*. A similar situation of antagonists (*streptomyces*) under NO_3-N has been postulated in the control of Poria and Armillaria root rot of pine.[10]

TABLE 5
Diseases Influenced by Phosphorus[4,12]

Disease-Pathogen	Effect
Bacterial Diseases	
Bacterial wilt of cotton—*Pseudomonas cryophylli*	Increase
Black fire of tobacco—*Ps. angulata*	Decrease
Blight of lima bean—*Ps. syringae*	Decrease
Fire blight of apple—*Erwinia amylovora*	Increase
Wilt of maize—*E. stewartii*	Increase
Nematode Diseases	
Pea nematode—*Pratylenchus penetrans*	Increase
Root knot of pea—*Meloidogyne incognita*	Increase
Virus Diseases	
Bean mosaic—tobacco mosaic virus	Decrease
Spinach virus—cucumber virus-1	Increase
Tomato/tobacco mosaic—tobacco mosaic virus	Increase
Fungal Diseases	
Bunt of wheat—*Tilletia* spp.	Decrease/Increase
Club root of cabbage-*Plasmodiophora brassicae*	Increase
Damping-off of pea—*Rhizoctonia solani*	Decrease
Downy mildew of cabbage—*Peronospora parasitica*	Decrease
Downy mildew of grape—*Plasmopara viticola*	Decrease
Downy mildew of lettuce—*Bremia taclucae*	Increase
Flag smut of wheat—*Urocystis tritici*	Decrease
Late blight of potato—*Phytophthora infestans*	Decrease/Increase
Powdery mildew of cereals—*Erysiphe graminis*	Decrease/Increase
Root rot of tobacco—*Thielaviopsis basicola*	Decrease
Root rot of cotton—*Phymatotrichum omnivorum*	Increase
Root rot of soybean—*R. solani*	Decrease
Take-all of wheat—*Gaeumannomyces graminis*	Decrease
Wilt of cotton—*F. oxysporum* f.sp. *vasinfecti*	Decrease/Increase
Wilt of tomato—*F. oxysporum* f.sp. *lycipersici*	Decrease/Increase

b. Phosphatic Fertilizers

Results of investigations reveal that effects of phosphates on plant diseases appear to be connected with one of the two situations: (1) either the crop grows on soil markedly deficient in P, and correction of the imbalance contributes to its health or (2) maturity of the crop is somewhat advanced by liberal supply of P, and this helps it escape from biotrophs (such as downy mildews) preferring younger tissues.[4] Effects of P on some important diseases are presented in Table 5.

c. Potassium

Usually high potassium (K) has been reported to reduce incidence of diseases. The mechanism by which K affects disease differ greatly, and includes both direct and indirect effects. Direct effects include reduction or stimulation of pathogen penetration, multiplication, survival, aggressiveness, and rate of establishment in the host. Indirect effects include promotion of wound healing, increase in resistance to frost injury and delay in maturity in some crops. Diseases suppressed by potassium are listed in Table 6.

TABLE 6
Plant Diseases Influenced by Postassium Nutrients[4, 12]

Disease—Pathogen	Effect
Bacterial Diseases	
Angular leaf spot of tobacco—*Pseudomonas tabaci*	D
Angular leaf spot of cucumber—*Ps. lachrymans*	D
Angular leaf spot of cotton—*Xanthomonas campestris* pv. *malvacearum*	D
Bacterial blight of lima-bean—*Ps. syringae*	D
Bacterial blight of rice—*X. campestris* pv. *oryzae*	D
Fire blight of pear—*Erwinia amylovora*	D/I
Soft rot of cabbage—*E. carotovora*	D
Wilt of tobacco—*Ps. solanacearum*	D
Fungal Diseases	
Ascochytosis of chickpea—*Ascochyta rabiei*	D
Blast of rice—*Pyricularia oryzae*	D
Brown spot of rice—*Helminthosporium oryzae*	D
Canker of potato—*Rhizoctonia solani*	D
Damping off of beet—*Pythium ultimum*	D
Downy mildew of lettuce—*Bremia lactucae*	D
Downy mildew of grape—*Plasmopara viticola*	D
Downy mildew of cauliflower—*Peronospora parasitica*	*D*
Early blight of tomato—*Alternaria solani*	D
Gray mold of grape—*Botrytis cinerea*	D
Late blight of potato—*Phytophthora infestans*	D
Northern blight of corn—*Drechslera turcica*	D
Powdery mildew of cereal—*Erysiphe graminis*	D
Root rot of pea—*Aphanomyces euteiches*	D
Root rot of pine apple—*Phytophthora cinnamomi*	D
Root rot of cotton—*Phymatotrichum omnivorum*	D
Root rot of jute—*R. solani*	D
Rust of cereals—*Puccinia* spp.	D
Sheath blight of rice—*Corticium sasaki*	D
Stalk rot of maize—*Fusarium moniliforme*	D
Diplodia zeae	D
Gibberella zeae	D
Wilt of cotton—*F. oxysporum* f.sp. *vasinfectum*	D
Wilt of melon—*F. oxysporum* f.sp. *melonis*	D
Wilt of tomato—*F. oxysporum* f.sp *lycopersici*	D
Nematode Diseases	
Pea nematode—*Pratylenchus penetrans*	I
Root knot of cucumber—*Meloidogyne incognita*	I
Root knot of lima bean—*M. incognita*	D
Sugarbeet cyst nematode—*Heterodera schachtii*	D
White tip of rice—*Aphelenchoides besseyi*	D
Viral Diseases	
Bean mosaic—Tobacco mosaic virus	D
Blotchy ripening of tomato—Tobacco mosaic virus	D
Potato leaf roll—Potato leaf roll virus	I
Spinach virus—Cucumber virus	I
Tobacco mosaic of tomato—Tobacco mosaic virus	I
Tobacco mosaic—Tobacco mosaic virus	D

Note: D = decrease, I = increase.

d. Calcium

The principal role of calcium in host-pathogen relationship is the formation of calcium pectate in the cell walls thus making them resistant to degradation by facultative parasites. These pathogens are *R. solani, S. rolfsii, Botrytis cinerea* and *Erwinia amylovora*. The vascular wilts caused by *F. oxysporum* are also reduced by calcium especially in conjugation with NO_3-N. However, calcium favors black shank of potato caused by *Phytophthora nicotianae* var. *nicotianae*. It also makes potato tubers prone to attack of common scab (*Streptomyces scabies*).

4. Adjustment of Sowing Dates

The choice of sowing dates in relation to crop diseases has one principal aim, viz., to reduce to a minimum the period over which infective agent (propagules, vectors) meets susceptible host tissue. This is also the aim of operations to influence the time of flowering and fruiting, especially by pruning and by breaking of dormancy. Since most crops grow under ranges of temperature and humidity wider than those favoring pathogen development, sowing or planting at seasons which give the crop an advantage is obviously good strategy.

Bunt (*Tilletia caries*) of winter wheat can be controlled by either planting before mid September or after mid October.[15] Cercosporella foot rot can be an important disease of winter wheat planted in the state of Washington before September 15, but almost no disease is present if planting is delayed by one month.[4]

5. Spacing

In crowded stand of crops the proper aeration is checked, humidity and temperature of the atmosphere as well as of the soil are likely to favor the survival and spread of pathogens. The plants remain tender and weak. The soil-borne fungi can easily reach and invade the roots of healthy plants. Damping off, late blight of potato, downy mildews, etc. are some of the diseases which spread fast in close space plantings. To avoid these conditions favorable for the pathogen proper spacing between plants must be maintained.

At 15 cm between rice plants, there is a higher percentage of seed-borne *Alternaria* spp., *Curvularia lunata, Drechslera oryzae* than at wider distances.[3] Soybean seeds from narrow row (25 cm) compared to wider rows (76 cm) spacing give higher recovery of total fungi and bacteria which adversely affect the quality of seeds.[3]

6. Water Management

Soil moisture is related to many diseases. As examples, wet soil favors club root (*Plasmodiophora brassicae*) of crucifers, silver scurf (*Helminthosporium solani*) of potatoes, and *Cercosporella* on wheat while dry soil increases severity of white mold (*Wetzelinia sclerotiorum*) of onion, common scab (*Streptomyces scabies*) on potato and Fusarium diseases of cereals.[16] Damping off diseases caused by *Pythium* spp. can be decreased by maintaining a dry soil surface.[17] Using irrigation to grow crops in an otherwise dry season allows avoidance of diseases such as potato late blight and wheat stem rust in Mexico. The charcoal rot fungus *Macrophomina phaseolina* attacks potato when there is a water stress. By irrigating the field stress is removed and the disease is suppressed.

Sprinkler irrigation has been studied in rather some detail with respect to foliar diseases. Generally, sprinkler irrigation increases diseases by increasing leaf wetness and by dispersing propagules of the pathogen by water splashes just like rain water. At the same time, it has some advantages also such as washing off of inoculum from the leaf surface.

In addition to amount of irrigation, the timing of irrigation is also important. The principles governing the timing of irrigation, its begining and its frequency, from the aspect of disease management are (1) providing the crop with as uniform a water supply as possible to obviate stresses of water deficit or excess, (2) timing of giving water in relation to periods of heightened susceptibility of host to disease, and (3) minimizing hours of contiguous leaf wetness, e.g., avoidance of sprinkling in continuation of dew periods.

C. SANITATION

Field and plant sanitation is a main part of disease control through cultural practices. This step is essential even if disease or pathogen free seed or propagating material has been used and other recommended cultural practices like crop rotation, alteration in date and method of sowing have been followed. The inoculum present on few plants in the field may multiply in soil or on the plant and in due course of time may be sufficient to nullify the effect of other cultural practices. Therefore, plants bearing such pathogens or plant debris introducing the inoculum in the soil should be removed as early as feasible.

In wilt disease of banana (*F. oxysporum* f. sp. *cubense*) it has been reported that so long as the dead roots and rhizomes of affected plants are present in soil the fungus continues its growth. When these plant remains are removed there is rapid decline in the population of the pathogen in the soil. Similarly, the wilt of cotton and root rot of bean, and Verticillium wilt of cotton are also reduced to some extent by removal of diseased plant debris.

1. Destruction of Crop Residue

Infected plant debris not only serves as source of perennation of pathogens, it also serves as a substrate for multiplication of inoculum. The destruction of this source of survival and multiplication of inoculum has been found to help in the control of many diseases. The fungi causing downy mildews of maize and pea, white blisters of crucifers, powdery mildews of pea and cereals are examples of pathogens which survive through their sexually produced oospores or perithecia in crop debris. In certain areas the linseed rust fungus, the rice blast and leaf spot fungi, and the fungus causing early blight of potato also perennate through dormant stages in diseased crop debris. Destruction of crop debris by burning immediately after harvest reduces the amount of inoculum surviving through debris.

Deep ploughing, especially during hot summer in the tropics, buries the debris to such depths where the pathogens are automatically destroyed. The pathogn-free subsoil is brought on the surface by deep ploughing. The beneficial effect of deep ploughing has been reported in diseases caused by *Sclerotinia sclerotiorum* and *Sclerotium rolfsii*.

Fallowing also helps in reduction of inoculum through sanitary effect of decomposition of crop debris in absence of a host. This has been demonstrated in Ascochyta blight of pea. Including fallow period in the rotation establishes types and number of microorganisms in the soil which are beneficial for the health of the crop. The principle of flood fallowing in which the land is left submerged in deep water for few weeks is also a method of cleaning the soil of pathogens.

2. Roguing

Regular removal of diseased plants or plant parts from a population has been found effective in reducing spread of many diseases. This method is one of the effective recommendations in the control of virus dieseases of crops.[18] The roguing of diseased plants not only checks spread of the disease, it reduces the amount of survival structures also.

In the production of virus-free potato seed tubers and in the control of soybean mosaic virus roguing is an important recommended practice. However, the procedure is effective only when the populations of the insect vectors of these pathogens are low or inactive. For the control of loose smut of wheat and production of disease free seed, roguing of diseased plants is always recommended in seed plots. In this disease also roguing is effective only if the smutted heads are removed before spores have been dispersed by wind. Red rot of sugarcane, wilt of cotton, smut of sugarcane, etc. are other diseases in which roguing of diseased plants is recommended for their control. Roguing is feasible and economical in small size fields and when the incidence of disease is not very high. In large size fields it is difficult to locate the diseased plants at the proper time unless a high percentage of plants is affected, and when very high percentage of plants is diseased roguing becomes uneconomical.

3. Crop-free Period and Crop-free Zone

The pathogens attacking crops of secondary importance and having a narrow host-range can be controlled by maintaining a crop-free period of definite duration. When the growers in an area agree not to grow the crops susceptible to the pathogen for a definite period, depending on longevity of the pathogen without its host, the pathogen is automatically starved out. Similarly, when the host crop is not grown in a zone surrounding the infested area, spread of the disease is checked. On this basis, for control of bunchy top of banana, it has been suggested that a crop-free belt around the area affected by the disease checks its spread provided diseased planting material is also quarantined and insect vectors are unable to cross the crop-free zone. This method is effective for those diseases which are not seed-borne and either there is no insect vector or if insect vectors spread them their flight range is limited.

4. Creating Barriers by Non-host or Dead Hosts

Many diseases depend for their spread on nearness of healthy roots to infected roots. Presence of roots of non-host crops between such roots may create barrier against spread of the pathogen. This aspect has been mentioned under mixed cropping.

One of the control measures for bacterial wilt of banana (*Pseudomonas solanacearum*) and spreading decline of citrus (*Radopholus similis*) is to destroy the healthy plants around the diseased plants. This checks the movement of the pathogen from diseased to healthy plants that are left in the field. In potato fields spread of late blight and many virus diseases is checked if the plants are detopped early. The procedure prevents production of inoculum for secondary spread or dispersal of virus by insect vectors.

5. Control of Weeds and Insects

Another aspect of field sanitation is keeping the field free from weeds and insects. Weeds reduce the amount of nutrients available for the plants and by lowering their vitality increase their disease proneness. Excess of weeds in the field also helps in increased humidity and low temperature. In addition, many weeds may be collateral or alternate hosts of pathogens attacking the crops grown in the field. The root knot nematode infects a large number of solanaceous weeds. Wild cucurbits help the continuous infection chain of many viruses attacking cultivated cucurbits. Therefore, as a sanitary precaution destruction of these weeds is necessary.

REFERENCES

1. **Singh, R. S.,** *Introduction to Principles of Plant Pathology,* 3rd ed., Oxford & IBH Publishing Co., New Delhi, 1984, 534.
2. **Clark, F. S.,** The development of an isolated area for the production of smut free barley seed, *Agric. Inst. Rev. (Canada),* 7, 37, 1952.
3. **Agrawal, V. K. and Sinclair, J. B.,** *Principles of Seed Pathology,* Vols. I and II., CRC Press, Boca Raton, FL, 1987, 176.
4. **Palti, J.,** *Cultural Practices and Infectious Crop Diseases,* Springer-Verlag, Berlin, 1981, 243.

5. **Curl, E. A.,** Control of plant diseases by crop rotation, *Bot. Rev.,* 29, 413, 1963.
6. **Kommedahl, T.,** The environmental control of plant pathogens using eradication, in *Handbook of Pest Management in Agriculture,* Vol. I., Pimental, D., Ed., CRC Press, Boca Raton, FL, 1981, 315.
7. **Garrett, S. D.,** *Pathogenic Root Infecting Fungi,* Cambridge Univ. Press, Cambridge, 1970, 294.
8. **Baker, K. F. and Cook, R. J.,** *Biological Control of Plant Pathogens,* W. H. Freeman, San Francisco, 1974, 433.
9. **Verma, M. K., Sharma, H. C., and Pathak, V. N.,** Efficacy of *Tagetes patula* and *Sepamum orientale* against root-knot of egg plant, *Plant Dis Rep.,* 62, 274, 1978.
10. **Huber, D. M. and Watson, R. D.,** Nitrogen form and plant disease, *Annu. Rev. Phytopathol.,* 12, 139, 1974.
11. **Henis, Y. and Katan, J.,** Effect of inorganic amendments and soil reaction on soil-borne plant diseases, in *Biology and Control of Soil-Borne Plant Diseases,* Bruehl, G. W., Ed., Am. Phytopathol. Soc., St. Paul, Minnesota, 1975, 100.
12. **Huber, D. M.,** The use of fertilizers and organic amendments in the control of plant disease, in *Handbook of Pest Management in Agriculture,* Vol. I., Pimental, D., Ed., CRC Press, Boca Raton, FL, 1981, 315.
13. **Huber, D. M.** Spring versus fall nitrogen fertilization and take-all of wheat, *Phytopathology,* 62, 434, 1972.
14. **Smiley, R. W.,** Forms of nitrogen and the pH in the root zone and their importance in root infection, in *Biology and Control of Soil-Borne Pathogens,* Bruehl, G. W., Ed., Am. Phytopathol. Soc., St. Paul., Minnesota, 1975, 52.
15. **Fischer, G. W. and Holtan, C. S.,** *Biology and Control of Smut Fungi,* Ronald Press, New York, 1957, 622.
16. **Calhoun, J.,** Effect of environmental factors on plant disease, *Annu. Rev. Phytopathol.,* 11, 343, 1973.
17. **Leach, S.,** Environmental control of plant pathogens using avoidance, in *Handbook of Pest Management in Agriculture,* Vol. I. Pimental, D., Ed., CRC Press, Boca Raton, FL, 1981, 315.
18. **Kelman, A. and Cook, R. J.,** Plant pathology in People's Republic of China, *Annu. Rev. Phytopathol.,* 15, 409, 1977.

Chapter 16

HOST RESISTANCE AND IMMUNIZATION

I. HOST RESISTANCE

A. INTRODUCTION

The prevention of epidemics and ultimately the reduction of losses in yield has been of great concern. The diseases may be controlled by using chemicals; but the chemicals create hazards to human health and produce undesirable side effects on nontarget organisms. Under the existing circumstances, the use of resistant plants is one of the most attractive approaches for suppression of plant diseases. Their use requires no particular action by the cultivators during the crop growth, is not disruptive to the environment, is generally compatible with other management practices, and is sometimes singularly sufficient to suppress disease to tolerable levels.

Resistant plant varieties developed jointly by pathologists and breeders are now grown on approximately 75% of U.S. land under cultivation. For small grains and alfalfa, 95 to 98% of the U.S. crop area is planted with varieties resistant to at least one pathogen.[1] For example, use of wheat cultivars resistant to stem rust (*Puccinia graminis* f.sp. *tritici*) has contributed immensely to the suppression of that disease. During the 20th century, in the midwestern U.S., the frequency of severe epidemics of stem rust has decreased as resistant varieties have been adopted.[2]

Plant resistance has attracted the attention of many scientists, and there are many excellent books and review articles devoted to it. For detailed information readers are advised to consult Van der Plank,[3] Day,[4] Eenink[5] Browning et al.,[6] Parlevliet and Zadoks,[7] and Nelson.[8]

B. CONCEPT

All the individuals of a plant species are not equally susceptible to a pathogen. The extent to which a plant prevents the entry or subsequent growth of the pathogen within its tissues or the extent to which a plant is damaged by a pathogen is used to measure the resistance or susceptibility[9] (Figure 1).

Resistance may be qualified by such words as high, intermediate, or low because there may be gradations between extreme resistance and extreme susceptibility. The magnitude of resistance can range from very small to very large. Even resistance that do not completely prevent pathogenesis can suppress disease adequately in populations of plants. If resistance has large enough effect to slow down pathogen reproduction rates to replacement levels, pathogen population will not increase. If resistance is of small magnitude disease can increase in the resistant plant population.

C. CLASSIFICATION OF RESISTANCE

1. Based on Effect of Genes

The terms major and minor genes are sometimes used instead of oligogenic and polygenic when referring to gene for resistance. However, neither all oligogenes are necessarily major genes, nor all polygenes are minor genes.

2. Based on Growth Stage of Host Plant

Several terms such as seedling resistance, post-seedling resistance, adult plant resistance, etc., have invariably been used to describe the exact stage of the host growth when it shows resistance. For example, resistance of wheat to leaf rust (*P. recondita*) may be expressed

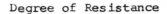

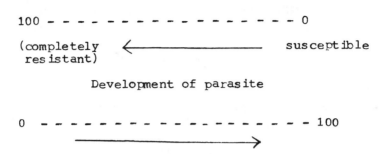

FIGURE 1.   Definition of resistance (Eenink, A. H., *Proc. Symp. Induced Mutat. Against Plant Dis. Int. At. Energy, Vienna.* With permission).

TABLE 1
Terms Frequently Used to Express Genetic Concept of Resistance[6]

Specific resistance	General resistance
Race specific	Race nonspecific
Vertical	Horizontal
Major gene	Minor gene
Monogenic	Polygenic
Multiple allele	Multiple gene
Qualitative	Quantitative
High	Low, moderate
Seedling	Adult plant
Nondurable	Durable
Hypersensitive	Nonhypersensitive
Protoplasmic	Generalized, uniform, field, partial, permanent, late resting,* slow rusting,* tolerance

* Applicable to only rust diseases.

at the "first leaf stage", is generally termed as "seedling resistance."[10] The term "post-seedling resistance" or "adult plant resistance" is used in the sense that a cultivar is susceptible at first leaf stage but at later stages of development it becomes resistant.[11]

3. Based on Mode of Inheritance

Types of resistance have been classified into those controlled by a single gene (monogenic) and those controlled by several genes (polygenic). Monogenic resistance is often sufficiently effective to qualify as immunity. It is stable under a wide range of environmental conditions, but is usually specific for certain race virulence gene of the pathogen. In contrast, polygenic resistance, which is more sensitive to environmental fluctuations, does not result in immunity, but is more uniformly effective against variants of the pathogen.

Browning et al.[6] used the terms "specific resistance" and "general resistance". Several interrelated and complementary terms have been employed to describe the genetic concept of resistance. They are summarized in Table 1.

4. Cytoplasmic Resistance

There are several plant diseases in which neither vertical resistance nor horizontal resistance is controlled by genetic material contained in the cytoplasm of the cell in which genes do not normally follow the Mendelian laws of inheritance. Such resistance is sometimes

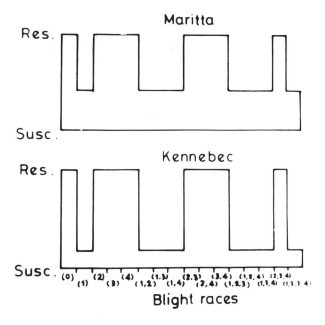

FIGURE 2. Vertical and horizontal resistance in two varieties of potato (for explanation see text).

referred to as cytoplasmic resistance. The two best known examples of cytoplasmic resistance occur in maize in which resistance to two leaf blights, the southern corn leaf blight caused by *Helminthosporium maydis* and the yellow leaf blight caused by *Phyllosticta maydis*, is conferred by characteristics present in normal cytoplasm of various types of corn but absent or suppressed in Texas male-sterile cytoplasm.

5. Resistance in Epidemiological Terms

When a host plant variety is more resistant to some races of a pathogen than to others, the resistance has been termed "vertical" or "differential", whereas resistance which is evenly spread against all races is "horizontal" or uniform.[3,12] Vertical and horizontal, as applied to resistance, are mathematically derived concepts, and refer respectively to the vertical and horizontal axes of a graph. This is shown in Figure 2 taken from Vanderplank,[12] which indicates the foliage resistance of two potato varieties to sixteen races of *Phytophthora infestans*. Both varieties show vertical (complete) resistance to races (0), (2), (3), (4), (2, 3), (2, 4), (3, 4), and (2, 3, 4), while resistance to the other eight races is horizontal and greater in Maritta than in Kennebec. Vertical resistance is probably always accompanied by some degree of horizontal resistance, as Van der Plank[3] points out, it is difficult to visualize a plant completely lacking in resistance to a pathogen. It is further suggested by Van der Plank[3] that vertical resistance operates mostly after the pathogen has penetrated the plant and that it may involve such active defense mechanisms as hypersensitivity, production of phytoalexins, etc. Although these mechanisms do not entirely prevent penetration they do prevent the subsequent spread of the pathogen in the tissue so that the plant is for most practical purpose immune. Vertical resistance is usually controlled by one or a few genes thereby the name monogenic or oligogenic.

Horizontal resistance protects against all races of a pathogen to varying extents, but the protection it affords is usually less than that given by vertical resistance against specific races. It may operate before or after infection through defense mechanisms which reduce or delay infection, colonization of the plant and/or production of spores by the pathogen. Horizontal resistance is controlled by many genes, thereby the name polygenic or multigenic

TABLE 2

**Effect of Discriminatory and Dilatory Resistance on the
Components of an Epidemic[6]**

Epidemic component	Discriminatory resistance	Dilatory resistance	Both combined
Incoming inoculum (Ii)	Decrease	No effect	Decrease
Rate of increase (r)	No effect	Decrease	Decrease
Delay of onset of epidemic (t)	No effect?	Increase	Increase
Amount of decrease (x)	No effect?	Decrease	Decrease
Outgoing inoculum (I_o)	No effect?	Decrease	Decrease

resistance. Each of these genes alone may be rather ineffective against the pathogen and may play a minor role in the total horizontal resistance.

Browning et al.[6] proposed two terms "discriminatory resistance" and "dilatory resistance" and emphasized that the use of these two terms should be restricted to epidemiological concepts of resistance. A population of host is defined as having discriminatory resistance or susceptibility if it affects the epidemic by discriminating among strains, i.e., by favoring or rejecting certain components of the pathogen population. A population of host plant is defined as having dilatory resistance if it affects the epidemic by reducing the rate of development of the pathogen population. Theoretical actions of discriminatory and dilatory resistance on the components of an epidemic are given in Table 2.

D APPARENT RESISTANCE

True resistance denotes incompatibility between the host and the parasite. It implies that so long as the incompatibility exists the host should ward off infection by the pathogen. However, under certain conditions, some very susceptible varieties or plants may remain free from infection or symptom and thus appear resistant. Apparent resistance to disease of plants known to be susceptible is generally a result of "disease escape" or "tolerance" to disease.

1. Disease Escape

The disease escaping varieties have no true genetic resistance as they are compatible with the pathogen, and can exhibit highly susceptible reaction if the pathogen can attack them at the proper time. However, due to their other genetic characters such as rapid growth, early maturity, etc. they avoid the period when the pathogen is in aggressive stage or when the environmental conditions are favorable for the pathogen. For example, varieties of pea which mature early (by January under Indian conditions) usually escape much damage from powdery mildew and rust. This disease normally becomes serious in January or later. If pods have developed before serious disease incidence the losses are considerably reduced. Varieties of raspberries susceptible to raspberry mosaic virus frequently escape the disease because the plant is a non-preferred host of the insect vestor.[13] Several leaf characters of rice cultivars like rolled, narrow, dark green, and slow senescence were associated with resistance to bacterial leaf blight (*X. campestris* pv. *oryzae*).[14]

2. Tolerance

A plant which is attacked by a pathogen to the same degree as other plants, but suffers less damage in terms of yield or quality as a result of the attack, is said to be tolerant.[15] Tolerance be defined as the inherent or acquired capacity to endure disease and give satisfactory return to the growers.[13,16] Browning et al.[6] in terms of epidemiological concepts of resistance said that "a population of host plants is defined as having tolerance if it is rated as susceptible visually, but is damaged less by the epidemic than another susceptible population''.

Tolerance results from specific heritable characters of the host plant that allow pathogen to develop and multiply in the host while the host, either by lacking receptor sites for, or by inactivating or compensating for the irritant excretion of the pathogen, still manages to produce a good crop. Vanderplank,[12] however, considers that tolerance is not a desirable character because it does not reduce either initial inoculum (X_o), infection rate (r) or infection time.

Tolerance to disease is most commonly observed in many plant-virus infections in which mild strains of viruses infect plants such as potato systemically and yet cause few or no symptoms and have little effect on yield. Russell[17] observed three different kinds of tolerance mechanisms for viral diseases:

1. The virus is able to multiply but symptoms do not appear.
2. The plants which develop disease symptoms which are as severe as those in other plants but which suffer less damage, for example sugarbeet yellows virus, such plants are referred to as disease tolerant.
3. The virus infected plants that do not show severe disease symptoms and are less damaged by infection than other plants, are true tolerant. For example, infected barley plants that are tolerant to barley yellow dwarf virus do not show pronounced yellowing or stunting of the leaves give an acceptable yield inspite of being infected.

C. DEVELOPMENT OF RESISTANT VARIETIES

The breeding for resistance to diseases is generally no way different than breeding for other traits. In breeding for other traits, the breeder deals with the variability in test material, while in breeding for disease resistance involvement of host plant and parasite/pathogen is taken into account. The first step in resistance breeding programme is the collection of natural variability followed by finding out the sources of resistance. Then next step is the incorporation of the resistance gene(s) from the donor parent(s).

1. In Self Pollinated Crops

a. Mass Selection

Plants with identical phenotypic characters are selected for resistance form a population and progenies are bulked to form the basis of variety. Production of such varieties is easy and their heterozygosity give them some advantage over pure line varieties.

b. Pure Line Selection

These are each derived from the progeny of a selfed homozygous plant selected from a land variety or commercial variety. The progeny is evaluated for resistance and other desirable character in succeeding generations and, if promising, it is multiplied to raise a new variety.

2. Hybridization

This method involves crossing of two pure line varieties for transferring resistance from donor parents and combining characteristics from each parent. It helps the breeder to combine resistance against several races/biotypes of the pathogen in a single variety. The F_1 plants from a cross are identical in their genetic constitution. Segregation occurs in F_2 generation onwards and homozygosity is retained in succeeding selfed generations. The transgressive segregants for resistance may occur with quantitatively inherited resistance, such segregants may be selected in the F_2 and in later generations. Selection after hybridization is based on such methods as pedigree method, bulk population method, mass pedigree method, and back cross method.

Exposure/treatment of seed with physical/chemical treatments

↓

Each M$_1$ plant is selfed. Screening of dominant mutations, if any may be carried out. Seeds from individual plant/spike/tiller may be harvested separately

↓

M$_2$ plant to progenies are grown. Selection of individual plants with moderate to high degree of resistance is carried under epiphytotic conditions.

↓

M$_3$ plants to progenies of selected M$_2$ plants are grown. Selection of plants/progenies under epiphytotic conditions.

↓

M$_4$ generation consists of progenies of individual selected M$_3$ plants and selected progenies. Preliminary yield trial of selected progenies alongwith parental variety and standard checks are conducted.

↓

Multilocation trial of the high yielding and disease resistant progenies alongwith checks.

↓

Release and seed multiplication of superior progeny.

FIGURE 3.   General outline of a mutation breeding procedure in autogamous crop plants, (courtesy, Dr. D. P. Singh)

a. Mutation

Development of resistant varieties by mutation depends on chance. Sometimes resistant gene(s) are not available in cultivated types and the wild germplasm is not easily crossable. Often the cultivated variety is high yielding and widely adapted, but is susceptible to a particular pathogen or a race/biotype and the breeder is interested to alter the susceptible alleles to resistance without changing its other traits. Under such conditions, mutation breeding for resistance is the obvious choice.

A general outline of a mutation breeding procedure in self-pollinated crop is presented in Figure 3. The same procedure may be followed in cross-pollinated and vegetatively propagated crops. In cross pollinated crops, instead of selfing individual plants in M$_1$ generation, a group of plants are selfed together. In asexually propagated crops cuttings are used in place of seeds.

3. In Cross-Pollinated Crops

Several methods can be employed in cross pollinated crops for development of varieties resistant to diseases.

a. Mass/Recurrent Selection

This is the most commonly used breeding method. In this method individual plants are selected from an artificially inoculated heterozygous population. The population from the selected plant is artificially reinoculated in the subsequent generation and those showing susceptible reaction are discarded before terminating for seed production. Because of high heritability of disease resistance, mass selection is an important breeding tool.

b. Polycross

This method consists of selecting resistant plants from a heterozygous population and intercrossing these or inbred lines derived from these, in all possible combinations. The progenies of a polycross can be bulked. Resistant plants are subsequently selected from the bulk population, or the progenies of individual lines can be tested separately.

c. Line Breeding

Selected plants are either selfed or interpollinated, and the resulting progenies/lines are individually tested for resistance; only the most resistant lines are retained for subsequent breeding. These are then interpollinated to produce a composite cross.

d. Synthetic/Hybrid varieties

Resistant lines obtained from recurrent selection or line breeding can be used to produce synthetic or hybrid varieties. A number of selected plants, lines, or clones are intercrossed to produce a synthetic variety. Hybrid varieties are produced by controlled pollination between the lines. The parent lines ought to be maintained separately so that synthetic or hybrid varieties may be reconstituted as required. Inbred lines may be improved for resistance by using the back cross method or convergent improvement. This may be done by using male sterile cytoplasm.

D. SOURCES OF RESISTANCE

In an on going resistance breeding programme, the availability of genes for resistance is the first concern. The primary and secondary gene centres of cultivated plants are the best places to find genuine resistance to common diseases.[18] The information on the origin and evolution of cultivated plants and their wild progenitors have been collected and summarized by several workers.[18]

The International Board of Plant Genetic Resources (IBPGR) established in 1973, has sponsored plant collections in over 20 countries or areas. All the major cereals, pulses, grasses, forage, legumes, groundnuts, vegetables, potatoes, cassava, and sweet potatoes have been included. In addition, other international research centers (ICRISAT, CIMMYT, ICARDA, ITTA, etc.), maintain germplasm collection.

Various sources exploited for collection of resistance genes include use of land races/cultivated types, wild/alien species, multiplasm, induced mutants, etc. Often genes for resistance are present in the varieties or species normally under cultivation in the area where the disease is severe, and in which the need for resistant variety is most pressing. With most diseases, a few plants remain virtually unaffected by the pathogen. These survivor plants are likely to remain disease-free because of resistance in them. Such materials are oftenly employed for breeding for disease resistance. If no resistant can be found within the local population, plants of the same species from other areas and plants of other cultivated or wild species are also checked, for resistance.

E. FAILURE OF RESISTANCE

There are very few varieties except some isolated cases (e.g., cabbage yellow-wilt caused by *F. oxysporum* f. sp. *conglutinans*) which could maintain their complete resistance for long. In most cases even complete resistance to a disease breaks down sooner or later. There are several reasons for it. Among avoidable causes are laxity in screening, inadequate backcrossing, and such other failures on the part of the breeder. Segregation in varieties is another cause of failure. However, the most important and unavoidable cause of failure of resistance in varieties is unchecked variability of the pathogen. Mutation and hybridization cause evolution of new races/biotypes or strains of the pathogen rapidly. During screening of varieties or breeding materials, if they are not exposed to all the existing races/biotypes of the pathogen or after the release of the variety new pathotypes have appeared in the area having gene for virulence which could counteract the genes for resistance in the variety, the variety is destined to become susceptible. The new races could develop in the same area or may be introduced from some external source. Prolonged cultivation of a single genotype in the area also helps in the development of new races. Soon the inoculum build up of these races may reach a density that could lead to an epidemic. Therefore, as far as possible, it is desirable to grow more than one genotype of the particular crop in a particular area to maintain their resistance for longer periods. This warrants continuous breeding programme to evolve new varieties as replacement for existing ones.

F. MANAGEMENT OF RESISTANT VARIETIES

In most cases the life of a resistant variety is limited. Complete reliance on resistant variety as the only method for reducing incidence and severity of disease is neither good for the crop variety nor for the growers. Resistance in a variety can be prolonged and efficient disease management achieved if such varieties are used as a part of an integrated control programme. In order to increase durability as well as stability of resistance and also to reduce disease damage, the use of cultural, biological and chemical control methods should be encouraged. In general, control measures are much more effective on resistant or partially resistant varieties than on susceptible ones. There must be varietal diversification in space and time using race specific or race nonspecific resistance. No single variety should be allowed to dominate cultivation. Practices like alteration of sowing dates, fertilization, irrigation, combined with fungicidal protection will prolong the life of a resistant variety by interfering with the capacity of the pathogen to develop races which overcome resistance.

II. HOST IMMUNIZATION

A. INTRODUCTION

Although plants do not possess the antigen-antibody system of man and animals, they can be systemically immunized against fungal, bacterial, and viral diseases by prior inoculation/treatment with mild strain or low doses of severe strain of the pathogens or elicitors produced by the pathogens. Mechanism of induced resistance has already been discussed in Chapter 4 (Table 3), the major aim of this chapter is to assess the potential of this phenomenon for the practical plant diseases control.

B. IMMUNIZATION AGAINST MICROBIAL DISEASES

Protection by immunization against several fungal and bacterial diseases has been demonstrated under field conditions. Caruso and Kuc[19] reported that in cucumbers, watermelons and muskmelons plants receiving restricted infection with *Colletotrichum lagenarium* prior to transplanting to the field had fewer and smaller lesions following challenge with high inoculum concentration of the pathogen in the field than plants that were not infected prior to transplant. The survival of protected watermelon plants was 98% as compared to 32% for controls. More recently Tuzun et al.[20] demonstrated that immunization of tobacco plants by stem injection of sporangial suspension of *Peronospora tobacina* protected the plants against blue mold as well as or better than most effective downy mildew fungicide, metalaxyl.

Above two examples and several others[21] prove that effectiveness is not a problem for the use of immunization for the control of plant diseases in the field. So why immunization is not popular? Kuc[21] has pointed out different advantages and disadvantages of plant immunization in relation to other plant disease control measures.

1. Advantages

1. Immunization is effective against fungal, bacterial, and viral diseases.
2. It is likely that immunization is dependent upon the activation of several different mechanisms, and therefore, is probably stable. Systemic fungicides with a single metabolic site of action are generally unstable with the development of new races of pathogens.
3. Immunization is systemic and persistent. In some cases it persists for the life of annual plants.
4. Since immunization utilizes mechanisms for resistance present in plants, it may be considered natural and as safe for man and the environment as disease-resistant plants.
5. Immunization in cucurbits and tobacco is graft transmissible from root stock to scion. This could be significant since many plants are propagated by grafting.
6. Immunization in tobacco appears nonreversible. Buds from immunized plants grafted onto susceptible root stock developed into fully grown immunized tobacco plants.
7. Immunization systemically sensitizes the plant to respond, but the major expenditure of energy and the expression of resistance are generally localized and occur in the presence of pathogens, that is, when and where needed.
8. The ability to immunize susceptible plants implies that the genetic potential for resistance is present in all plants.
9. Plants can be immunized by chemicals extracted from immunized plants. This suggests the possibility of seed treatments.

2. Disadvantages/Drawbacks

1. The natural chemical signals for systemic immunization have not been characterized.
2. Immunization is not economically competitive with our present technology in modern agriculture although it appears as effective as available systemic pesticides.
3. Immunization has not received sufficient field testing to determine its stability and persistence under high natural pathogen pressure.
4. People have difficulty in accepting the reality that plants can be systemically immunized against disease.

Instead of using plant pathogens directly for immunization, possibility of using chemical elicitors, where known, produced by these pathogens to induce the host defense should be explored. At least in one case they have been found to be very effective, practical and economically sound.

Metlitsky et al.[22] reported that a lipoglycoprotein (LGP complex) elicitor isolated from *Phytophthora infestans*, when applied as a tuber treatment before sowing at the extremely low concentration of 0.0005%, was as affective as fungicide treatment against late blight and early leaf mold, and more effective against brown patch and scab. The acquired immunity was retained not only throughout the life span of the plants but also during storage of potatoes.

B. IMMUNIZATION AGAINST VIRAL DISEASES
1. Cross Protection

In viruses, phenomenon of cross protection although first demonstrated in 1929 by McKinney,[23] could be utilized for the disease control under field condition only after Rast[24] produced a mild mutant (MII-16) in 1972 form a common tomato strain of TMV by nitrous acid mutagenesis. Since then this mutant has been used commercially and has been applied to a high proportion of glasshouse grown tomato crops in the Netherlands and the United

Kingdom.[25] Successful control of TMV with an attenuated mutant (L11A), isolated from the tomato strain of TMV in plants treated with high temperature, was also reported in Japan.[26] Cross protection has been used on a large scale to control citrus tristeza virus, a closterovirus that is important worldwide. Naturally occurring mild strains of CTV have been demonstrated to offer protection in the field.[27] In Brazil, the number of protected sweet orange trees exceeded 8 millions in 1980, and no breakdown in protection was observed.[28] Recently mild strain generated from severe strain of papaya ring spot virus (PRV) by nitrous acid mutagenesis was used to protect papaya plants against infection by severe strain of PRV under field condition and a high degree of protection was obtained.[29]

Although cross protection is a general phenomenon of plant viruses, not all plant diseases caused by viruses can be controlled by using mild strain for preimmunization. Careful attention to the selection of the best protective isolates of virus and to their introduction into the crop to be protected is essential.[30]

For practical application of cross protection, the mild protective virus should:[29] (1) not cause severe damage to the protected plants, (2) be stable for a long period, (3) protect plants against the effects of severe strains, (4) be suitable for infecting large number of plants, (5) not affect other crops in the vicinity of the crop protected, and (6) have no synergistic reaction with other viruses.

2. Genetically Engineered Cross Protection

By using recombinant DNA techniques, plant viral genes for coat protein have been isolated, characterized, cloned and transferred to host-plant cells using Ti-plasmid as a vector. Coat protein (CP) gene got integrated with host's genome and was expressed in transgenic plants. For some viruses it has been found that transgenic plants expressing a nuclear integrated coat protein gene show a significant degree of resistance to infection by homologous virus. In several ways this phenomenon resembled cross protection: the term "genetically engineered cross-protection" was coined by the Tumer et al.[31] to describe this phenomenon.

Transgenic tobacco plants that accumulate the CP of the U1 strain of tobacco mosaic virus (TMV) showed delay in disease development upon inoculation with the U1 strain.[32] Besides resistance to U1 strains, transgenic plants also showed reduced susceptibility to more severe PV 230 strain of TMV.

As a second example of genetically engineered cross-protection it was shown that transgenic plants containing alfalfa mosaic virus (AMV) coat protein gene resist infection by AMV.[31] Only coat protein gene is capable of providing protection to the transgenic plants. Viral genes, other than coat protein, although expressed in transgenic plants, failed to provide protection against the viral infection.[33] Recent studies have shown that coat protein gene has multifunctional role during infection that induce encapsidation, symptom expression, and differential elicitation of resistance gene.[34] During last 3 years coat protein genes from a number of viruses have been characterized, isolated and cloned. In next few years it is expected that a number of transgenic plants containing viral coat protein gene would be generated.

Although cross protection may sometimes operate through coat-protein it cannot do so in all cases because viroids, which have no coat protein, also exhibit cross protection. So another hypothesis to explain cross protection involves antisense mRNA (see Table 3 of Chapter 4), first expounded by Palukaitis and Zaitlin.[35] Replication of many viruses involves the synthesis of a negative ($-$) strand of RNA from a postive ($+$) strand contained in the virus particle. The newly synthesized ($-$) strand then acts as a template for the synthesis of many ($+$) strands. In cross protection, it is proposed that, the ($-$) strand synthesized by the challanger virus will anneal with the many positive strands of the first virus to form double-stranded RNA that is inactive. Tobacco plants have been genetically engineered by inserting in their genome reverse oriented cDNA of cucumber mosaic virus (CMV) using Ti-plasmid as a vector with the hope that it would directly produce the $(-)$ strand viral RNA which should be able to anneal with the $(+)$ strand of inoculated virus, and making transgenic plants resistant to CMV.[36] Results of inoculated tests are eagerly awaited.

At present "genetically engineered cross protection" looks potentially sound. It is still difficult to predict its practical utility in plant disease management in the near future. None of the transgenic plants generated so far has been tested in field.

REFERENCES

1. **Young, R. A.**, *Plant Disease Development and Control*, Publ. 1596, Nat. Acad. Sci., Washington, D.C., 1968.
2. **Fry, W. E.**, *Principles of Plant Disease Management*, Academic Press, New York, 1987, 378.
3. **Vanderplank, J. E.**, *Disease Resistance in Plants*, Academic Press, New York, 1968, 206.
4. **Day, P. R.**, *Genetics of Host-Parasite Interaction*, Freeman, San Francisco, 1974, 238.
5. **Eenink, A. H.**, Genetics of host-parasite relationships and uniform and differential resistance, *Neth. J. Plant Pathol.*, 82, 133, 1976.
6. **Browning, J. A., Simous M. D., and Torres, E.**, Managing host genes: epidemiological and genetic concepts, in *Plant Diseases: An Advanced Treatise*, Vol. I, Horsfall, J. G. and Cowling, E. B. Eds., Academic Press, New York, 1977, 191.
7. **Parlevliet, J. E. and Zadoks, J. C.**, Integrated concept of disease resistance—new view including horizontal and vertical resistance in plants, *Euphytica*, 26, 5, 1977.
8. **Nelson, R. R.**, Genetics of horizontal resistance in plant diseases, *Annu. Rev. Phytopathol.* 16, 359, 1978.
9. **Eenink, A. H.**, Genetics of host-parasite relationship and the stability of resistance, in *Proc. Symp. Induced Mutat. Against Plant Dis.* Int. Atomic Energy, Agency, Vienna, 1977, 47.
10. **Bartos, P., Fleishchman, G., Samborski, D. J., and Shipton, W. A.**, Studies on asexual variation in the virulence of oat rust, *Puccinia coronata* f. sp. *avenae* and wheat leaf rust, *Puccinia recondita, Can J. Bot.*, 43, 1383, 1969.
11. **Anderson, R. G.**, Studies on the inheritance of resistance to leaf rust of wheat, *Proc. Symp. 2nd Wheat Genet Lund, Hereditas, Suppl.*, 2, 144, 1966.
12. **Vanderplank, J. E.**, *Plant Diseases, Epidemics and Control*, Academic Press, New York, 1963, 349.
13. **Nelson, R. R.**, the meaning of disease resistance in plants, in *Breeding Plants for Disease Resistance: Concepts and Application* Nelson, R. R., Ed., Pa State Univ. Press., University Park, 1973, 13.
14. **Singh, C. B. and Rao, Y. P.**, Association between resistance to *Xanthomonas oryzae* and morphological and quality characters induced mutants of *indica* and *japonica* varieties of rice. *Ind. J. Genet. Plant Breed.*, 31, 369, 1971.
15. **Robinson, R. A.**, Disease resistance terminology, *Rev. Appl. Mycol.*, 48, 593, 1969.
16. **Rao, M. V.**, Control of plant diseases-some possible approaches, *Ind. J. Genet. Plant Breed.*, 28, 128, 1968.
17. **Russell, G. E.**, *Plant Breeding for Pest and Disease Resistance*, Butterworth, London, 1978, 485.
18. **Singh, D. P.**, *Breeding for Resistance to Diseases and Insect Pests*, Narosa Publishing House, New Delhi, 1986, 222.
19. **Caruso, F. and Kuc, J.**, Field Protection of cucumber, watermelon and muskmelon against *Collectotrichum lagenarium* by *Colletotrichum lagenarium, Phytopathology*, 67, 1290, 1977.
20. **Tuzun, S. W., Nesmith, W., and Kuc, J.**, The effect of stem infection with *Peronospora tabacina* and metalaxyl treatment on growth of tobacco and protection against blue mold in field, *Phytopathology*, 74, 304, 1984.
21. **Kuc, J.**, Plant immunization and its practicability for disease control, in *Innovative Approaches to Plant Disease Control*, Chet, L., Ed., John Wiley & Sons, New York, 1987, 225.
22. **Metlitsky, L., Ozeretskovskaya, O. Chalova, L., Ivanyuk, V., Chalenko, G., and Platonova, T.**, Immunization of plants with the aid of biogenic inducers of systemic protective reactions, *Doklady AN SSSR*, 283, 1, 1986.
23. **McKinney, H. H.**, Mosaic diseases in the canary islands, West Africa, and Gibraltor, *J. Agric. Res.*, 39, 557, 1929.
24. **Rast, A. T. B.**, MII-16, an artificial symptomless mutant of tobacco mosaic virus for seedling inoculation of tomato crops, *Neth. J. Plant Pathol.*, 78, 110, 1972.

25. **Fletcher, J. T.,** the use of avirulent virus strains to protect plants against the effects of virulent strains, *Ann Appl. Biol.* 89, 110, 1978.

26. **Oshima, N.,** The control of tomato mosaic disease with attenuated virus of tomato strain of TMV, *Rev. Plant Prot. Res.,* 8, 126, 19

27. **Muller, G. W. and Costa, A. S.,** Tristeza control in Brazil by preimmunization with mild strains, *Proc. Int. Soc. Citric,* 3, 868, 1977.

28. **Costa, A. S. and Muller, G. W.,** Tristeza control by cross protection: A U.S. Brazil cooperative success, *Plant Dis.,* 64, 538, 1980.

29. **Yeh, S. D., Gonsalves, D., Want H. L., and Namba, R., Chiu, R. J.,** Control of papaya ringspot virus by cross protection, *Plant Dis.,* 72, 375, 1988.

30. **Fulton, R. W.,** Practices and precautions in the use of cross protection for plant virus disease control, *Ann. Rev. Phytopathol.,* 24, 67, 1986.

31. **Tumer, N. E., O.'Connell, K. M., Nelson, R. S., Sanders, P. R., Beachy, R. N., Fraley, R. T. and Shah, D. N.,** Expression of alfalfa mosaic virus coat protein gene confers cross-protection in transgenic tobacco and tomato plants. *EMBO* J., 6, 1181, 1987.

32. **Abel, P. P., Nelson, R. S., De, B., Hoffmann, N., Rogers, S. G., Fraley, R. T. and Beachy, R. N.,** Delay in disease development in transgenic plants that express the tobacco mosaic virus coat protein gene, *Science,* 232, 738, 1986.

33. **Van Dun, C. M. P., Van Vlotten-Doting, L., and Bol, J. F.,** Expression of alfalfa mosaic cirus cDNA 1 and 2 in transgenic tobacco plants *Virology,* 163, 572, 1988.

34. **Dawson, W. O., Butrick, P., and Grantham, G. L.,** Modification of the tobacco mosaic virus coat protein gene affecting replication, movement, and symptomology, *Phytopathology,* 78, 783, 1988.

35. **Palukaitis, P. and Zaitlin, M.,** A model to explain the "cross-protection phenomenon shown by plant viruses and viroids, in *Plant Microbe Interactions—Molecular and Genetic Perspectives,* Kosuge, T. and Nester, E. W., Eds., Macmillan, New York, 1984, 420.

36. **Rezaian, A.,** unpublished data, 1988.

Chapter 17

CHEMICAL CONTROL

I. FUNGICIDES

A. INTRODUCTION

The word fungicide has originated from two latin words: viz. *fungus* and *caedo* (to kill). So literally speaking a fungicide would be any agency (physical or chemical) which has ability to kill a fungus. However, the word is restricted to chemicals. Hence, the word fungicide should mean a chemical capable of killing fungi. However, there are a number of compounds which do not kill the fungus. They simply inhibit fungal growth or spore germination temporarily. If the fungus is freed from such substances, it would revive. Such a chemical is called a "fungistat" and the phenomenon of temporary inhibition of growth is called "fungistasis". Some other chemicals, like certain phenanthrene derivatives and Bordeaux mixture,[1] may inhibit spore production without affecting the growth of vegetative hyphae. These are called "antisporulants". There are other groups of chemicals which exhibit very poor or no antifungal activity *in vitro* condition but provide protection to the plants against the disease either by inhibiting the penetration of host surface by the fungi or by inducing the host defense system. The former type of chemicals are termed as "*antipenetrants*" and latter as "*antipathic agents*". Even though fungistats, antisporulants, antipenetrants and antipathic agents do not "kill" fungi, they are included under the broad term fungicide because by common usage, the word fungicide has been defined as a chemical substance which has ability to prevent damage caused by fungi to plants and their products.[1] Some workers prefer the use of term "fungitoxicants" instead of fungicide but technically it is also not appropriate for "antipenetrants" and "antipathic agents".

Fungicide which is effective only if applied prior to fungal infection is called "protectant". On the other hand, fungicide which is capable of eradicating a fungus after it has caused infection, and thereby "curing" the plant, is called "therapeutant". Most of the nonsystemic fungicides are protectants in nature and systemic fungicides usually exhibit both protective and curative actions. There are few nonsystemic funcicides, like dodine, organomercurials, etc., which, due to their limited ability to penetrate host surface, can eradicate the fungal pathogens upto some extent from infected area. These compounds are called as eradicants. However, this classification does not hold much ground as compounds listed as eradicants are basically effective in plant disease control only when used as a protectant.

B. USE OF FUNGICIDES

Fungicides alone accounts for about 18% (approx. 2.8×10^9 U.S. dollars)[2] of the total world market of the pesticides[3] (Figure 1). Most of the fungicides are used in areas with highly developed agriculture: Western Europe (39%), the Far East (28%), and the U.S. (12%). Only 21% is used in the rest of the world.[4] Fungicides are used primarily on susceptible crops of high value (Table 1) where losses due to disease are likely to exceed costs of application. Thus, relatively few crops and a relatively small acreage consume most fungicides. Fewer than a dozen diseases create demand for most of the fungicides used in the world[5] (Table 2). Among different fungicides, dithiocarbamates are still the most popular followed by benzimidazoles (Table 3).

C. FORMULATIONS

Formulation is an art which is mainly concerned with methods of presenting the active ingredient (pure fungicidal compound), in the most effective physical form—effective, that

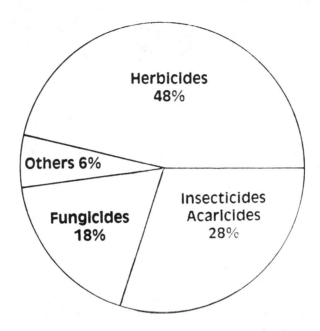

FIGURE 1.   World market of plant protection chemicals according to product group 1983 (sales to third parties: 25.7 billion SFr. = 100%) (Courtsey Schwinn, F.).

TABLE 1
Proportion of Fungicides Used on
Different Crops[2]

Crop	Percentage share of fungicide market
Fruit and vegetables (including vines)	46
Rice	15
Wheat	14
Soya	2
Maize	1.4
Cotton	1.4
Sugarbeet	1.4
Others	19

is, with regard to storage, application, and ultimate biological activity. Actually the amount of fungicide that is biologically effective at the plant surface is so minute that for economic use the chemical must be diluted, either with a solid (dust, powder, etc.) or liquid (concentrates, emulsion, suspension, etc.), before application. Water provides a cheap and effective dilution medium and, with few exceptions is used as the carrier for agricultural sprays. As majority of the fungicides are of very low water solubility they must be formulated to make them compatible with water, hence surface active agents are required to prepare water-dispersible powders, stock emulsions, or emulsifiable concentrates. In addition surface-active agents may be needed to improve the suspending, spreading, and wetting properties of the sprays. Other supplements that may be incorporated with sprays include stickers to improve the weather resistance of the deposit, and materials to improve storage stability, deposition on plant surface and penetration of plant surfaces and/or fungal cells. The type

TABLE 2
Crops of Worldwide Importance for Which Fungicides are Used Intensively[5]

Crop	Disease	Pathogen
Apples	Scab	*Venturia inaequalis* (Cooke) Wint.
Bananas	Leaf spot (Sigatoka)	*Mycosphaerella musicola* Leach
Cereals	Smuts, seed rots, seedling blight	Various fungi
Cocoa	Black pod	*Phytophthora palmivora* (Butl.) Butl.
Coffee	Leaf rust	*Hemileia vastatrix* Berk & Br.
	Berry disease	*Colletotrichum coffeanum* Noack
Cotton	Seed rots and seedling blight	Various fungi
Grape	Powdery mildew	*Uncinula necator* (Schw.) Burr.
	Downy mildew	*Plasmopara viticola* Berl. and de Toni
Potato	Late blight	*Phytophthora infestans* (Mont.) de Bary
Rice	Blast	*Pyricularia oryzae* Cav.
Tobacco	Blue mold	*Peronospora tabacina* Adam.

TABLE 3
Fungicides Used in World Agriculture[2]

Chemical type	Percentage share of fungicide market
Nonsystemic fungicides	21
Dithiocarbamates	
Inorganic compounds	16
(mainly copper and sulfur)	
Phthalimides	9
Other	
(Chlorothalonil, iprodione, dodine and many others)	9
Total	55
Systemic fungicides	
Benzimidazoles	13
Triazoles	8
Other	24
(many small products)	
Total	45

of formulation chosen for a fungicide is determined by a variety of factors of which cost and biological efficiency are the most important.[6] Different formulations of the pesticides (including fungicides) are listed in Table 4.

D. SYSTEMICITY

A fungicide when applied on the plant surface either may remain on the surface (non-systemic) or absorbed by the plant. The latter may remain there in treated plant part (lo-cosystemic); may move in the direction of evapotranspiration stream (apoplastic) or may move with photosynthates to "sinks" (symplastic) or in both directions (ambimobile) (Table 5). The symplast is the living part of the plant which is enclosed by membranes, i.e., protoplasts and plasmodesmata, including the phloem sieve cells. Long-distance transport in phloem is symplastic. The apoplast is the nonliving part of the plant, i.e., cell walls and cuticle, including xylem vessels and tracheids. Long distance movement in the transpiration stream is apoplastic.

Information concerning movement of systemic fungicides indicate that majority of them are translocated apoplastically. Only few fungicides (e.g., fosetyl-Al, metalaxyl, pyroxy-chlor, fenapanil etc.) are ambimobile and there too, except fostel-Al, apoplastic translocation

TABLE 4
Different Formulations of the Pesticides

Formulations	Abbreviations
Wettable powder	WP (formerly w.p.)
Dustable powder	DP (formerly d.p.)
Water soluble powder	SP (formerly s.p.)
Powder for seed treatment	DS
Solution for seed treatment	LS
Flowable concentrate for seed treatment	FS
Water soluble powder for seed treatment	SS
Water dispersible powder for slurry treatment (of seed)	WS
Emulsifiable concentrate	EC (formerly e.c.)
Electrochargeable liquid	ED
Emulsion, water in oil	EO
Emulsion, oil in water	EW
Oil miscible liquid	OL
Hot fogging concentrate	HN
Ultra-low volume liquid	UL
Aerosol generator	AE
Gas generating product	GS
Granules	GR
Water soluble granules	SG
Water dispersible granules	WG
Suspension concentrate (flowable)	SC (formerly s.c.)
Soluble concentrate	SL

TABLE 5
Terminologies Related with Transport of Fungicides in Plants[7-9]

General

Octanol number: (log P or log Kow)	Measure of lipophilicity of a compound. At equilibrium $$= \frac{\text{concentration in Octan-1-ol phase}}{\text{concentration in aqueous phase}}$$
Root/seed concentration factor: (RCF/SCF)	Measure of accumulation of a compound in root/seeds $$= \frac{\text{concentration in roots/seeds}}{\text{concentration in ambient solution}}$$
Transpiration stream: concentration factor (TSCF)	Measure of efficiency of xylem translocation $$= \frac{\text{concentration in transpiration stream}}{\text{concentration in external solution}}$$

Transport in Tissues

Apoplastic:	Transport in apoplast, the coherent network of free space, cell walls, and nonliving cells
Euapoplastic:	Apoplastic movement without any passage through protoplasts
Pseudoapoplastic:	Apoplastic movement with occasional passage through or retention in protoplasts
Symplastic:	Transport in the symplast, the coherent network of protoplasts connected by plasmodesmata

Long Distance Transport

Xylem-mobile (apoplastic):	Apoplastic transport in vessels and tracheids of xylem by means of the transpiration stream (xylem-systemic)
Phloem-mobile (symplastic):	Symplastic transport in the sieve tubes of phloem by means of the mass flow from source to sink (phloem-systemic)
Ambimobile:	Transport in xylem and phloem (ambisystemic)
Locally mobile	Transport within the organ of application (locosystemic)
Amobile	No long-distance transport from site of application (nonsystemic)

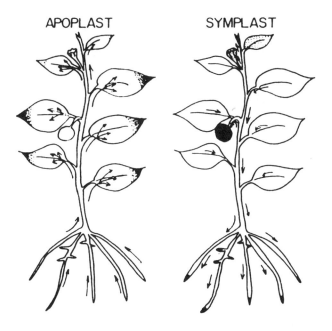

APOPLAST   SYMPLAST

FIGURE 2.   Translocation patterns in the apoplast and symplast. Arrows indicate direction of transport, black area indicates accumulation of systemic chemicals. Stippling indicates areas of lesser accumulation (from Edgington and Peterson[11]).

is predominant. Fosetyl-Al is the only fungicide which exhibits predominantly symplastic translocation. So far no fungicide is known which is truely symplastic (without any apoplastic movement).

1. Apoplastic Translocation

Apoplastic translocation, within the plant, is usually directed from roots to transpiring areas, especially leaves (Figure 2). Fungicides are absorbed by the roots, mainly in root hair zone, along with water. Root absorption is usually a passive phenomenon, i.e., do not involve expenditure of metabolic energy[10]. Radial movement through the cortex zone occurs either symplastically or apoplastically (Figure 3). Symplastically moving substances cross the plasmalemma and are transported via protoplasts and plasmodesmata of cortex cells and then through the endodermis cells to the vessels of the xylem (Figure 3b). Apoplastic movement is supposed to be accomplished in the free space of the cell walls. Since this route is blocked by lipophilic (suberized) incrustations of the casparian stripes at the endodermis level, fungicide should enter the symplast at this site to be transferred to xylem.[7] Entrance into the symplast requires lipophilic properties therefore, extremely hydrophilic compounds (log P < 0.5) (Table 5) are retained in the free space of the cortex and are usually not translocated in the long distance stream of the xylem to the leaves. At the same time too lipophilic chemicals (log P > 3.5) are absorbed not only by the soil organic matter resulting in poor root uptake (following soil application) but also by the constituents of plant cell wall (lignin, suberin etc.) after uptake. These chemicals may fail to enter into the xylem or transpiration stream. Partitioning into the xylem or transpirational stream (transpirational stream concentration factor, TSCF) (Table 5), is favored by water solubility.[12] Once the fungicide enters the roots, systemic movement occurs over the range of log P values from −0.5 to 3.5.

Accumulation of the fungicide inside the roots or seeds is estimated by root/seed concentration factor (Table 5); RCF/SCF value of more than unity indicates accumulation against concentration gradient. It is not only dependent upon the log P of the fungicide but also

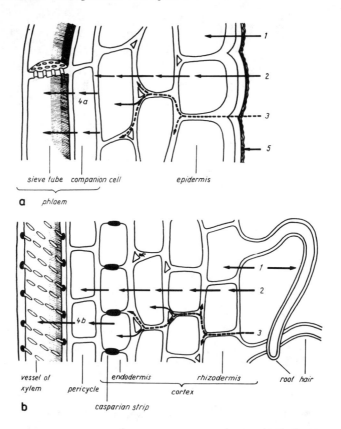

FIGURE 3.   The pathway of fungicide movement in plant tissue by penetrating (a) the leaf or (b) the root surface. 1 uptake in outer cells, 2 symplastic transport, 3 apoplastic transport, 4a absorption in sieve tubes, 4b transfer to vessels, 5 absorption in cuticular layers (Jacob and Neumann[7]).

plant species, particularly chemical composition of roots/seeds. Lignin is one such compound which binds with the fungicides[13] resulting in its accumulation in roots and at the same time poor apoplastic translocation to aerial plant parts.

Fungicides with typical apoplastic translocation display following properties:[14]

1.   Upward movement within the plant following seed, root, or stem application.
2.   Movement into various plant organs is dependent on their transpiration rate.
3.   Accumulation at tips and margins of leaves. For fungicides, particularly ambimobiles, which enters the cytoplasm, intraleaf distribution also depends on venation of leaves. In monocotyledons, where venation is convergent palmate parallel type, metalaxyl is drawn much strongly through veins resulting in its accumulation at tips and margins[9],[15] (Figure 4a). In dicotyledons because of reticulate venation, driving forces toward the periphery get drastically reduced, thereby, allowing metalaxyl to enter the cytoplasm. In such cases metalaxyl is either uniformly distributed in entire lamina (e.g., pea)[16] (Figure 4c) or accumulation at margins is quite delayed (e.g., pigeonpea,[17] cowpea,[18] french bean,[19] etc.) (Figure 4b). In cowpea, when applied through roots or seeds, metalaxyl exhibits delayed accumulation at margin of primary leaves but remained uniformly distributed in trifoliates.[18] However, majority of the fungicides with typical apoplastic translocation are accumulated at tips and margins of the leaves irrespective of the plant species involved.[14]

FIGURE 4. Autoradiograph showing distribution of ^{14}C-metalaxyl in (A) maize, (B) frenchbean and (C) pea when applied as a root treatment (root dip).

Above-mentioned properties of a apoplastically translocable fungicide may result in following problems situations:[14]

1. A lack of downward translocation means that chemical applied to above ground parts of the plant does not move to the roots.
2. Plant organs (e.g., fruits and flowers) which normally have a negligible amount of transpiration are largely bypassed by fungicides moving in the xylem.
3. (a) A lack of basipetal movement means that fungicide cannot be exported from leaves into other part of the plant. (b) Chemicals may buildup to phytotoxic concentrations at tips and margins of leaves. (c) The basal and central area of the leaf may not retain

enough fungicide to prevent new infections several days after foliar application. (d) Foliage developing subsequent to foliar treatment will not be protected from disease.

Direction of movement of the transpiration stream imposes basic limitations on the direction of fungicide movement. Movement of water in the xylem may be very rapid and fungicide movement therefore be expected to be very rapid unless the fungicide is bound to the xylem walls. However, in addition to long-distance transport of apoplastic fungicides in the xylem, a concomitant slower distribution by diffusion in the wall also occurs, allowing the substance to spread laterally within the plant. Since major driving force in apoplastic translocation is transpirational pull, factors affecting transpiration also affect fungicide translocation. In order to decelerate the undesirable marginal tip accumulation of fungicides in leaves, rates of transpiration have been reduced using abscisic acid which partially closes stomata.[14]

2. Symplastic Translocation

Mode of the symplastic translocation is well worked out for the sugars. Sugars are actively "loaded" into the phloem (i.e., sieve tube-companion cell complex) in plants. It is generally agreed that accumulation of sugars results in water influx and subsequent hydrostatic pressure which moves the phloem assimilates (including sugars) to sinks. Fully expanded photosynthesizing leaves serve as a source and roots, flowers, young growing leaves and fruits serve as a sink. A fungicide with symplastic translocation, after entry into the phloem follows the same source to sink pathway as followed by the phloem assimilates (Figure 5).

Metalaxyl in photosynthesizing leaves gets loaded inside the phloem as indicated by its accumulation in minor veins. This phloem loading was inhibited by 2,4-DNP and restored by further application of ATP but only in the presence of sugar. It indicates that loading of metalaxyl inside the phloem is coupled with active sugar loading. After phloem loading metalaxyl moves from source to sink with respect to sugar. Application of sucrose to metalaxyl treated leaves improved and removal of phloem by girdling drastically reduced, downward translocation of fungicide to roots.[19] Unlike sugar, metalaxyl was not accumulated in roots particularly at growing tips. It may be due to diffusion of metalaxyl from symplastic route to apoplastic route and subsequently retransport to leaves. Probably same is the true for other ambimobile fungicides.

Since in majority of the ambimobile fungicides, except fosetyl-Al, predominant mode of transolocation is apoplastic, they cannot be expected to provide protection to the roots on foliar application. However, due to their better redistribution inside the plant, they provide protection to the young growing host tissues thereby minimizing the number of sprays required to protect the crop.

Structure-systemicity relationship studies have demonstrated what weak acids (with carboxylic, sulfonic or phosphoric groups) are more phloem mobile.[11] However, nonacidic compounds like oxamyl also show phloem mobility probably due to their ability to enter phloem by solubilization into plasma membrane.

3. Factors Affecting Translocation

Different factors which may affect uptake, translocation and/or distribution of fungicides in plant are as follows.[11,18]

a. *Membrane Permeability and Selective Toxicity*

Since systemic fungicides enter the protoplasm, they must be nontoxic to host plant cells. It is due to this reason that nonselective fungicides particularly those acting as a metal

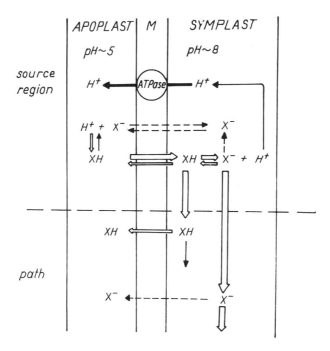

FIGURE 5.   Diagram of symplastic transport of fungicides. Uptake of phloem-and ambimobile, as well as xylem-mobile compounds into sieve tube - companion cell complex by simple diffusion of the un- dissociated molecules (XH). Weak acids get ionized at alkaline pH of phloem and hydrophilic anions (X^-) are retained in phloem by ion trap mechanism and move along with phloem assimilates (particularly sugar) from source to sink. Along the pathway the relatively lipophilic, un- dissociated molecules XH are leaked out very quickly from the sieve tube, while the relatively hydrophilic anions X^- escape to a lesser extent.

chelator, an oxidative uncoupler, or as an thiol alkylating agent or heavy metals are not systemic. So selectivity is a prerequisite for the systemicity.

b. Metabolic Stability and Cuticular Penetration

Cuticle, which covers all primary above ground organs of higher terrestrial plants, offers first formidable barrier to fungicides applied to the foliage, to flowers, or fruits. Plant cuticle has not only high sorptive capacity for lipophilc chemicals but also a large number of binding sites for reactive chemicals. Due to this they can bind and accumulate pollutants and agro- chemicals including fungicides. A systemic fungicide sprayed on the surface of the leaf must partition into the cuticle, diffuse through and repartition into the apoplast underneath the spray droplets. If a fungicide is quite lipophilic, penetration is rapid. However, if too lipophilic, the fungicide may be retained in the cuticle, retarding the transcuticular diffusion and release to the apoplast. Very hydrophilic molecules, e.g. oxamyl (nematicide), may partition less rapidly into the cuticle, but they continue to be partitioned from the cuticle into the apoplast for a long period and thus have greater total uptake.[11] As in root, uptake of fungicides by aerial plant parts and seeds is by simple diffusion and does not involve any expenditure of metabolic energy. It has been postulated that uptake of metalaxyl and pyr- oquilon into apparent free space (AFS) of roots is through simple diffusion and their passive entry into cytoplasm involves solubilization into lipoprotein plasma membrane.[16,20]

c. Plant Species

Uptake, translocation and inter- and intraplant part distribution of a fungicide is also influenced by plant species involved and this in turn may influence the *in vivo* efficacy of the fungicide.[24] Roots of pearl millet and pea (Figure 4c) retain much higher quantity of metalaxyl than roots of maize (Figure 4a) and cowpea. Unlike in any other crop distribution of metalaxyl, following seed treatment, was uniform in entire pea seedling and this coupled with higher seed uptake may be probably responsible for the prolonged protection provided by metalaxyl seed treatment to pea seedlings against *Pythium* and downy mildew.[16]

E. SELECTIVITY

No fungicide can be used in practice to control all fungal dieseases. Except probably strong biocidal agents like mercury compounds no fungicide can kill or is inhibitory to an equal extent to all the fungi. Moreover fungicides must be much more toxic to the fungal pathogen than the host plant. So in true sense every fungicide shows at least some degree of selectivity. Nevertheless, there are fungicide which are effective against a wide spectrum of fungi (e.g., dithiocarbamates, phthalimdies, organotin compounds, etc.) while others exhibit moderate (e.g., benzimidazoles, EBIs, etc.) to high degree of selectivity (e.g., melanin biosynthesis inhibitors are used only against *Pyricularia oryzae;* phenylamides are effective only against oomycetes) in their antifungal spectrum. The fungicides belonging to first category are classified as nonselective and those belonging to latter two groups as selective.

Selectivity to the fungicides is probably regulated by same biochemical factors which decide fungicide resistance in otherwise sensitive fungi. These are:[22] (1) differences in the accumulation of a fungicide in the cell, (2) different structures of receptor or target site, (3) differences in ability to toxify (conversion from inactive to active form) a compound, (4) differences in ability to detoxify a compound, and (5) different degrees of importance of a receptor or target site for survival of the fungus.

F. CLASSIFICATION

Ever since the introduction of systemic fungicides, systemicity has been the most popular criterion for the broad classification of the fungicides. They were classified as (1) nonsystemic and (2) systemic. However, in this book selectivity has been preferred over systemicity and based on that fungicides are broadly classified as (1) nonselective and (2) selective. Selectivity is the prime requirement for the systemicity. All the systemic fungicides are selective but all the selective fungicides are not systemic. So selectivity not only gives the idea about antifungal spectrum and mode of action of a fungicide and associated risk for the development of fungicide resistant strains but also reflects upon its systemicity. However, like any other classification system, it also suffers from the same problem. There are compounds (e.g., anilazine, fenitropan, chinomethionate, dichlofluanid, etc.) whose mode of action is same as non-selective fungicides, still they exhibit some degree of selectivity with respect to their antifungal spectrum. Whether to group these compounds as nonselective or selective? In this book they are grouped as selective, however, there can be differences in opinion in such cases.

G. NONSELECTIVE FUNGICIDES
1. Sulfur Fungicides

Sulfur fungicides are the oldest but still very popular. Several inorganic (elemental S, lime S) and organic (dithiocarbamates) (Figure 6) sulfur fungicides are in agricultural use for the control of plant diseases (Table 6). McCallan[24] stated in 1967 that "the dithiocarbamates are, without question, the most important and versatile group of organic fungicides yet discovered". If we consider the relative use and market of different fungicides (Table

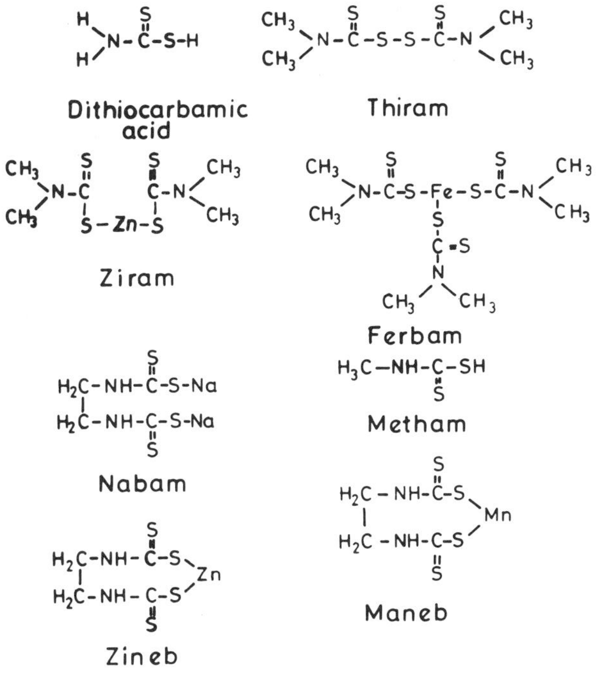

FIGURE 6. Structure formula for dithiocarbamate fungicides.

3), this statement looks still valid, however, relative significance of dithiocarbamates has declined considerably as other types of fungicides have come into use.

All sulfur fungicides are nonsystemic, and except lime-sulfur, are nonphytotoxic (excluding "sulfur shy" varieties) and compatible with most of the pesticides. As far as their mode of action is concerned elemental sulphur interferes with energy production by intercepting electron on the substrate side of cytochrome C in the mitochondrial electron transport system.[25] The dialkyldithiocarbamates are known to inhibit a multitude of enzymes; therefore, fungitoxicity probably involves concurrent inhibition of enzymes at several sites.[25] The pyruvate dehydrogenase reaction is particularly highly sensitive to dialkyldithiocarbamates.[26]

2. Copper Fungicides

Ever since the discovery of Bordeaux mixture in 1885, copper fungicides predominated the field of fungicidal plant disease control for more than 50 years until synthetic organic fungicides invaded the market. Even today some of the copper compounds are used widely in many countries. All such copper compounds which are currently in use are shown in Figure 7 and are described in Table 7. Copper is a multisite biochemical inhibitor (probably interact with −SH groups of enzymes) with little biological specificity. Therefore, phytotoxicity is always a potential problem. Nevertheless, when the level of dissolved cupric ion is carefully regulated, copper fungicides can be used to control fungal diseases with little or no deleterious effect on the plant.

3. Mercury Fungicides

Mercury is a general biocide. Hiltner was probably the first person to use murcuric chloride successfully for the control of *Fusarium* disease of rye[1] as early as 1910. Since then several organomercurial compounds were introduced as a fungicide. However, most of them are now withdrawn or discontinued because of their high mammalian toxicity. Only two organic and three inorganic mercury compounds (Figure 8) are in use and that too in

TABLE 6
Sulfur Fungicides

Common name	Chemical name	Year of introduction (by)	Uses	Acute oral toxicity (LD$_{50}$ rat mg/kg)	Formulations
Sulfur Sulphur	Sulfur	19th century	Acaricide and fungicide, applied as dust or spray against powdery mildews	Non-toxic	'Elosal', 'Kumulus S', 'Flotox', 'Fortho', 'Kolthior', 'Sodil B', 'Solfa', 'Ultranix', WP (800 g/kg); 'Imber', 'Sulfospor', SC, 'Sandotox', DP *Mixtures*: with carbendazim, maneb, dichlone, nitrothal-isopropyl, parathion-methyl, dicofol, rotenone etc.
Lime Sulfur	Calcium polysulfide (Ca Sx)	19th century	Powdery mildew, also for control of scale insects and mites; phytotoxic; limited compatibility with other pesticides		Prepared by dissolving S (15 lbs) in aqueous suspension of calcium hydroxide (20 lbs rock lime/50 gallons water); formulation is diluted to 10 ml/l
Thiram TMTD	Tetramethyl-thiuram disulfide	1931? (du Pont)	Foliar spray to control: *Botrytis* spp. on lettuce, ornamentals, soft fruits and vegetables, rusts on ornamentals; *Venturia pirina* on pears; seed treatment: damping off of vegetables, ornamentals and maize etc; at high doses bird, rodent and dear repellent	780— 865	'Arasan', 'Tersan', 'Pomarsol', 'Fernasan', WP ($\leq$ 800 g/kg), SC, seed treatment *Mixtures*: several in combination with different fungicides and insecticides like benomyl, carbendazim, gamma-HCH, permethrin, petroleum oil, fonofos, rotenone, thiophanatemethyl, bendiocarb, thiabendazole, vinclozolin, ziram, carboxin etc.
Ziram	Zinc bis (dimethyl-dithiocarbamate)	1944 (du Pont)	Foliar spray on fruit and vegetable crops—*Alternaria, Septoria* spp. etc.; also used as repellent to birds and rodents	1400	'Crittam', WP, 'Aaprotect' repellent paste (370 g/kg) *Mixtures*: 'Ramedit', WP (with copper oxychloride)
Ferbam	Iron tris (dimethyl-	1942 (du Pont)	Protection of foliage against fungal	>4000	'Fermate' WP (760 g/kg)

Common name	Chemical name	Year (Company)	Uses	LD_{50}	Trade names / Mixtures
	dithiocarbamate)		crops—*Alternaria, Septoria* spp. etc.; also used as repellent to birds and rodents		paste (370 g/kg) *Mixtures:* 'Ramedit', WP (with copper oxychloride)
Ferbam	Iron tris (dimethyl-dithiocarbamate)	1942 (du Pont)	Protection of foliage against fungal pathogens including *Taphrina deformans* (peaches)	>4000	'Fermate' WP (760 g/kg) *Mixtures:* 'Trimanzone' WP (ferbam + maneb + zineb)
Metham	Sodium Methyl-dithiocarbamate	1954 (Stauffer Chem. Co.)	Soil fungicide, nematicide and herbicide with a fumigant action; phytotoxic, applied about 14 day before planting	1700–1800	'Polefume', 'Vapam' SL (382 g/l), 'Vapam B', SL (334 g/l); 'Sistan', SL
Nabam	Disodium ethylenebis (dithiocarbamate)	1943? (du Pont) Rohm & Hass)	Now superseded by zineb and maneb; too phytotoxic for general use on foliage; soil application exhibit systemic action on *Phytophthora fragariae*; used as an algicide in paddy field	395	'X-Spor', SL
Zineb	Zinc ethylenebis (dithiocarbamate)	1943 (Rohm & Hass, du Pont)	To protect foliage and fruit of a wide range of crops against diseases such as potato and tomato blight, *Botrytis* spp. downy mildews and rusts	>5200	'Dithane-Z-78', 'Tiezene', WP (700 or 750 g/kg); 'Dithane' dust, DP *Mixtures:* with copper oxychloride ('Cupro-Phynebe', 'Super Mixy', WP), or mancozeb ('Turbair Dicamate')
Naneb	Manganese ethylenebis (dithiocarbamate)	1950 (du Pont, Rohm & Hass)	Foliar spray to control many fungal diseases of field crops, fruits, nuts, ornamentals, turf and vegetables, especially blights of potatoes and tomatoes, downy mildew of lettuce and vine; frequently used in combination with other fungicides	6750	'Farmaneb', 'Dithane M-22', 'Manzate', WP (800 g/kg) *Mixtures:* two or three way mixtures with several fungicides like carbendazim, fentin acetate, sulfur, tridemorph, fentin hydroxide, thiophanate methyl, fenarimol, nuarimol, metalaxyl etc.
Maniozeb	Manganese ethylenebis (dithiocarbamate) complex with zinc salt (maneb containing 20% of Mn + 2.55% of Zn)	1961 (Rohm & Hass, du Pont)	Wide range of foliar fungal diseases including those caused by *Phytophthora infestans* (on potatoes), *Fulvia fulva* (on tomatoes), *Venturia* spp. (on apples and pears), *Dilocarpon rosae* and *Phragmidium mucronatum* on roses; also used in combination with some systemic fungicides to increase duration of protection	>8000	'Blecar MN', DS (480 g/kg); 'Dithane LF', SC (455 g/l) 'Acarie M', 'Critox MZ'; 'Dithane M-45', 'Dithane 945', 'Manzate', WP (600 or 800 g/kg); Dithane M 45 Poudrage', DP (60 g/kg). *Mixtures:* with carbendazim, fenpropimorph, cymoxanil, metalaxyl, benalaxyl, zineb, maneb, copper oxychloride, tridemorph, oxadixyl, cyprofuram, gamma-HCH, fosetyl-Al, Bordeaux mixture etc.

Note: Data collected from Nene and Thapliyal[1] and Worthing.[23]

Cu$_2$Cl(OH)$_3$

Copper oxychloride

Cu$_2$O

Cuprous oxide

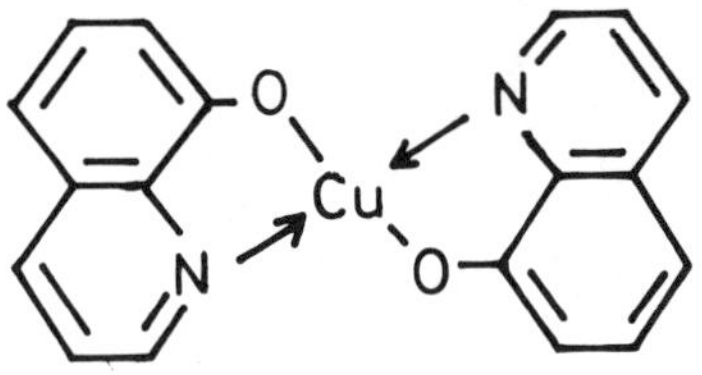

Oxine-copper

FIGURE 7. Structure formula of copper fungicides.

very limited cases (Table 8). Like copper, mercury also exhibits multisital action due to its interaction with the −SH group of the susceptible enzymes.

4. Organotin Fungicides

Vander Kirk and Luijten[27] in 1954 reported the strong antifungal activity of triphenyltin derivatives. Two such compounds are now in use as fungicides (Figure 9, Table 9). Organotin fungicides are locosystemic and also exhibit antibacterial activity. They act as antifeedant for certain insects (e.g., *Diacretia obliqua*). The primary mechanism of toxicity of triphenyltins in fungi is interference with oxidative phosphorylation.[26]

5. Phthalimide and Quinone Fungicides

One of the quinone derivatives, chloranil (2,3,5,6-tetrachloro-1,4,benzoquinone), was introduced as a fungicide as early as in 1940. However, now it has been replaced by dichlone, another quinone derivative introduced in 1943. Properties of dichlone are listed in Table 10. Quinone fungicides are multisital in their mode of action. The two mechanisms which are thought to be most likely for dichlone and chloronil are:[1] (1) binding of the quinone nucleus to −SH and −NH$_2$ groups in fungal cell, and (2) interference with the electron transport system.

Amongst the phthalimide compounds, captan was introduced first and is still being used widely as a spray, dust, or seed dressing (sometimes as soil drench also) against a number of diseases. Basically phthalimide fungicides (Figure 10; Table 10) are protective in nature but limited systemicity has been reported for captan[1] and captafol.[1] Like quinone, phthalimides are highly reactive against thiol (−SH) groups of proteins (enzymes) and low molecular weight metabolites (cystein, glutathione etc.).[25]

H. SELECTIVE FUNGICIDES
1. Aromatic Hydrocarbon Fungicides

Aromatic hydrocarbon fungicides (AHF) include a heterogenous group of compounds (Figure 11) which have been in use since long. Actually AHF group was formed after the general cross resistance between these fungicides was recognized, which is indicative of their common site of action.[28] Although AHFs are reported to affect several metabolic processes in fungal cell, the primary toxic effect is on induced lipid peroxidation. The development of AHFs was stimulated by their relatively simple structure (Figure 11), low production cost and low mammalian toxicity combined with good but selective antifungal activity (Table 11). The AHFs have gained more popularity as soil fungicides because of their high volatility, low UV light stability and high activity against some soil-borne fungi. They are also used against seed borne pathogens and post-harvest decay fungi (Table 11). Except chloroneb and etridiazole, where poor systemicity has been reported, all other AHFs are non-systemic. However, because of their selective action AHFs result in rapid buildup of resistant strains which are cross resistant not only to all members of this artificially

TABLE 7
Copper Fungicides

Common name	Chemical name	Year of introduction (by)	Uses	Acute oral toxicity (LD$_{50}$ rat mg/kg)	Formulations
Bordeaux mixture	Mixture, with or without stabilizing agents, of calcium hydroxide and copper (II) Sulfate	1885 (A. Millardet)	Protective fungicide for foliar application, major uses includes *Phytophthora infestans* (potato), *Venturia inaequalis* (apple), *Pseudoperonospora humuli* (hops); high tenacity		WP, SC, can be prepared as a tank-mix using (CuSO$_4$ · 5H$_2$O (1.0 kg) with Ca(OH)$_2$ (1.25 kg) in water (100 l) for high volume application or (4 kg + 2 kg + 100 l) for low volume application
Copper oxychloride	Dicopper chloride trihydroxide	early 1900's (Sandoz, Bayer)	Mainly for control of *Phytophthora infestans* (potatoes) and *Pseudoperonospora humuli* (hops)	1440	'Recop', 'Cupravit', 'Fernacot', 'Pere-col'. Mixtures: with propineb, dichlorofluanid, zineb, benalaxyl, mancozeb, metalaxyl, oxadixyl, captafol, cymoxanil, copper sulphate etc.
Cuprous oxide	Copper (I) oxide	1932 (Sandoz, ICI)	Mainly for seed treatment; for foliar application against blight, downy mildews and rusts; nonphytotoxic except to brassicas and 'Copper shy' varieties	470	'Cacobre', 'Copper-Sandoz', 'Perenox', 'Yellow Cuprocide' WP or WG (500 g Cu/kg); a stabilizer is necessary to delay oxidation and formation of carbonate
Oxine-copper	Cupric 8-quinolinoxide		Seed treatment: in cereals (against *Drechsleria graminea, Fusarium nivale, Septoria nodorum, Tilletia caries*), flax, oilseed rape, sugarbeet and sunflowers (against *Alternaria, Botrytis, Cercospora, Phoma, Pythium* and *Sclerotinia* spp.) and beans and peas (*Ascochyta* spp.); mixture with other insecticides and fungicides: mixture with bitumen for sealing wounds and pruning cuts on trees	4700	'Quinolate 400', SC (400 g/l); 'Quinolate Semences WK', DS (600 g/kg). Mixtures: for seed treatment with anthraquinone, gamma-HCH, endosulfan, carboxin etc.

Data collected from Nene and Thapliyal[1] and Worthing[23]

$$Hg\,Cl_2$$

Mercuric chloride

$$Hg\,O$$

Mercuric oxide

$$Hg_2Cl_2$$

Mercurous chloride

$$CH_3OCH_2CH_2\,Hg\text{-}silicate$$

2- methoxyethyl mercury silicate

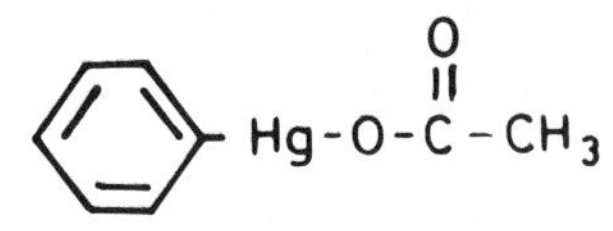

$$\langle \text{phenyl ring} \rangle - Hg - O - \overset{\displaystyle O}{\overset{\displaystyle \|}{C}} - CH_3$$

P M A

FIGURE 8. Structure formula of mercury fungicides.

composed group but also against dicarboximide fungicides. However, the problems of resistance have never been a widespread one in soil fungi probably due to their low fecundity in natural conditions as compared to foliar pathogens.

2. Guanidine Fungicides

This group is comprised of three non-systemic fungicides (Figure 12). Dodine and guazatine are in use since long; iminoctadine which is a major component of mixture of reaction products comprising guazatine, was introduced only recently. These fungicides interfere with the membrane permeability. Long hydrocarbon chain of dodine is reported to get incorporated in membrane resulting in loss of permeability. Primary antifungal action of guazatine and iminoctadine has been suggested as inhibition of lipid biosynthesis. Guanidines are broad spectrum fungicides. They are used in combination with other fungicides (Table 12). *In vitro* resistance was detected against dodine and guazatine in several fungi and fungal strains resistant against these two fungicides exhibited negatively correleated cross resistance against fenarimol, a sterol biosynthesis inhibitor.

3. Dicarboximides

Dicarboxymide fungicides (DCOF) represent a group of highly active and selective fungicides (Figure 13; Table 13). These compound are very effective against the diseases caused by *Botrytis cinerea*, *Sclerotinia* spp., *Monilia* spp., and *Phoma* spp. in cereals, fruits, vegetables, and ornamental crops.[30] Although DCOFs are classified as contact fungicides which, as a rule, are applied prophylactically, certain reports have appeared indicating their (particularly in vinclozolin and procymidone) uptake via the roots and leaves and apoplastic translocation, though poor, inside the plant.[30] Primary site of action of these compounds is yet to be worked out induction of lipid peroxidation in treated fungi seems to be one of the major effect of DCOFs. These compounds are considered as of "moderate risk" from resistance point of view. Fungicide resistant strains are frequently detected in laboratory and field. Failure of disease control due to resistant strains are also recorded from fields and glass houses particularly in *Botrytis cinerea* but these failures are not as spectacular as in the case of benzimidazoles and phenylamides. Resistant strains developed against one DCOF are cross resistant to not only other DCOFs but also to aromatic hydrocarbon fungicides.

4. 2-Aminopyrimidine Fungicides

This group includes three fungicides, dimethirimol, ethirimol and bupirimate, which are effective against powdery mildews (Figure 14; Table 14). Dimethirimol and ethirimol are

TABLE 8
Mercury Fungicides

Common name	Chemical name	Year of introduction (by)	Uses	Acute oral toxicity (LD_{50} rat mg/kg)	Formulations
Mercurous chloride	Mercury (I) chloride	1929	Preplant soil application to control root maggots, club root of brassicas, white rot of onions; as a fungicide and moss-killer on turf; phytotoxic	210	'Cyclosan', 'M-C Turf Fungicide,' DP (40 g/kg) *Mixtures:* 'Merfusan' 'Mersil' (mercurous chloride + mercuric chloride)
Mercuric chloride	Mercury (II) chloride		Used only in Canada in combination with mercurous chloride on turf; phytotoxic	1-5	Mixtures: 'Merfusan', 'Mersil'
Mercuric oxide	Mercury oxide	(Sandoz)	As paint against apple canker, bark injuries, pruning cuts on fruits, ornamentals (shrubs and trees)	18	'Santar' paint (30 g/kg)
2-methoxy-ethylmercury silicate		(Bayer)	Seed treatment against various seed borne diseases of cereals	1140	'Soprasan', DS (15 g/kg); *Mixtures:* with anthraquinone, gamma-HCH, endosulfan
PMA	Phenylmercury acetate	1932 (Bayer)	Powerful eradicant fungicide used as a treatment for cereal seed, often in combination with insecticides and fungicides, also used as selective herbicide to control crabgrass in lawns		'Agrosan D', Ceresol, DS, WS (100-300 g mercury/kg) *Mixtures:* with carboxin, gamma-HCH, or 2-(1, naphthyl) acetic acid

Information collected from Worthing.[23]

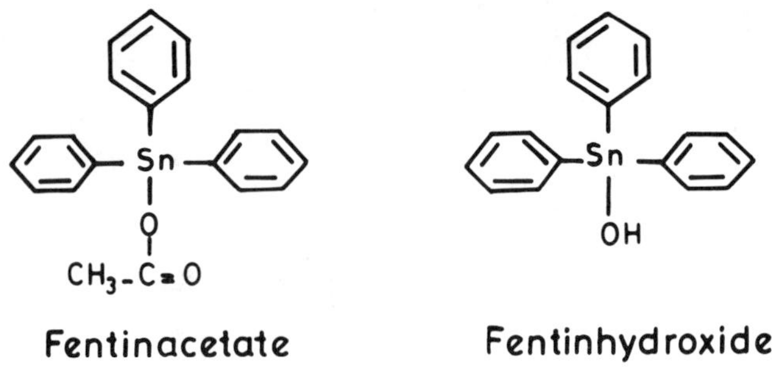

Fentinacetate   Fentinhydroxide

FIGURE 9.   Structure formula of organo-tin fungicides.

effective largely against powdery mildews of herbaceous plants; bupirimate was developed to control these diseases in woody plants and ornamentals. They are applied either as a spray, seed treatment (e.g., barley) or soil drench. On soil application they are absorbed by the roots and translocated apoplastically to leaves.[31] Uptake is poor if soil is acidic or having high organic matter content. These fungicides mainly inhibit appressoria and haustoria formation. They inhibit nucleic acid synthesis by inhibiting enzyme adenosine deaminase (ADA-ase). Although ADA-ase is present in many fungi, but only the enzymes from powdery mildew fungi are sensitive to 2-aminopyrimidine fungicides.[31] Problem of resistance development was encountered particularly in glass houses against dimethirimol and ethirimol, but it did not attain serious proportion in field. It may be either due to the less fitness of the resistant strains or replacement of these fungicides with morpholines and triazoles in late seventies avoiding their extensive field use. However, problem of development of resistance in powdery mildews against sterol biosynthesis inhibitors has renewed the interest in 2-aminopyrimidines. Development of resistant strains are not yet reported against bupirimate. It shows excellent protective and eradicative efficacy; may be due to its good vapor phase activity.

5. Organophosphorus Fungicides

Chemicals with direct conjunction of the phosphorus atom and carbon-atom and derivatives of phosphoric acid without C-P-conjunction are all counted as organophosphorous compounds.[32] At present only four such compounds are in use (Table 15) as fungicides. They include compounds especially active against blast of rice, e.g., phosphorothiolates (iprobenos [IBP] and edifenphos), while others are primarily active against diseases caused by powdery mildew fungi, e.g., phophorothianates (e.g., pyrazophos and toclofos-methyl). Toclofos-methyl has already been described under aromatic hydrocarbon fungicides, rest of the organophosphorus fungicides are shown in Figure 15 and described in Table 15. These fungicides display protective as well as curative activity; this is partly due to their systemic properties. The most systemic of these compounds is IBP. It can be applied through irrigation water as it is readily absorbed by the roots and translocated apoplastically to leaves. Pyrazophos is absorbed by the roots and translocated to leaves, its activity against powdery mildew is restricted to the vein (xylem) system probably due to poor distribution of compound from veins to mesophyll cells.[32] However, on foliar application pyrazophos penetrates leaf surface and exhibits lateral movement.

The metabolic conversion of IBP and edifenphos (mainly by cleavage at P-S linkage), probably into fungitoxic metabolites, by sensitive strains of *Pyricularia oryzae* has been demonstrated.[32,33] It has been suggested that pyrazophos is converted by fungal oxidases to PP-pyrazophos (2-hydroxy-5-methyl-6-ethoxycarbonyl-pyrazolo (1,5a) pyrimidine), which is considered as actual fungitoxic principle of pyrazophos. With regard to biochemical mode of action IBP and edifenphos interfere with the synthesis of phospholipids by inhibiting specific conversion of phosphat dylethanolamine to phosphatidylcholine by transmethylation

TABLE 9
Fentin Fungicides

Common name	Chemical name	Year of introduction (by)	Uses	Acute oral toxicity (LD_{50} rat mg/kg)	Formulations
Fentin acetate	Triphenyltin (IV) acetate	1954 (Hoechst)	*Ramularia* spp. (celery, sugarbeet), *cercospora beticola* (sugarbeet), *Phytophthora infestans* (potato); as an algicide in rice	140—298	'Brestan', WP (190 or 540 g/kg); 'Brestan 60' WP (540 g fentin acetate + 160 g maneb/kg)
Fentin hydroxide	Triphenyltin (IV) hydroxide	1954 (Duphar)	Early and late blight of potato, leaf spot on sugarbeet, rice blast, coffee berry disease, brown spot of tobacco; act as antifeedant against insects	110—171	'Du-Ter' WP (190 g/kg); 'Du-Ter Extra', WP (475 g/kg), 'Du-Ter Forte', WP (600 g/kg); 'Farmatin' 50, WP (500 g/kg) Mixtures: with maneb ('Du-Ter M' WP) or metoxuron ('End Spray')

Information collected from Worthing.[23]

TABLE 10
Phthalimide and Quinone Fungicides

Common name (Code No.)	Chemical name	Year of introduction (by)	Main uses	Acute oral toxicity (LD_{50}) rat mg/kg	Formulations
			Phthalimides		
Captan (SR 406)	N (trichloromethylthio) cyclohex-4-ene-1,2-dicarboximide	1952 (Standard Oil Development Co.)	Diseases of many fruit, ornamental and vegetable crops including *Venturia* spp. (on apple and pear); also used as a spray, root dip or seed treatment to protect young plants against rots and damping off; not compatible with oil sprays	9000	'Orthocide 50 WP', 'Phytocape' WP (500, 800 or 830 g/kg), DP (50 or 100 g/kg), DS (600-750 g/kg) and WS Mixtures: with nitrothalisopropyl, fosetyl-Al, metalaxyl, thiabendazole, dicloran, carboxin, gamma-HCH, bendiocarb, penaconazole, etaconazol, etc.
Captafol (Ortho-5865)	N (1,1,2,2-tetrachloroethylthio) cyclohex-4-ene-1, 2- dicarboximide	1961 (Chevron)	Widely used to control foliage and fruit diseases of tomatoes, coffee berry disease, potato blight, tapping panel disease of rubber and many other diseases; also used in lumber and timber to check wood rotting fungi	5000— 6200	'Captaspor', SC (480 g/l); 'Kenofol', 'Ortho Difolatan 80 W', WP (800 g/kg), 'Ortho Difolatan 4 Flowable', 'Sanspor', SC (480 g/l); 'Sanseal', LA (sealing pruning cuts) Mixtures: with carbendazim, triadimefon, flutriafol, folpet, diclobutrazole, propiconazole or cymoxanil etc.
Folpet	N-(trichloromethylthio) phthalimide	1952 (Standard Oil Dev. Co.)	Foliar application against *Alternaria, Botrytis, Pythium, Rhizoctonia, Venturia* spp., leaf spot, downy and powdery mildews - mainly on apples, citrus, cucumbers, soft fruits, lettuce, melons, onions, ornamentals, tomatoes etc.	>10000	'Acryptan', 'Phaltan', WP (500 g/kg) Mixtures: with metalaxyl, ofurace, captafol, oxadixyl, cyprofuram, fosetyl-Al etc.

Quinone

Dichlone	2,3-dichloro-1,4-naphth-oquinone	1943 (Uni-royol)	As a seed treatment in corn, peas, rice, sorghum, beet, pepper, tomato against damping off, anthracnose, grain smut etc.; foliar spray against apple scab, leaf curl of peach etc.	1300	'Phygon' WP (500 g/kg) Mixtures: 'Kolo-100', DP (35 g dichlone + 754 g S/kg)

Data compiled from Worthing[23] and Nene and Thapliyal.[1]

TABLE 11

Aromatic Hydrocarbon Fungicides in Commercial Use

Common/trivial name (Code no.)	Chemical name	Year of introduction (by)	Mode of application	Active against	Acute oral toxicity (LD$_{50}$ rat mg/kg)	Formulations
				Chlorobenzenes		
Hexachlorobenzene/HCB Quintozene/PCNB	Hexachlorobenzene Pentachloronitrobenzene	1945 1930 (IG Furben now Bayer)	seed, soil seed, soil	Wheat bunts and *Urocystis agropyri Rhizoctonia, Sclerotium rolfsii, Sclerotinia, Botrytis, Tilletia caries*	10000 >12000	DS, WS 'Brassicol', 'Tritisan' (200 g/ kg; 'Cryptonol Special E' DP, GP (300 g/kg); 'Terrachlor', WP (750 g/kg), EC (240 g/l) Mixtures: with cycloheximide ('Actidione RZ'); etridiazole ('Terrachlor Super X'; 'TerraCoat L 205'), or fuberidazole ('Voronit Special')
Tecnazene/TCNB	1,2,4,5-tetrachloro-3-nitrobenzene	1946 (Bayer)	soil	*Fusarium coeruleum* (potato tubers), *Botrytis* spp.	7500	'Arena', 'Fusarex' DP (60 g/ kg); 'Arena', 'Fusarex G', GR (100 g/kg) 'Hytec', 'New Hystore', 'Turbostore', GR (50 g/ kg) Mixtures: with thiabendazole ('Fusarex T', 'Storite SS')
Chloroneb (Soil Fungicide 1823)	1,4-dichloro-2,5 dimethoxy-benzene	1967 (du Pont)	seed, foliar	*Rhizoctonia solani, Corticium rolfsii, Sclerotinia sclerotiorum, Botrytis, Ustilago maydis, Pythium, Typhula*	>11000	'Demosan' 65 W, WP; 'Tersan SP'; Turf fungicide (650) g/kg)
Dichloran or Dicloran (RD 6584)	2,6-dichloro-4-nitroaniline	1960 (Boots Co., now FBC)	Seed, fruit (dip), foliar	*Botrytis, Rhizopus, Mucor, Sclerotinia, Penicillium, Monilia*	4000	'Botran' WP (500 or 750 g/kg), DP (40-80 g/kg) Mixtures: with captan ('Botec' SD) or thiram ('Turbair Botryticide', UL)

TABLE 11 (continued)
Aromatic Hydrocarbon Fungicides in Commercial Use

Common/trivial name (Code no.)	Chemical name	Year of introduction (by)	Mode of application	Active against	Acute oral toxicity (LD$_{50}$ rat mg/kg)	Formulations
				Biphenyls		
Biphenyl Diphenyl	1,1'-biphenyl	1944	Fruit	*Penicillium* spp. (citrus fruit)	3280	Used to impregnate citrus fruit wraps
Phenyl-2-phenol or 2 phenyl-phenol	Diphenyl-2-ol	1936	Disinfectant, fruit	*Penicillium, Diplodia, Botrytis, Nectria galligena*	2480	'Dowicide', (2-phenylphenol) 'Dowicide A' (sodium salt); 'Nectryl' canker paint
				Others		
Tolclofos-methyl (53349)	*O*-2,6-dichloro-*p*-tolyl *O,O*-dimethyl phosphorothioate	1983 (Sumimoto)	Seed, foliar	*Rhizoctonia solani, Sclerotium rolfsii, Sclerotinia sclerotiorum, Botrytis, Ustilago maydis, Penicillium, Typhula*	5000	'Rizolex' EC (200 g/l), DP (50, 100 or 200 g/kg); WP (500 g/kg), SC (250 g/kg)
Etridiazole or Echlomezol (OM 2424)	Ethyl 3-trichloromethyl-1,2,4-thiadiazole-5-yl ether	1969 (Uniroyl)	Seed, soil	*Pythium, Phytophthora Rhizoctonia, Fusarium*	2000	'Terrazole' WP (350 g/kg), EC (250 or 440 g/l); 'AAterra' 35 WP; 'Koban' 25 EC; 'Pansoil' DP (40 g/kg) 'Truban' 30WP Mixtures: with thiophanate-methyl ('Ban-rot', WP), quintozene ('Terraclor Super X'; Terra-coat L 250'), or chlorothalonil ('Terradactyl')

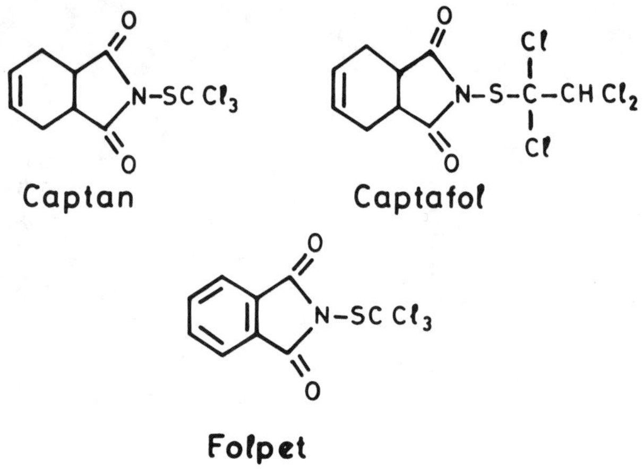

FIGURE 10.   Chemical structures of phthalimide fungicides.

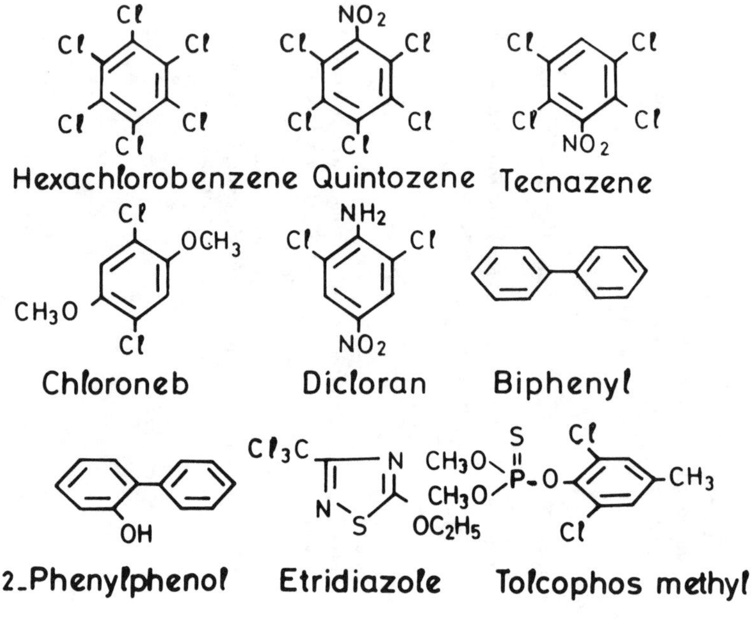

FIGURE 11.   Structure formula of aromatic hydrocarbon fungicides.

of S-adenosylmethionine.[34] They are also reported to interfere with the chitin biosynthesis. Exact mode of action of pyrazophos is not yet known. Alteration of membrane permeability due to an accumulation of free fatty acids and reduction of synthesis of phospholipids, inhibition of oxygen uptake or respiration, and reduced protein synthesis are reported to be associated with pyrazophos or PP pyrazophos.[32] Because of the numerous sites of action of the organophosphorus fungicides, only a few report on the development of resistance are known, e.g. in *Pyricularia oryzae* against IBP and edifenphos in field (after about 10 years of use) and in *Spaerotheca fuliginea* against pyrazophos in glasshouse.[32]

6. Benzimidazoles and Related Fungicides

Benzimidazoles and thiophanates (Figure 16) which are transformed to benzimidazoles represent a group of highly effective broad spectrum systemic fungicides (Table 16) which

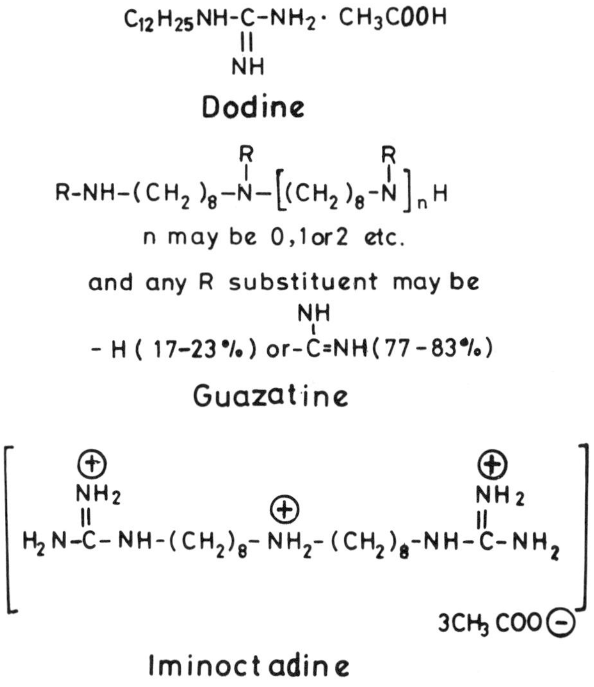

$$C_{12}H_{25}NH-\overset{\displaystyle \underset{\displaystyle NH}{\|}}{C}-NH_2 \cdot CH_3COOH$$

Dodine

$$R-NH-(CH_2)_8-\overset{\displaystyle \underset{}{R}}{\underset{}{N}}-\left[(CH_2)_8-\overset{\displaystyle \underset{}{R}}{\underset{}{N}}\right]_n H$$

n may be 0, 1 or 2 etc.

and any R substituent may be

$$-H\,(17-23\%)\ or\ -\overset{\displaystyle \underset{}{NH}}{C}=NH\,(77-83\%)$$

Guazatine

$$\left[H_2N-\overset{\displaystyle \underset{}{\overset{\oplus}{N}H_2}}{\overset{\|}{C}}-NH-(CH_2)_8-\overset{\oplus}{N}H_2-(CH_2)_8-NH-\overset{\displaystyle \underset{}{\overset{\oplus}{N}H_2}}{\overset{\|}{C}}-NH_2 \right] \quad 3CH_3COO^{\ominus}$$

Iminoctadine

FIGURE 12. Structure formula of guanidine fungicides.

are widely used for efficient plant disease control. Most Ascomycetes, some of the Basidiomyces and Deuteromycetes, and none of the Phycomycetes are sensitive to these fungicides (Table 17). Their mild cytokinin-like effects on some plants tend to retain chlorophyll and in some cases increase yield and delay maturity.[35] Effects of nontarget organisms are minimal because of selective toxicity and strong adsorption to plants and soil. Benzimidazoles are readily absorbed by the different plant parts and exhibit typical apoplastic translocation inside the plants. Their translocation is poor in woody plants due to their strong binding with the root and stem lignins.[13] At acidic pH in soil carbendazim gets protonated, and strongly adsorbed on soil particles/organic matter, resulting in poor uptake by plant roots. The major limitation of these compounds is high resistance risk. Exclusive use of these fungicides have resulted in serious control failures in fields in several cases due to development of resistant strains, which are as virulent and aggressive as wild type sensitive strains and show cross resistance to all benzimidazole and thiophanate fungicides.[35] In order to tackle the resistance problem these fungicides are now being used in various combinations with other fungicides (Table 16). Compatibility of benzimidazole wettable powder formulation with most other agricultural chemicals is good except with highly alkaline pesticides such as Bordeaux mixture or lime sulfur.

Benzimidazoles bind with the β-tubulin subunit of the microtubules of sensitive fungi and thereby inhibit formation of spindle and subsequently chromosomal separation during nuclear division. β-tubulins of resistant strains show less affinity for binding with these fungicides.[36] N- phenylcarbamate compounds like methyl N-(3,5-dichlorophenyl) carbamate (MDPC) and isopropyl N-(3,4-diethoxyphenyl) carbamate (diethofencarb) which bind with the α-tubulin subunit of microtubules exhibit negatively correlated cross resistance with benzimidazoles and thereby offer good promise for the use as a combination product for the management of benzimidazole resistant strains.

TABLE 12
Guanidine Fungicides in Commercial Use

Common name (Code No.)	Chemical name	Year of introduction (by)	Uses	Acute oral toxicity (LD$_{50}$ rat mg/kg)	Formulations
Dodine (CL 7521, AC 5223)	1-dodecylguanidinium acetate	1956 (American Cyanamid Co.)	a number of major fungal diseases of fruit, nut and vegetable crops like scab (apple, pear), leaf spot (*Mycosphaerella* spp.), blossom brown rot and leaf blight	1000	'Cyprex', 'Guanidol', 'Melprex', WP (650 or 800 g/kg), liquid (200 or 250 g/l), DP (750 g/kg) *Mixtures*: with dodemorph 'Badilin Rosenfluid) or nitrothal-isopropyl ('Pummel')
Guazatine (EM 379, MC 25)	A mixture of reaction-products comprising mainly octamethylene-diamine, iminodi (octamethylene) diamine, octamethylenebis (imino-octamethylene) diamine and carbamonitrile	1968 (Evans Medical Co.)	cereal seed treatment against *Septoria nodorum, Tilletia caries, Fusarium* and *Helminthosporium* spp.; dip for seed potato, sugarcane, citrus fruit; and spray against *Pyricularia oryzae* (rice), *Septoria* (wheat) *Cercospora* spp. (peanut and soybeans)	300	'Kenopel' SL, 'Mist-O-Matic Murbenine', LS; 'Panoctine', LS, DS. Mixtures: with imazalil 'Mist-O-Matic Murbenine plus' 'Panoctine Plus'; fenfuram ('Panoctine Super'); or imazalil + fenfuran ('Panoctine Universal', LS)
Iminoctadine (DF-125)	1,1'-iminodi (octamethylene) diguanidinium triacetate	1986 (Dainippon)	used as seed dressing for *Fusarium* spp., *Septoria nodorum* and *Tilletia caries*; post-harvest fruit dip in citrus against *Geotrichum candidum* and *Penicillium* spp.; foliar spray in cereals against *S. nodorum*	300—326	'Befran' liquid (25%), paste (3%)

Data collected from Worthing[23] and Gasztonyi and Lyr.[29]

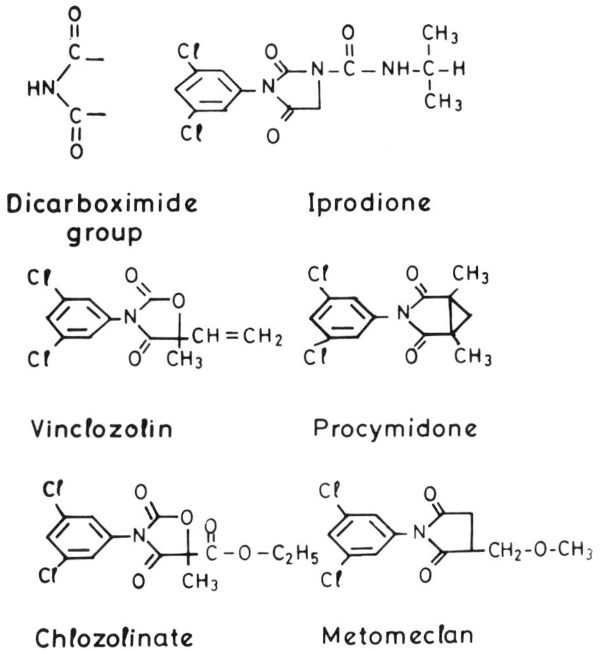

FIGURE 13. Chemical structures of dicarboxymide fungicides.

7. Carboxins and Related Fungicides

Amongst systemic fungicides carboxin and oxycarboxin (Figure 17) were first to be discovered and introduced for plant disease control.[37] These fungicides are readily absorbed by the seeds, roots and leaves and translocated apoplastically inside the plant. They are very effective mainly against Basidiomycetes-smuts, bunts and rusts of cereal grains, and soil fungus-*Rhizoctonia solani* (Table 18). The treatment of seeds or plants with these chemicals results in higher yield of crops not only because of disease control but also because of growth stimulation. Both carboxin and oxycarboxin have low animal toxicity and are quickly degraded in soil, plants and animals and leave no residue in crops.[38] Their mode of action is very specific. They interact with succinate-ubiquinone reductase complex (complex II) resulting in inhibition of oxidation of succinate via electron transport chain. So far, no widespread development of resistance of fungal pathogens to carboxin and oxycarboxin has been observed in the field.

Discovery of the fact that 1,4-oxathiin ring in carboxin is not essential either for fungitoxicity or systemicity prompted the development of several other systemic fungicides, with antifungal spectrum similar to carboxins, by substituting 1,4-oxathiin ring with dihydropyran, furan or benzene (Figure 17). All such compounds, which are currently in commercial use as fungicides, except probably pyracarbolid,[23] are listed in Table 18.

8. Sterol Biosynthesis Inhibiting Fungicides

This group includes numerous fungicides from chemically heterogenous classes. However, all these compounds inhibit ergosterol biosynthesis in fungi and therefore are classified as "ergosterol biosynthesis inhibitors" (EBIs) or "sterol biosynthesis inhibitors" (SBIs). They are effective against a wide range of fungal pathogens except Oomycetes which do not need sterol for growth and reproduction. Based on their exact site of action in ergosterol biosynthetic pathway SBIs are divided into two groups: (1) Δ^8-Δ^7 sterol isomerization and/or Δ^{14}-sterol reduction inhibitors. This group includes morpholines and piperidine fungicides,

TABLE 13
Dicarboximide Fungicides in Commercial Use

Common name (Code No.)	Chemical name	Year of introduction (by)	Acute oral toxicity LD_{50} rat mg/kg	Formulations
Iprodione (26 019 RP)	3-(3,5-dichlorophenyl)-N-isopropyl-2,4-dioxoimidazolidine-1-carboxamide	1974 (Rhone-Poulenc Phytosanitaire)	3500	'Rovral', WP (500 g/kg), 'Rovral' HN
Vinclozolin (BAS 352 F)	(RS)-3-(3,5-dichlorophenyl)-5-methyl-5-vinyl-1,3-oxazolidine-2,4-dione	1975 (BASF AG)	10000	'Ronilan', WP (500 g/kg) 'Ronilal FL', SC (500 g/l); Mixture: 'Silbos T', WP (100 g vinclozolin + 640 g thiram/kg)
Procymidone (S-7131)	N-(3,5-dichlorophenyl)-1,2-dimethylcyclopropane-1,2-dicarboximide	1976 (Sumitomo Chemical Co.)	6800	'Sumisclex', WP (500 g/kg); 'Sumilex', WP (500 g/kg), 'Sumisclex', HN (300 g/kg); 'Sumilex', HN (300 g/kg), Sumiclex, SP (250 g/kg)
Chlozolinate (M 8164)	Ethyl (±)-3-(3,5 dichlorophenyl)-5-methyl-2,4-dioxo-1,3-oxazolidine-5-carboxylate	1980 (Farmoplant S.p.A.)	>4500	'Serinal' 20 WP (200 g/kg), 'Serinal', 50 WP (500 g/kg)
Metomeclan (CO-6054)	1-(3,5-dichlorophenyl)-3-methoxymethyl-2,5-pyrrolidindione	1984 (Wacker)	10000	'Drawifol'

Data from Worthing[23] and Pommer and Lorenz.[30]

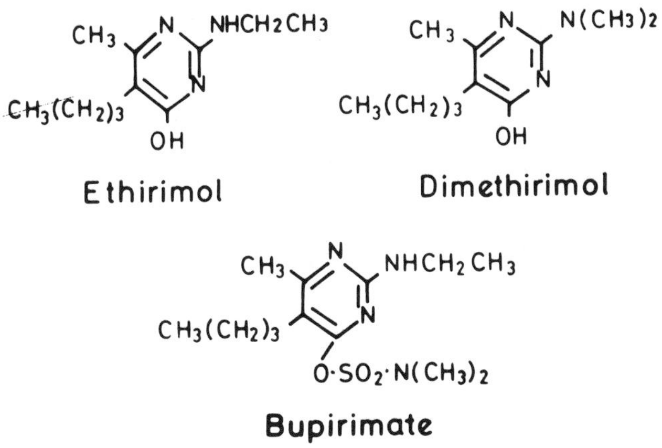

FIGURE 14. Chemical structures of 2-aminopyrimidine fungicides.

and (2) Δ^{14}- demethylation inhibitor (DMIs). This group includes chemically diverse compounds like piperazine, pyridine, pyrimidine, and azole fungicides.

a. Morpholine and Piperidine Fungicides

There are five morpholine and one piperidine (fenpropidin) compounds currently registered as fungicides (Figure 18). These fungicides were primarily introduced to control powdery mildew and/or rusts (Table 19). However, tridemorph and fenpropimorph exhibit a wide range of activity against pathogens belonging to the classes Ascomycetes, Basidiomycetes and Deuteromycetes.[39] Except fenpropimorph where only *cis* isomer is toxic rest of the morpholine fungicides are used as mixture of *cis-trans* stereoisomers. Among different morpholine fungicides, fenpropimorph shows good persistence, and eradicative, curative, and vapor phase activity. It exhibits good stop effect against powdery mildews. Fenpropidin behaves similar to fenpropimorpholine and exhibits strong secondary dispersion through the vapor phase.[39] As discussed earlier primary site of action of these compounds is the sterol biosynthesis; more specifically inhibition of enzymes, Δ^8-Δ^7 sterol isomerase and/or Δ^{14} sterol reductase. All the fungicides of this group are absorbed by the roots and are transported acropetally, through transpirational stream to the leaves. Morpholines are considered "safe" or "low-risk" fungicide from resistance point of view. Fungal strains cross resistant to DMIs show normal or even increased sensitivity towards morpholines. Potential of fenpropimorph and fenpropidin to check the build-up of resistant strains against DMIs particularly in cereal rusts and powder mildews, is being assessed under field conditions.

These fungicides have no adverse effect on soil microflora. However, morpholine fungicides may be phytotoxic because of their inhibitory effect on sterol biosynthesis in plants. Sometimes, phytotoxicity decides use of these fungicides on a particular crop.

b. Demethylation Inhibitors

This is the largest group of systemic fungicides, which includes compounds of diverse chemical structures (Figures 19 and 20) but surprisingly with similar mode of action. They block the enzyme cytochrome P-450 mixed function oxygenase involved in the removal of the sterol C-14 methyl group.[46] Since cytochrome P-450 is also involved in gibberellin biosynthesis, DMIs sometimes result in growth retardation of host plant. Different DMIs currently in use as fungicides are described in Table 20 and major plant diseases controlled by them are listed in Table 21.

Among different DMIs, triazoles are by far the most rapidly growing class of fungicides. A breakthrough in this field was the introduction of triadimefon in early 1970s. This com-

TABLE 14
2-Aminopyrimidine Fungicides

Common name (Code No.)	Chemical name	Year of introduction (by)	Used against	Acute oral toxicity (LD_{50} rat mg/kg)	Formulations
Ethirimol (PP 149)	5-butyl-2-ethylamino-6-methylpyrimidin-4-ol	1969 (ICI)	Powdery mildews of cereals	6340	'Milstem', LS (280 g/l), 'Milgo E' SC (280 g/l) 'Milcurb Super', EC (250 g/l) Mixtures: with captafol ('Junospor;, 'Milcap' SC), or flutriafol + thiabendazole ('Ferrax' seed treatment)
Dimethirimol (PP 675)	5-butyl-2-dimethyl-amino-6-methyl-pyrimidin-4-ol	1969 (ICI)	Powdery mildews of chrysanthemum, cineraria and cucurbits	2350	'Milcurb', SL (125 g/l) as hydrochloride
Bupirimate (PP 588)	5-butyl-2-ethylamino-6-methylpyrimidin-4-yl dimethylsulphamate	1975 (ICI)	Powdery mildews of woody (apple, vine, mango, peach, rose, currants, apricot, gooseberry) and herbaceous (hop, strawberry, cucurbits, peas, sugarbeet, etc.) plants	4000	'Nimrod', EC (250 g/l), WP (250 g/kg) Mixtures: 'Nimrod T' EC (62.5 g bupirimate + 62.5 g triforine/l)

Data from Worthing[23] and Hollomon and Schmidt.[31]

TABLE 15
Organophosphorus Fungicides in Commercial Use

Common name (Code No.)	Chemical name	Year of introduction (by)	Used against	Acute oral toxicity (LD$_{50}$) rat mg/kg	Formulations
			Phosphorothiolates		
Iprobenfos/IBP	S-benzyl O, O-di-isopropyl phosphorothioate	1966 (Kumiai)	Rice blast	600 (mice)	'Kitazin P', EC (480 g/l), DP (20 g/kg), GR (170 g/kg)
Edifenphos (Bayer 78418)	O-ethyl S,S-diphenyl phosphorodithioate	1965 (Bayer)	Rice blast	150—340	'Hinosan', EC (300, 400 or 500 g/l), DP (15, 20 or 25 g/kg)
			Phosphorothionates		
Pyrazophos (Hoe 02 873)	Ethyl 2-diethoxyphosphino-thioyloxy-5-methylpyrazolo [1,5-α] pyrimidine-6-carboxylate	1977 (Hoechst)	Powdery mildews of apple, cucumber, ornamentals, hops, vineyards and cereals	151—778	'Afugan', 'Curamil', WP (300 g/kg); 'Afugan', 'Missile' EC (295 g/l)
Toclofos-methyl	See aromatic hydrocarbon fungicides for detail				

Data from Worthing[23] and Schreiber.[32]

TABLE 16
Benzimidazoles and Related Fungicides

Common name (Code No.)	Chemical name	Year of introduction (by)	Acute oral toxicity (LD_{50} rat mg/kb)	Formulations
Benomyl (du Pont 1991)	methyl 1-(butylcarbamoyl) benzimidazol-2-ylcarbamate	1966 (du Pont)	>10000	'Benlate', WP (500 g/kg), 'Tersan' 1991 turf fungicide (500 g/kg) Mixture: 'Benlate T', WP (200 g benomyl + 200 g thiram/kg)
Carbendazim (BAS 346 F, Hoe 17411)	methyl benzimidazol-2-ylcarbamate	1967 (BASF, Hoechst, du Pont)	>15000	'Battal FL', 'Bavistin FL', SC (500 g/l); 'Bavistin' WP, 'Agrozim', WP (500 g/kg); 'Carbate', 'Derosal', WP (594 g/kg), SC (188 g/l); 'Derroprene FL', 'Focal', 'Maxim', SC (510 g/l); 'Stempor DG', WG (500 g/kg); 'Supercarb', 'Fungicide BLP' SL (7 g arbendazim phosphate/l) Mixtures: with maneb ('Bavistin M', 'Delsene M', 'Septal', WP), triadimefon ('Bayleton BM', 'Bayleton Total'), triadimefon + captafol ('Bayleton triple'), and with several other fungicides like Sulfur, triforine, chlorothalonil, fenpropimorph, fenarimol, mancozeb, diclobutrazol, propiconazole, flutriafol, etc.
Fuberidazol (Bayer 33172)	2-(2-furyl) benzimidazole	1968 (Bayer)	1100	'Voronit' seed treatment Mixtures: for seed treatment with triadimenol ('Baytan'), triadimenol + imazalil hydrogen sulphate ('Baytan IM'), nabam ('Neo-Voronit'), quintozene ('Voronit special') and anthraquinone
Thiabendazole (MK360)	2-(1,3-thiazol-4-yl) benzimidazole	1968 (Merck)	3300	'Mertect', 'Tecto', 'Storite' WP (400, 600 or 900 g/kg); SC (450 g/l); FT (7 g a.i.). Mixture: two-or three-way mixtures with several fungicides like fosetyl-AL, captan, metalaxyl, carboxin, ethirimol, flutriafol, tecnazene, thiram, bendiocarb, etc.

TABLE 16 (continued)
Benzimidazoles and Related Fungicides

Common name (Code No.)	Chemical name	Year of introduction (by)	Acute oral toxicity (LD_{50} rat mg/kb)	Formulations
Thiophanate (NF 35)	Diethyl 4,4'-(o-phenylene) bis (3-thioallophanate)	1970 (Nippon Soda)	>15000	'Topsin', 'Cercobin', 'Nemafax', WP (500 g/kg)
Thiophanate-methyl (NF 44)	Dimethyl 4,4'-o-phenylene) bis (3-thioallophanate)	1970 (Nippon Soda)	6640—7500	'Topsin M', 'Cercobin M', 'Mildothane', 'Cytosin' WP (500 or 700 g/kg) Mixture: with etridiazole ('Ban-rot'), thiram ('Homai'), maneb ('Labilit') and imazalil ('Mist-O-Matic Muridal') (all WP)

Data from Worthing[23] and Delp.[35]

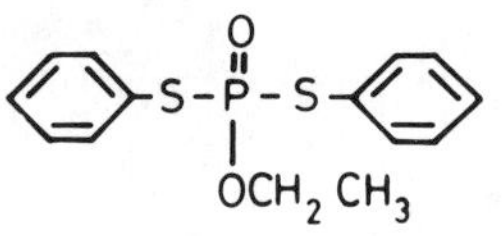

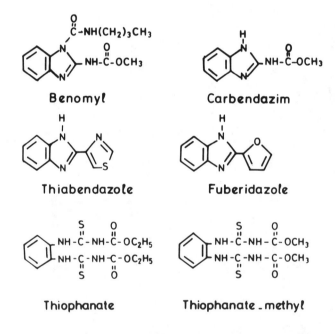

Edifenphos

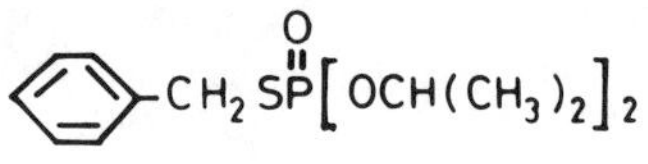

Iprobenfos

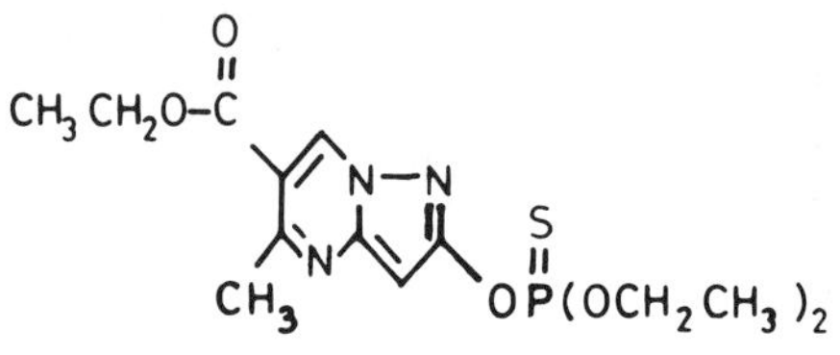

Pyrazophos

FIGURE 15.   Chemical structures of organophosphate fungicides.

Benomyl

Carbendazim

Thiabendazole

Fuberidazole

Thiophanate

Thiophanate - methyl

FIGURE 16.   Chemical structures of benzimidazoles and related fungicides.

pound combines a broad spectrum of fungicidal activity against important pathogens or cereals with excellent systemic properties and low application rates. Several imidazole compounds are being used in medical field as antimycotics.

All DMIs penetrate the plant cuticle (except PP 969, an experimental triazole fungicide), seed coat and root to at least some extent and therefore, are curative in action. However, their long distance apoplastic translocation may be absent (triflumizole), limited (buthiobate, fenarimol, prochloraz, bitertanol, etc.) or good (triforine, fenapanil, triadimefon, triadimenol, propiconazole, etaconazole, etc). Fenapanil exhibits ambimobility in soybean.[41] A marked vapor phase activity is exhibited by triadimefon and propiconazole. DMIs resistant strains of different powdery mildew fungi have been isolated from both *in vitro* culture and

TABLE 17
Major Practical Applications of Benzimidazoles and Related Fungicides[35]

Crops	Pathogens
Almond *(Prunus)*	*Monilinia* spp.
Banana *(Musa)*	*Colletotrichum musae, Fusarium* spp., *Mycosphaerella musicola, M.fijiensis* var. *difformis*, and *Penicillium* spp.
Beans *(Phaseolus)*	*Botrytis cinerea, Colletotrichum* spp., *Sclerotinia sclerotiorum* and *Cercospora* spp.
Carrot *(Daucus)*	*Cercospora carota*
Chestnut *(Castanea)*	*Colletotrichum castanea*
Citrus	*Botrytis cinerea, Colletotrichum gloeosporioides, Diplodia natalensis, Guignardia citricarpa, Mycosphaerella citri, Penicillium* spp., *Phomopsis citri*, and *Oidium tingitaninum*
Clove *(Eugenia)*	*Cylindrocladium quinqueseptatum*
Coffee	*Cercospora coffeicola, Colletotrichum coffeanum*, and *Pellicularia*
Cole crops, canola or rape *(Brassica)*	*Fusarium oxysporum, Mycosphaerella brassicola, Phoma lingam*, and *Sclerotinia sclerotiorum*
Cucurbits and melons *(Cucumis)*	*Cladosporium cucumerinum, Colletotrichum gossypii, Colletotrichum lagenarium, Erysiphe cichoracearum, Fusarium* spp., *Mycosphaerella citrullina, Oidium* spp., *Rhizoctonia solani*, and *Sclerotinia sclerotiorum*
Grape *(Vitis)*	*Botrytis cinerea, Gloeosporium ampelophagum, Guignardia bidwellii, Melanconium fuligenum, Pseudopeziza* sp., and *Uncinula necator*
Lettuce *(Lactuca)*	*Botrytis cinerea*, and *Sclerotinia sclerotiorum*
Mango *(Mangifera)*	*Colletotrichum gloeosporioides*, and *Oidium*
Mulberry *(Morus)*	*Phyllactinia* sp.
Olive *(Olea)*	*Cycloconium oleaginum*
Onion *(Allium)*	*Botrytis allii, Colletotrichum, gloeosporiodes*, and *Fusarium oxysporum*
Ornamentals and trees	*Ascochyta, Botrytis cinerea, Ceratocystis* spp., *Cercospora* spp., *Cylindrocladium, Oidium* spp., *Penicillium* spp., *Phomopsis* spp., *Phyllostictina* spp., *Ramularia* spp., *Rhizoctonia* spp., *Sclerotinia* spp., *Sphaerotheca pannosa.*
Papaya *(Carica)*	*Oidium caricae*
Peanut *(Arachis)*	*Ascochyta* spp., *Aspergillus* spp., *Cercospora arachidicola, Cerocosporidium personatum*, and *Mycosphaerella arachidicola*
Pea *(Pisum)*	*Mycosphaerella pinoides*, and *Oidium* spp.
Pepper *(Capsicum)*	*Botrytis cinerea, Fusarium piperi*, and *Fusarium solani*
Pineapple *(Ananas)*	*Fusarium moniliforme, Rhizoctonia* sp., and *Thielaviopsis paradixa*
Pome fruit *(Malus* and *Pyrus)*	*Botrytis cinerea, Cladosporium* spp., *Gloeodes pomigena, Gymnosporangium* spp., *Marssonina mali, Microthyriella rubi, Mycosphaerella pomi, Penicillium* spp., *Phyllactinia pyri, Physalospora* spp., *Podosphaera leucotricha, Rosellinia necatrix, Valsa ceratosperma*, and *Venturia* spp.
Potato *(Solanum)*	*Fusarium solani, Oospora* sp., *Phoma* sp., and *Rhizoctonia solani*
Rice *(Oryza)*	*Acrocylindrium oryzae, Cercospora oryzae, Gibberella fujikuroi, Pyricularia oryzae, Thanatephorus cucumeris*, and *Rhizoctonia solani* (stem rot)
Rubber *(Hevea)*	*Ceratocystis fimbriata, Microcyclus ulei*, and *Mycosphaerella* spp.
Soybean *(Glycine)*	*Cercospora kikuchii, C.sojina, Colletotrichum truncatum, Diaporthe* spp., *Phomopsis sojae*, and *Septoria glycinea*
Stone fruit *(Prunus)*	*Cladosporium carphophilum, Coccomyces hiemalis, Cytospora leucostoma, Fusarium* spp., *Monilinia* spp., *Oidium* spp., *Phomopsis persicae*, and *Sphaerotheca pannosa*
Strawberry *(Fragaria)*	*Botrytis cinerea, Cercospora* sp., *Fusarium oxysporum, Mycosphaerella fragariae, Odium* spp., *Sphaerotheca humuli*, and *Verticillium* spp.
Sugar beet *(Beta)*	*Cercospora beticola*
Sugar cane *(Saccharum)*	*Ceratocystis paradoxa*, and *Cercospora* spp.
Tea *(Thea)*	*Gloeosporium theae-sinensis, Elsinoe leucospila*, and *Rosellinia necatrix*
Tobacco *(Nicotiana)*	*Ascochyta nicotianae, Cercospora nicotianae, Helicobasidium mompa*, and *Pellicularia filamentosa*
Tomato *(Lycopersicon)*	*Botrytis cinerea, Cercospora* spp., *Cladosporium* sp., *Colletotrichum phomoides, Corynespora melongenae, Fusarium oxysporum, Odium* spp., *Phoma destricitiva, Sclerotinia sclerotiorum*, and *Septoria lycopersici*
Wheat *(Triticum)*	*Erysiphe graminis, Fusarium* spp., *Pseudocercosporella herpotrichoides, Rhynchosporium* sp. *Septoria avenae*, and *Ustilago tritici*

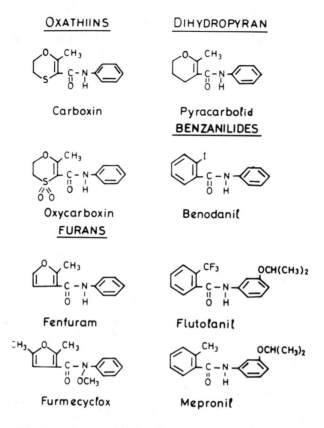

FIGURE 17.   Chemical structures of carboxins and related fungicides.

field. However, resistant strains usually exhibit poor parasite fitness and resistance associated serious control failures are yet to be encountered from field. Fungal strains resistant to one DMI-fungicide generally are cross- resistant to other DMI-fungicides but not to morpholines. Resistance to imazalil, fenarimol and triazoles is reported to be associated with energy dependent efflux of fungicides resulting in their low intracellular concentration.

9. Anti-Oomycetes Fungicides

The past decade has witnessed the introduction of five classes of fungicides controlling diseases caused by oomycetes: (1) the carbamates (e.g., prothiocarb, propamocarb), (2) the isoxazoles (e.g., hymexazol), (3) the cyanoacetamide-oximes (e.g., cymoxanil), (4) ethyl phosphonates (e.g., fosetyl-AS), and (5) phenylamides. The last class, phenylamides, covers three groups of compounds: (1) acylalanines (e.g., furalaxyl, metalaxyl, benalaxyl), (2) acylamino butyrolactones (e.g., ofurace), and (3) acylamino-oxazolidinones (e.g., cyprofuram, oxadixyl) (Figure 21, Table 22). All these fungicides show comparatively high water solubility and all are selective against Oomycetes except hymexazol and fosetyl which also show activity against pathogens other than Oomycetes (Table 23). Main practical uses and mode of application of these fungicides are listed in Table 24. So far metalaxyl is the most active, versatile and broadly used compound of this class. Its main biological features can be summarized as[47] (1) high inherent fungitoxicity, (2) protective and curative activity against all Peronosporales, (3) rapid uptake, high acropetal and limited basipetal systemicity, (4) protection of new growth, (5) good persistence in plant tissue (extended spray interval), (6) control of systemic seed- and soil-borne diseases, and (7) weak on old plant tissue (senescence).

The antioomycetes fungicides show a wide variation in their systemicity which is illustrated by cymoxanil, prothiocarb, propamocarb and hymexazol with only locally systemic activity, by the phenylamide with excellent apoplastic and limited symplastic (e.g., metalaxyl)[9,15,18-19] transport and by fosetyl-AL with excellent symplastic and good apoplastic transport.[47] The symplastic translocation or the activity of fosetyl-AL appears to be based on the transport of its metabolite H_3PO_3.[48]

Phenylamides interfere with the activity of RNA polymerase I; thereby inhibit the synthesis of vRNA.[36] Hymexazol and cymoxanil also inhibit RNA synthesis but exact site of action is not yet clear. No information is available regarding mode of action of other fungicides.[47]

Of all the antioomycetes fungicides only phenylamides have met with serious resistance problems in the field and resistance development was due to altered target site. Field resistance to phenylamides has developed shortly after their introduction in cases of intensive, continuous, and exclusive use of the straight product, mainly in *Phytophthora infestans, Plasmopara viticola* and *Peronospora tabacina*. For all practical purposes cross- resistance exists among phenylamides even though the resistance factors can vary in some strains.[36] Parasite fitness of metalaxyl resistant strains is as good as wild type strains. Cymoxanil, fosetyl-AL, and a number of protectant fungicides are used in mixture with phenylamides to control resistance problem. Cymoxanil is reported to act synergistically with metalaxyl.[49]

Application of metalaxyl to control white rust and downy mildews of *Brassica campestris* and *B. juncea* enhanced the infection of *Alternaria brassicae* probably due to stimulatory effect of fungicide on spore germination and growth of the pathogen.[50] Fosetyl-AL and metalaxyl are reported to stimulate host resistance by inducing the phenolic and phytoalexin synthesis and accumulation.[51]

10. Some Miscellaneous Selective Fungicides

There are several miscellaneous fungicides (Figure 22, Table 25) belonging to diverse chemical groups which show at least some degree of selectivity in their antifungal spectrum. This selectivity is surprising particularly for the compounds like anilazine, fenitropan, chinomethionate, dichlofluanid, and chlorothalonil, which interact with $-NH_2$ and/or $-SH$ groups and hence, their mode of action is multisital. Probably in such compounds selectivity is caused by differences in degradation or penetration in the fungal cell. After enzymatic hydrolysis dinocap and binapacryl give rise to 2,4-dinitrophenol which is probably responsible for their activity as uncoupler of oxidative phosphorylation.[29] All the fungicide listed in Table 25 and nonsystemic except pyroxychlor, which is ambimobile inside the plant system. Anilazine and chinomethionate are sensitive to photodegradation. Anilazine is rapidly degraded in soil. It acts synergistically with Cu^{++} and Zn^{++}. Drazoxolon is incompatible with lime S and dodine and binapacryl with organophosphorus and alkaline preparations. Development of resistant strains are reported against anilazine (*Sclerotinia homoeocarpa*) and butylamine (*Penicillium digitatum*) but only after several years of their use.[29]

11. Antipenetrants

There are several compounds which show no or poor fungitoxicity under *in vitro* conditions but prevent infection by interfering with the penetration process. These are termed as 'antipenetrants'. The action of antipenetrants might be directed at one or more of the following.[53,54]

1.   Inhibition of synthesis or activity of penetration enzymes, particularly cutinases. Certain organophosphorus fungicides are potent cutinase inhibitors. Apart from its antimitotic activity benomyl also acts as cutinase inhibitor. It is probably its breakdown product, butylisocyanate, rather than carbendazim, which act as anticutinase. As it

TABLE 18
Carboxins and Related Fungicides in Commercial Use

Common Name (Code No.)	Chemical name	Year of introduction (by)	Main uses	Acute oral toxicity (LD_{50} rat mg/kg)	Formulations
Oxathiines					
Carboxin (D 735)	5,6-dihydro-2,-methyl-1,4-oxathi-ine-3-carboxanilide	1966 (Uniroyal)	seed treatment of cereals against bunts and smuts; against *Rhizoctonia* spp. on cotton, groundnut and vegetables; *Exobasidium vexans* (tea), *E. vaccinii* (blueberry); *Helminthosporium* spp. (barley); combination product with other fungicides for most other seed-borne and soil-borne seedling diseases	3820	'Vitaflow', FS (400 g/l); WP or SC; 'Vitavax', WP (750 g/kg), LS (340 g/l), dust (100 g/kg). Mixtures: with thiabendazol ('Cerevax'), imazalil + thiabendazole ('Cerevax Extra'), phenyl-mercury acetate ('Mist-O-Matic Murganic'; 'Murganic'), anthroquinone + oxine-copper ('Quinolate'), gamma-HCH ('Vitavax 750L'), thiram, or captan + maneb
Dihydropyran					
Oxycarboxin (F 461)	5,6,dihydro-2-methyl-1,4-oxathiine-3-carboxanilide 4,4-dioxide	1966 (Uniroyal)	rust diseases of cereals, ornamentals, vegetables, coffee, sunflower, bean, safflower, flax etc.	2000	'Plantvax' EC (200 g/l), WP (750 g/kg)
Benzanilides					
Pyracarbolid (Hole 13764)	3,4-dihydro-6-methyl-2 *H*-pyran-5-carboxanilide	1970 (Hoechst)	effective against rusts, smuts and *Rhizoctonia solani*, blister blight of tea etc.; currently not much in commercial use		'Sicarol', WP (500 g/kg)
Benodanil (BAS 3170 F)	2-iodobenzanilide	1974 (BASF)	*Puccinia striiformis* (wheat and barley), *P. hordei* (barley), rust diseases of coffee, ornamentals and vegetables	>6400	'Calirus', WP (500 g/kg)

TABLE 18 (continued)
Carboxins and Related Fungicides in Commercial Use

Common Name (Code No.)	Chemical name	Year of introduction (by)	Main uses	Acute oral toxicity (LD_{50} rat mg/kg)	Formulations
Flutolanil (NNF-136)	α,α,α-trifluoro-3'-isopropoxy-o-toluanilide	1982 (Nihon Nohyaku)	against some basidiomycetes	>10000	'Moncut'
Mepronil	3'-isopropoxy-o-toluanilide	1981 (Kumiai)	against basidiomycetes, foliar application-*Rhizoctonia solani* (rice), *Gymnosporangium fusicim* (pears) and *Puccinia chrysanthemi* (chrysanthemums); soil or seed treatment-*Thanatephorus cucumeris* (potato) and *R. solani* (vegetables)	10000	'Basitac', WP (750 g/kg), DP (30 g/kg), SC (400 g/kg)
			Furans		
Fenfuram (WL 22361)	2-methyl-3-furanilide	1974 (Shell Research Ltd.)	as seed dressing against smuts and bunts of temperate cereals	12900	'Pano-ram', DS, LS Mixtures: with guazatine acetates ('Panoctine Super', LS), guazatine acetate + imazalil ('Panoctine Universal', LS), guazatine acetate + imazalil + gamma-HCH ('Panoctine AT Universal') or imazalil + anthraquinone + gamma-HCH
Furmecyclox (BAS 389F)	methyl N-cyclohexyl-2,5-dimethylfuran-3-carbohydroxamate	1977 (BASF)	seed treatment for cereals, cotton (against *Rhizoctonia solani*), potato and other crops; also to protect wood against fungal decay	3780	'Campogran', 'Xyligen B', wet treatment (500 g/l)

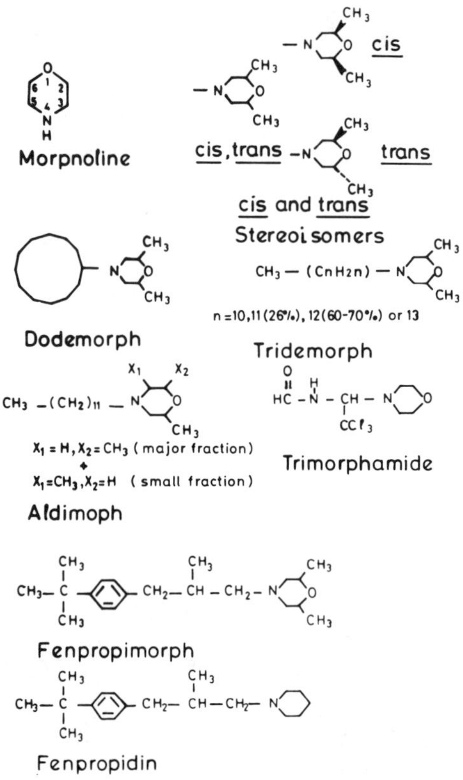

Morpholine

cis

cis, trans   trans

cis and trans
Stereoisomers

$CH_3 - (C_nH_{2n}) - N$

n = 10,11 (26%), 12 (60-70%) or 13

Dodemorph

Tridemorph

$CH_3 - (CH_2)_{11} - N$

$X_1 = H, X_2 = CH_3$ (major fraction)
+
$X_1 = CH_3, X_2 = H$ (small fraction)

Aldimoph

Trimorphamide

Fenpropimorph

Fenpropidin

FIGURE 18.   Structure formula of morpholines and related fungicides.

has been shown that fungicide benomyl, but not carbendazim, protects pea stem from infection by *Fusarium solani*, a cutinase producer.[52] Other potent cutinase inhibitor, DFP (diisopropylflurophosphate), which protects papaya fruits against infection by cutinase producing pathogen, *Colletotrichum gloeosporioides*, is not registered as a fungicide.[53]

2.   Inhibition of the induction of appressorial development. Although solid surfaces, plant exudates, or nutrients like K^+ are reported to induce/interfere with appressoria development, no specific chemical is available which can be used to block this process.

3.   Interference with appressorial structure or function. Melanin biosynthesis inhibitors (MBI) do the same. Although several MBI-compounds are known (Figure 23) only three (fthalide, tricyclazole, and pyroquilon) are registered as fungicides (Table 26). Chlobenthiazone (4-chloro-3-methylbenzothiazol-2 (3H)-one) (Figure 23) is recently developed antipenetrant having activity resembling that of tricyclazol and pyroquilon, but it is not yet in practical use. These three compounds are effective against rice blast when used as submerged (seed soil, or root) or foliar application, while fthalide is effective only when used as foliar application (Table 26). All three MBIs registered as fungicides are systemic in nature with predominantly apoplastic translocation. Pyroquilon is readily absorbed by rice roots and seed against concentration gradient.[20] It is translocated to leaves where it gets accumulated at tips and margins. It may also get diffused/excreted out to the leaf surface alongwith leaf exudates.[55] Presently all MBIs are being used against rice blast only however, they also show good activity against *Colletotrichum* spp.[54] In field MBIs provide protection to the plant for a very long period. Seed treatment with pyroquilon protects rice seedlings for about 50 d. Similarily only two sprays of tricyclazole protect rice crop for whole duration.

TABLE 19
Morpholine and Related Fungicides in Commercial Use

Compound (Code no.)	Chemical name	Year of introduction (by)	Main use	Acute oral toxicity (LD_{50} rat mg/kg)	Formulations
Dodemorph (BAS 238F)	4-cyclododecyl-2,6-di-methylmorpholine	1965 (BASF)	powdery mildew on ornamentals	1800	used as dodemorph acetate in commercial EC formulations e.g. 'Meltatox', EC (400 g/l). Mixture: 'Badilin Rosenfluid' (280 g dodemorph + 58 g dodine /l)
Tridemorph (BAS 200F, BAS 220F)	2,6-dimethyl-4-tridecyl-morpholine	1969 (BASF)	powdery mildew on cereals (barley), sigatoka disease of banana	825	'Bardew', 'Calixin', EC (750 g/l). Mixture: 'Cosmic', WP (94 g tridemorph + 38 g MBC + 400 g maneb/kg); 'Multi-B', pack for tank mix (tridemorph + mancozeb + maneb); 'Tilt Turbo 375', EC (250 g tridemorph + 125 g propiconazole/l)
Fenpropimorph (BAS 421F)	($\pm$) - cis-4-[3-(4-tert-butylphenyl)-2-methyl-propyl] 2,6 dimethyl-morpholine	1980 (BASF/Maag)	powdery mildew and rust on cereals (wheat and barley)	3515	'Corbel', 'Mistrol', EC (750 g/l). Mixture: 'Corvet CM', WP (188 g fenpropimorph + 50 g MBC + 400 g mancozeb/kg)
Aldimorph	N-n-dodecyl-2,5/2,6-di-methylmorpholine	1980 (Fahlberg-List)	powdery mildew on cereals (barley)	3500	'Fahmorph', EC (662 g/l)
Trimorphamide (VUAg T-866, BAS 463F)	N-2,2,2-trichloro-1-(4-morpholinyl)-ethylformamide	1981 (Vysk. Utsav Agrochem. Technol.)	powdery mildew on grapes and apple	2820	'Fademorph' EC (200 g/l)
Fenpropidin (Ro 12-3049)	(RS)-1-[3-(4-tert-butyl-phenyl)-2-methylpro-pyl] piperidine	1986 (ICI/Maag)	powdery mildew and rust on cereals (wheat and barley)	1800	'Patrol', EC (750 g/l)

Data from Worthing,[23] Pommer,[39] and Bohnen et al.[40]

TABLE 20

Sterol-Biosynthesis Inhibiting Fungicides in Commercial Use

Common name (Code No.)	Chemical name	Year of introduction (by)	Acute oral toxicity (LD_{50} rat mg/kg)	Formulations
			Piperazines	
Triforine (Cela W524)	*N,N'*-[piperazine-1,4-diyl-bis[(trichloromethyl) methyl-ene]] diformamide	1969 (Cela-merck)	16000	'Saprol', EC (200 g/l); 'Funginex', EC (190 g/l) Mixtures: with carbendazim ('Brolly', EC), bupirimate ('Nimrod T', EC)
			Pyridines	
Buthiobate (S-1358)	butyl 4-*tert*-butyl-benzyl *N*-(3-pyridyl) dithiocarbonimidate	1975 (Sumitomo)	3200—4400	'Denmert', EC (100 g/l), WP (200 g/kg)
Pyrifenox (Ro 15-1297)	2',4'-dichloro-2-(3-pyridyl)-aceto-phenone *O*-methyloxime	1986 (Maag)	2900	'Dorado', EC (200 g/l), WP (250 g/kg) Mixtures: with captan ('Rondo', WP), sulphur or man-cozeb
			Pyrimidines	
Fenarimol (EL-222)	(±)-2,4'-dichloro-α-(pyrimi-din-5-yl) benzhydryl alcohol	1975 (Elanco)	2500	'Bloc', 'Rimidin', 'Rubigan' as EC, SC and WP Mixtures: 'Rumidine Plus', WP (16 g fenarimol + 80 g carbendazim + 640 g maneb/kg)
Nuarimol (EL-228)	(±)-2-chloro-4'-fluoro-α-(py-rimidin-5-yl) benzhydryl alco-hol	1975 (Elanco)	1250	'Trimidal', 'Triminol', 'Gauntlet', 'Murox' as EC, SC and SL Mixtures: 'Trimisem Total', WS (65 g nuarimol + 165 g anthraquinone + 165 g gamma − HCH + 265 g maneb); 'Trimidal GT'
			Imidazoles	
Imazalil ('R 23979' for base, 'R 27180' for hydrogen sulphate, 'R	allyl 1-(2,4 dichlorophenyl)-2-imidazol-1-ylethyl ether	1972 (Janssen)	320	'Florasan R' EC (200 g/l); 'Fungaflor'; EC (200, 500 or 700 g/l); imazalil hydrogen sulphate, SP (750 g base/kg)

18531' for nitate)

				Mixtures: 'Baytan IM' (33 g imazalil (as hydrogen sulphate) + 30 g fuberidazole + 250 g triadimenol/kg) 'Sisthane'
Fenapanil (RH-2161)	(±)-2-(imidazol-1-yl-methyl)2-phenyl-hexanenitrile	1978 (Rohm & Hass)	1600—2400	
Prochloraz (BTS 40542)	1-N - propyl-N-[2-(2,4,6-trichlorophenoxy) ethyl] carbamoylimidazole	1977 (FBC/Schering)	1600	'Sportak', EC (400 or 450 g/l), WP (500 g/kg); seed treatment and liquid (200 g/l) Mixtures: with manganese ('Sporgon') or carbendazim ('Sportak Alpha')
Triflumizole (NF-114)	(E)-4-chloro-α,α,α-trifluro-N-(1-imidazol-1-yl-2-propoxy-ethylidene)-o-toluidine	1983 (Nippon Soda/Uniroyal)	695—715	'Trifmine', WP (300 g/kg), 'Trisosol'; 'Trifludol'

Triazoles

Fluotrimazole (BAY BUE 0620)	1-(3-trifluoromethyltrityl)-1H-1,2,4-triazole	1973 (Bayer)	>5000	'Persulon', WP (500 g/kg), EC (125 g/l)
Triadimefon (BAY MEB 6447)	1-(4-chlorophenoxy)-3,3-dimethyl-1-(1H-1,2,4-triazol-1-yl) butanone	1973 (Bayer)	363—568	'Bayleton', WP (50 or 250 g/kg), EC (100 g/l), DP (10 g/kg) Mixtures: with carbendazim ('Bayleton BM'; 'Bayleton Total', WP), captafol ('Bayleton CF'), captafol + carbendazim ('Bayleton triple') or cymoxanil + propineb (Diametam B)
Triadimenol (BAY KWG 0519)	1-(4-chlorophenoxy)-3,3,dimethyl-1-(1H-1,2,4-triazol-1-yl)butan-2-ol (mixture of two diastereoisomers)	1977 (Bayer)	700—1500	'Baytan 15', WP (150 g/kg); 'Bayfidan' (250 g/l) Mixtures: with fuberidazol ('Baytan'), fuberidazol + imazalil ('Baytan I', seed treatment
Bitertanol (BAY KWG 0599)	all-rac-1-(biphenyl-4-yloxyl)-3,3-dimethyl-1-(1H-1,2,4-triazol-1-yl) butan-2-ol	1978 (Bayer)	>5000	'Baycor' WP (250 g/kg), EC, AE and dry seed treatment
Propiconazole (CGA 64250)	(±)-1-[2-(2,4-dichlorophenyl)-4-propyl-1,3-dioxolan-2-ylmethyl]-1H-1,2,4-triazole	1979 (Ciba-Geigy)	1517	'Tilt', EC (100 or 250 g/l), 'Tilt', SL (125 g/l) Mixture: with carbendazim ('Hispor' 45; 'Tilt CB' 45, WP,; 'Tilt C' 275, SC), captafol ('Tilt CF' 72.5, WP) or tridemorph ('Tilt Turbo 375', EC)
Etaconazole (CGA 64251)	(±)-1-[2-(2,4-dichlorophenyl)-4-ethyl-1,3-dioxolan-2-ylmethyl] 1H, 1,2,4-triazole	1979 (Ciba-Geigy)	1343	'Sonax 1', 'Vangard 1', WP (100 g/kg) Mixture: 'Sonax C 52', WP (20 g etaconazole + 500 g captan/kg)

TABLE 20 (continued)
Sterol-Biosynthesis Inhibiting Fungicides in Commercial Use

Common name (Code No.)	Chemical name	Year of introduction (by)	Acute oral toxicity (LD_{50} rat mg/kg)	Formulations
Penconazole (CGA 71819)	1-(2,4-dichloro-β-propylphenethyl)-1H-1,2,4-triazole	1983 (Ciba-Geigy)	2125	'Award', WP (100 g/kg) 'topas' 100 EC (100 g/l), 'Topas' 1 EC Mixture: with captan ('Onmex', 'Topas C 50', WP) or mancozeb ('Topas MZ', WP)
Diclobutrazol (PP 296)	(2 RS, 3RS)-1-(2,4-dichlorophenyl)-4,4-dimethyl-2-(1H-1,2,4-triazol-1-yl) pentan-3-ol	1979 (ICI)	4000	'Vigil', SC (125 g/l) Mixtures: with carbendazim ('Vigil K', SC) or captafol ('Vigil T', SC)
Flutriafol (PP 450)	(RS)-2,4'-difluoro-α-(1H-1,2,4-triazol-1-ylmethyl) benzhydryl alcohol	1983 (ICI)	1140—1480	'Impact' (125 g/l) Mixtures: with captafol ('Impact T', liquid; 'Impact TP', WP), carbendazim ('Early Impact', 'Impact R'; liquid), or with ethirimol + thiabendazole ('Ferrax' liquid)
Flusilazol (DPX H 6573)	bis (4-fluorophenyl) methyl-(1H-1,2,4-triazol-1-ylmethyl) silane	1984 (du Pont)	674—1110	'Nustar', 'Punch'
(BAS 45 406F)	1-(2,4-dichlorophenyl)-2-(1H-1,2,4-triazol-1yl) ethanon-O-(phenylmethyl)-oxim	1983 (BASF)	4640	
Diniconazol (S-3308L,XE-779L)	(E-(±)-p[(2,4-dichlorophenyl)-methylene]-α-(-(1,1-dimethylethyl)-1H,1,2,4-triazol-1-ethanol	1983 (Sumitomo/Chevron)	474—639	'Sumi-8', 'Spotless,' ortho spotless
(PP-969)	(5 RS, 6 RS)-6-hydroxy-2,2-7,7-tetramethyl-5-(1H-1,2,4-triazol-1-yl) octan-3-one	1983 (ICI)		
Myclobutanil-(RH-3866)	2-p- chlorophenyl-2-(1H-1,2,4-triazol-1-ylmethyl) hexanenitrile	1987 (Rohm & Hass)	1600—2229	'Systhane', WP (400 g/kg), EC (250 g/l)
Hexaconazole (PP-523)	(RS)-2-(2,4-dichlorophenyl)-1-(1H-1,2,4-triazol-1-yl) hexan-2-ol	1986? (ICI)	2189—6071	'Anvil', SC (50 g/l), SG (50 g/kg)

| (BAY HWG 1608) | (RS)-1-(4-chlorophenyl)-4-4-di-methyl-3-(1H-1,2,4-triazol-1-ylmethyl) pentan-3-ol | 1986 (Bayer) | >3933 | 'Raxil', DS; 'Folicur', WS, FS, WP, EC, EW |
| (SAN 619 F) | (2 RS, 3RS)-2-(4-chloro-phenyl)-3-cyclopropyl-1-(1H-1,2,4-triazol-1-yl) butan-2-ol | 1986 (Sandoz) | 1020—1330 | 100 SL, 10 WP, 10 WG, 40 WP, 40 WG and also mixtures with other fungicides |

Data from Worthing,[23] Scheinpflug and Kuck,[41] Gisi et al.,[42] Reinecke et al.,[43] and Shephard et al.[44]

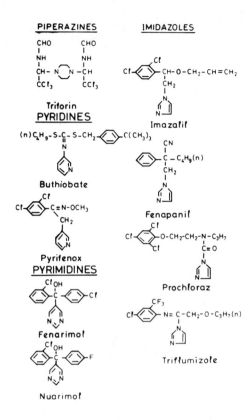

FIGURE 19.   Chemical structures of piperazine, pyridine, pyrimidine, and imidazole fungicides.

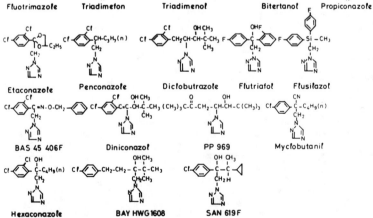

FIGURE 20.   Chemical structures of triazole fungicides.

TABLE 21
Activity Spectrum of Sterol Biosynthesis
Inhibitors for Pathogens of Major Crops[45]

Crop	Pathogens
Cereals	
Stems, leaves	*Erysiphe graminis*
	Puccinia spp.
	Rhynchosporium secalis
	Septoria spp.
	Pyrenophora teres
	P. tritici-repentis
	Typhula incarnata
	Pseudocercosporella spp.
Seed	*Ustilago* spp.
	Tilletia spp.
	Gerlachia nivalis
	Pyrenophora teres
	Septoria spp.
Apples, pears, stone fruits	*Venturia inaequalis*
	V. pirina
	Podosphaera leucotricha
	Gymnosporangium spp.
	Monilinia spp.
	Taphrina deformans
Grapes	*Uncinula necator*
	Guignardia bidwellii
Bananas	*Mycosphaerella* spp.
	Guignardia musae
Peanuts	*Mycosphaerella* spp.
	Puccinia arachidis
Coffee	*Hemileia vastatrix*
Tea	*Exobasidium vexans*

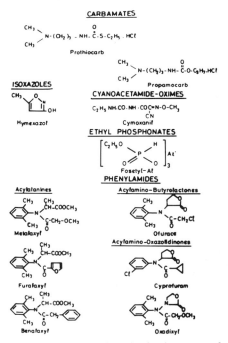

FIGURE 21. Structure formula of anti-oomycetes fungicides.

TABLE 22
Anti-Oomycetes Fungicides in Commercial Use

Common name (Code no.)	Chemical name	Year of introduction (by)	Acute oral toxicity (LD_{50} rat mg/kg)	Formulations
Carbamates				
Prothiocarb (SN 41703)	S-ethyl (3-dimethyl-amino-propyl) thio-carbamate	1974 (Schering AG)		'Previcur' s 70
Propamocarb (SN 66 752)	Propyl 3-(dimethyl-amino) propyl carbamate	1978 (Schering AG)	2000—8550	'Previcur N', SL (aqueous) (772 g propamocarb hydrochloride/kg); 'Ba ol' (665 g propamocarb hydrochloride/kg); Filex
Isoxazoles				
Hymexazole (F-319,SF-6505)	5-methylisoxazol-3-ol	1974 (Sankyo)	3909—4678	'Tachigaren', EC (30 g/l), SD (700 g/kg), DP (40 g/kg)
Cyanoacetamideoximes				
Cymoxanil (DPX-3217)	1-(2-cyano-2-methoxyimi-noacetyl)-3-ethylurea	1976 (du Pont)	1196—1390	'Curazole' WP *Mixtures:* 'Curazole M', Fytospore' (cymaxanil + mancozeb), 'Pulsan', Ripost M' (32 g cymoxanil + 560 g mancozeb + 80 g oxadixyl); several other formulations with mancozeb, captafol, folpet, zineb, and copper fungicides
Ethyl phosphonates				
Fosetyl (LS 74783)	Ethylhydrogen phosphonate, aluminum salt	1977 (Rhone-Poulenc)	5800	'Aliette', WP (800 g/kg) *Mixtures:* 'Aliette Extra', 'Hy-Cote', WS (fosetyl-Al + Captan + thiabendazole); 'Hy-Tona' WS (fosetyl-Al + bendiocarb + captan + thiabendazole); 'Mikal', WP (500 g fosetyl-Al + 250 g folpet/kg); fosetyl-Al + mancozeb
Phenylamides Acylalanines				
Metalaxyl (CGA 48988)	Methyl N-(2-methoxy-acetyl)-N-(2,6-xylyl)-DL-alaninate	1977 (Ciba-Geigy)	669	'Ridomil' 5G, GR (50 g/kg); 'Apron SD 35', (350 g/kg); 'Apron FW 350', FS (350 g/l). *Mixtures:* with folpet ('Acylon Super F', 'Ridomil Combi'), Captan ('Apron 70 SD', WS), mancozeb ('Fubol 58', 'Ridomil MZ', WP), maneb ('Ridomil

M'), carbendazim (Ridomil mbc', WP), thiabendazole ('Apron T 69', WS) and copper oxychloride ('Ridomil plus')

Common name (code)	Chemical name	Year (Company)	LD50	Formulations / Mixtures
Furalaxyl (CGA 38140)	Methyl N-(2-furoyl)-N-(2,6-xylyl)-DL-alaninate	1977 (Ciba-Geigy)	940	'Fongarid', WP (250 or 500 g/kg), GR (50 g/kg)
Benalaxyl (M 9834)	Methyl N-phenylacetyl-N-2,6-xylyl-DL-alaninate	1981 (Farmoplant S.P.A.)	4200	'Galben' 2E, EC (240 g/l), 25 WP (250 g/kg), 5 G, GR (50 g/l kg); *Mixtures:* with mancozeb ('Galben M') copper oxychloride ('Galben C', WP), folpet ('Golben F', WP) and zineb ('Galben Z', WP)
Acylaminobutyro lactones				
Ofurace (Ortho 20615)	(±)-α-2 chloro-N-2,6-xylyl-acetamido-γ-butyrolactone	1978 (Chevron)	2600—3500	*Mixtures:* with folpet ('Caltan', SC, FS), captafol + folpet ('Caltan C'), mancozeb + Zineb (Patafol Plus', WP)
Acylamino-Oxazolidinones				
Cyprofuram (SN 78314)	(±)-α-[N-(3-chlorophenyl) cyclopropanecarboxamido]-γ-butyrolactone	1982 (Schering AG)	174	*Mixtures:* with mancozeb ('Stanza', 'Vinicur M-SC'), folpet ('Vinicur F 50-SC')
Oxadixyl	2-methoxy-N-(2-oxo-1,3-Oxazolidin-3-yl) acet-2', 6'-xylidide	1983 (Sandoz)	1860	'Sandofan' 7·5 FG, GR (75 g/kg) *Mixtures:* with cymoxanil + mancozeb ('Pulsan', 'Ripost M'), mancozeb ('Pulsan T', 'Recoil', 'Sandofan M' and 'Sandofan M8', folpet ('Sandofan F') and copper fungicides (Sandofancopper)

Data from Worthing[23] and Schwinn and Staub.[47]

TABLE 23
Activity Spectrum of New Oomycetes Fungicides[47]

Common name	Pathogen on root/stem	Foliar pathogens	Additional activity against nonoomycete pathogens
Hymexazol	*Aphanomyces* *Pythium*		*Corticium sasaki* *Fusarium* spp.
Propamocarb/ Prothiocarb	*Aphanomyces* *Pythium* *Phytophthora*	*Bremia* *Peronospora* *Pseudoperonospora*	
Cymoxanil	—	*Phytophthora* *Plasmopara* *Peronospora* *Pseudoperonospora*	
Fosetyl	*Phytophthora* (*Pythium*)	*Bremia* *Plasmopara* *Pseudoperonospora*	*Phomopsis viticola* *Guignardia bidwellii* *Pseudopeziza tracheiphila*
Phenylamides	*Peronosclerospora* *Phytophthora* *Pythium* *Sclerospora* *Sclerophthora*	*Albugo* *Bremia* *Peronospora* *Peronosclerospora* *Phytophthora* *Plasmopara* *Pseudoperonospora* *Sclerospora* *Sclerophthora*	

TABLE 24
Main Practical Uses of The New Oomycetes Fungicides

Compound	Main uses against	Main crops	Application method
Prothiocarb/propanocarb	Diseases of roots and stem	Ornamentals vegetables	Drench
Hymexazol	Diseases of roots and stem in seedling stage	Rice, sugarbeets carnations	Drench, seed dressing, dust
Furalaxyl	Diseases of roots and stems	Ornamentals	Drench
Cymoxanil	Foliar diseases	Grapes, potato	Spray
Fosetyl	Foliar, stem and root diseases	Grapes, avocado, pineapple, citrus, ornamentals	Spray, drench, dip, injection
Metalaxyl and related compounds	Foliar, stem and root diseases	Grapes, potatoes, avocado, pineapples, citrus, tobacco, hops, maize, sorghum, millet, brassicas, lettuce, sunflowers	Spray, drench, dip, granules, seed dressing

Modified from Schwinn and Staub.[47]

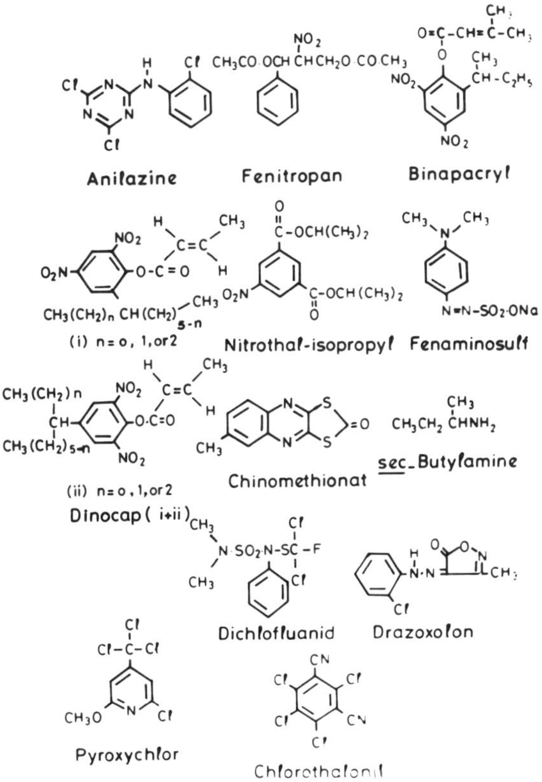

FIGURE 22. Chemical structures of some miscellaneous selective fungicides.

The antibiotic validamycin is effective primarily for the control of diseases caused by *Rhizoctonia* type fungi. It is widely used in Japan to control sheath blight of rice, caused by *R. solani*. Trinci[56] suggested that it may control disease by preventing the pathogen from penetrating the host or by reducing the rate of spread of the pathogen in the host so that enough time is available for the host to mobilize its defense system. As such it is not fungitoxic to *R. solani*.

12. Host-Defense Inducing Compounds

Various chemicals which have been found to trigger host resistance mechanism and by doing so provide protection against diseases on prior or sometimes simultaneous application with the pathogen. They include both chemical constituents of plants or microbes (biomolecules or metabolites) and synthetic compounds which are foreign to the living system (*xenobiotics*). Biomolecules like cell wall glucans, glycoproteins, certain enzymes, chitosan, fatty acids, etc., are reported to act as elicitor for phytoalexin synthesis and hypersensitive response in various host-parasite systems. They render compatible host parasite associations incompatible. Plant metabolites capable of inducing host-defense system include certain sugars, several amino acids and their derivatives, hormones like ethylene and indole acetic acid (IAA), and dicarboxylic acids like succinate and malate. Each metabolite is effective only against certain host-parasite systems. They operate by inducing synthesis of lignin, callose, suberin, phytoalexins, phenolics, or phenol oxidizing enzymes which may form a sort of physical and/or chemical barrier for advancing pathogen.[57]

Xenobiotics with host immunizing ability include fungicides, herbicides and antipathic agents. Among fungicides most effective ones are certain systemic fungicides like fosetyl-

TABLE 25
Miscellaneous Selective Fungicides

Common name Code No.	Chemical name	Year of introduction (by)	Main uses	Acute oral toxicity LD_{50} rat mg/kg	Formulations
Anilazine (B-622)	4,6-dichloro-N-(2-chloro-phenyl)-1,3,5-triazin-2-amine	1966—68 (Bayer)	*Botrytis, Septoria, Colleto-trichum, Helminthospor-ium, Fusarium, Alternaria* in turf grasses, cereals, coffee, vegetables and ornamentals	2710	'Dyrene', 'Direz', 'Kemate', WP (500 g/kg)
Fenitropan (EGYT 2248)	(1RS, 2RS)-2-nitro-1-phenyl-trimethylene di (acetate)	1981 (EGIS)	As seed treatment for cereals, maize, rice and sugarbeet	3237	'Volparox' (200 g/l) for seed treatment
Dinocap (CR-1693)	Isomeric reaction mixture mainly 2-(1-methylheptyl) 4,6 dinitrophenyl crotonate	1946 (Rohm & Hass)	Powdery mildews on various fruits and ornamentals	980—1190	'Karathane' WP (250, 500 or 830 g/kg), Sialite, WP 250g/kg) *Mixtures:* 'Tedane Combi,' WP (dinocap + dicofol + tetradifon), dinocap + gamma HCH
Binapacryl	2-sec-butyl-4,6-dinitrophenyl 3-methylcrotonate	1960 (Hoechst)	Powdery mildew of apples, citrus, cotton pears and top fruit; also used as acaricide	150—225	'Acricid', 'Endosan', 'Moro-cide', 'Morrocid' EC (384 g/l), SC (500 g/l)
Nitrothalisopropyl (BAS 30000 F)	Di-isopropyl 5-nitroiso-phthalate	1973 (BASF)	Powdery mildews	>6400	*Mixture:* with S ('Kumulan), captan ('Alcap'), dodine (Pummel') or zinc am-mmoniate ethylenebis (di-thio-carbamate)-poly (ethylenethiuram disulphite)) ('Pallinal')
Chinomethionate (Bayer 36205)	6-methyl-1,3-dithio-lo[4,5,b]quinoxalin-2-one	1962 (Bayer)	Powdery mildews of fruits ornamentals and vegetables; also used as acaricide	2500—3000	'Morestan' WP (250 g/kg); 'Morestan 2', DP (20 g/kg), FU

279

Name	Chemical name	Year (Company)	Uses	Value	Products
Fenaminosulf (Bayer 22555, Bayer 5072)	Sodium 4-dimethyl-aminobenzenediazo-sulfonate	1955 (Bayer)	Seed and soil fungicide for *Phythium*, *Phytophthora*, and *Aphanomyces*	60	'Lesan', WP (700 g/kg)
Butylamine	(RS)-sec-butylamine	1962 (Univ. of California)	Mainly fruit rotting fungi like *Penicillium digitatum*, *P. italicum*, *P. expansum*, *Phoma exigua*	380	Free amine, concentrated aqueous solution of appropriate salt
Dichlofluanid (Bayer 47531, Ku 13-032-C)	N-dichlorofluoro methylthio-N',N'-dimethyl-N-phenyl-sulphamide	1965 (Bayer)	*Venturia* spp. (apple and pear), *Botrytis* spp., downy mildews (*Plasmopara viticola*, *Pseudoperonospora humuli*, *Sphaerotheca humuli*), *Phytophthora infestans*, *Alternaria solani*	500—2500	'Euloaren', 'Elvaron', WP (500 g/kg), DP (75 g/kg) *Mixtures*: with copper oxychloride ('Cupro-Euparene, WP)
Drazoxolon (PP 781)	4-(2-chlorophenyl hydrazono)-3-methyl-isoxazol-5 (4H)-one	1967 (ICI)	Powdery mildews on black currants, roses and other crops, *Pythium*, *Fusarium* spp., *Ganoderma* (on rubber)	126	'Mil-Col', 'SAISan', SC (300 g/l), Ganocide GS (100 g/kg) *Mixtures*: 'Pirimicid', seed treatment (drazoxolon + pirimiphos-ethyl)
Pyroxychlor (Dowco 269)	2-chloro-6-methoxy-4 trichloromethyl-pyridine	(Dow Chem. Co.)	Peronosporales		Experimental fungicide
Chlorothalonile	Tetrachloroisophthalonitrile	1963 (Dimond Alkali Co.)	Broad range of fungal pathogens in vegetable and agricultural crops specially used against *Botrytis cinerea*; also acts as algicide	>10000	'Bravo'W-75, 'Daconil 2787 W-75 Fungicide', WP (750 g/kg); 'Bravo' 500, 'Daconil Flowable' 'Faber', Repulse', SC (500 g/l) *Mixture*: with etridiazole (Terradactyl)

Data from Worthing[23] and Gasztonyi and Lyr.[29]

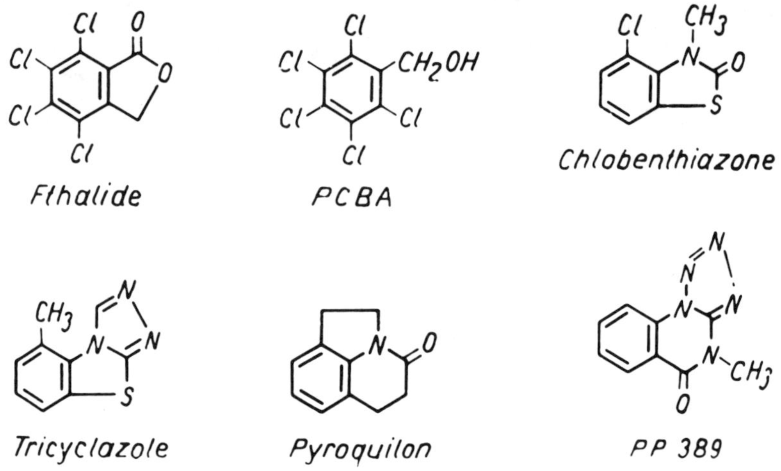

FIGURE 23.   Chemical structure of melanin biosynthesis inhibitors.

Al, metalaxyl, etc. These compounds, probably due to their ability to enter symplasm produce dramatic effects on the plant's physiological processes. They condition the plant in such a way that it responds immediately by producing toxic secondary metabolites as soon as pathogen tries to attack it (Table 27). There are other types of effects too. Carbendazim provides protection against certain viral diseases due to its ability to protect disintegration of chloroplasts, a sort of cytokinin like activity. Host immunizing ability of these fungicides may act synergistically with their antifungal activity in combating plant diseases.

Seedlings of egg plant and tomato raised in soil containing the herbicide, trifluralin, were resistant against *Rhizoctonia, Fusarium oxysporum* f. sp. *lycopersici* or *Verticillium dahliae*. Similar was the effect of alkenyl hydrazinium salts particularly (*E*)-3-(1,1-dimetheylhydrazin-1-io) propenoate against *F. oxysporum*. These herbicides as such are nontoxic to the pathogens involved. Probably due to their injurious effect on host metabolism, they induce the production of antifungal compounds when treated plants are exposed to the pathogens (Table 27).

Antipathic agents are the compounds which protect the plants against infection but as such or their metabolites are nontoxic to the pathogen. Recently, a few such compounds are introduced into the arsenal of plant protection chemicals in the name of fungicides. These are probenazole, WL 28325 and phenylthiourea. Details of these compounds and their effect on plant host are given in Table 27.

Certain miscellaneous compounds like polyacrylic acid, salicylic acid, aspirin etc. have been found to induce the acquired resistance in tobacco and other plant species against viral infections. These compounds induce the synthesis of pathogenesis related (PR) proteins whose exact role is not yet understood.

Although studies are limited and conclusive evidences are available only for a few compounds, they certainly suggest possibility of immunizing host by using chemicals. It may offer effective and economic control for various diseases including those incited by viruses and bacteria for which we still do not have any answer. These compounds might be able to overcome several limitations of present day systemic fungicides and antibiotics like development of resistant strains of the pathogen, nonaccumulation in desired plant part and environmental pollution.[57]

An ideal host immunization chemical should be (1) nontoxic to both host as well as pathogen, (2) water soluble, (3) ambimobile, (4) effective at low concentration, (5) it must

TABLE 26
Antipenetrant Fungicides:Validamycin and Melanin Biosynthesis Inhibitors

Common name Code No.	Chemical name	Year of introduction (by)	Main uses	Acute oral toxicity LD_{50} rat mg/kg	Formulations
Validamycin	1L-(1,3,4/2,6)-2,3-dihydroxy-6-hydroxymethyl-4-[(1S, 4R, 5S, 6S)-4,5,6-trihydroxy-3-hydroxy-methyl-cyclohex-2-enylamino]cyclohexyl β-D-glucopyranoside	1975 (Takeda)	Narrow range of fungi particularly *Rhizoctonia solani* (rice, potato, vegetables)	<20000	'Validacin', 'Valimon' SL (30 g/l); DP (3g/kg)
Phthalide (KF-32, Bay 96610)	4,5,6,7-tetrachlorophthalide	1975 (Bayer)	*Pyricularia oryzae*-rice (mainly used as spray)	<10000	'Rabcide', WP (300 or 500 g/kg), DP (25 g/kg), SC (200 g/l)
Tricyclazole (EL-291)	5-methyl-1,2,4-triazolo [3,4-b] [1,3] benzothiazole	1976 (Eli Lilly)	*Pyricularia oryzae*—rice (applied as flat drench, root soak or foliar spray)	314	'Beam', 'Bim', 'Blascide', WP (200 or 750 g/kg), DP (10 g/kg), GR (30 g/kg)
Pyroquilon (CGA 49 104)	1,2,5,6,-tetrahydropyrrol [3,2,1-*ij*] quinolin-4-one	1979 (Ciba-Geigy)	*Pyricularia oryzae*-rice (seed, soil, root and foliar applications)	321	'Coratop'2, 'Coratop'5, GR (20 or 50 g/kg); 'Fungorene' 50 WP (500 g/kg)

Data from Worthing,[23] Sisler and Rigsdale,[53] Sisler.[54]

TABLE 27
Host-Defense Inducing Compounds

Chemical	Effective against	Direct fungitoxicity	Treated plant's response following infection
Probenazole or Oryzemate (3-allyloxy-1,2-benzisothiazole-1,1-dioxide)	*Pyricularia oryzae*— rice *Xanthomonas campestris* pv. *oryzae* - rice	No appreciable *in vitro* toxicity of probenazole or its break down products	• Augmentation of peroxidase (PO), phenyl ammonia lyase (PAL) and catechal-*O*-methyltransferase • Induced lignification • Accumulation of antimicrobial compounds like α-linolenic acid; 13-hydroxy-*cis-9,trans*-11, *cis*-15 octadecatrienoic acid etc.
DDCC or WL 28325 (2,2-dichloro-3,3-dimethylcyclopropane carboxylic acid)	*P. oryzae* — rice	Not yet recorded	• Hypersensitive reaction • Accumulation of phytoalexins, momilactones A and B in the tissue surrounding invasion site • Probably interferes with host-parasite recognition
PTU (Phenylthiourea)	*Cladosporium cucumerinum* — cucumber	Protects plants at concentration exhibiting little or no *in vitro* fungitoxicity	• Enhanced lignification around penetration site • Increased PO activity
Fosetyl-Al	Peronosporales	Metabolic product H_3PO_3 is fungitoxic	*Phytophthora capsici* — tomato • Inhanced phenonic accumulation and necrotic reaction around infection on leaves *Phytophthora nicotianae* var *parasitica*/*P. nicotianae* var *nicotianae* - capsicum/tobacco • Hypersensitive reaction • Accumulation of phytoalexin (capsidiol) • Papillae formation *Plasmopara viticola* — grapes • Accumulation of phytoalexins-resveratrol, viniferin and flavanoids

| Metalaxyl | Peronosporales | Fungitoxic | *Phytophthora megasperma* var *sojae* — soybean
• Hypersensitive response
• Accumulation of phytoalexin-glyceollin |
| Trifluralin | Herbicide (Protect tomato and egg plant against vascular wilts) | No fungitoxicity | Tomato/egg plant — *Fusarium/Verticillium* spp.
• Accumulation of fungitoxic compounds |

For references see Lazarovits[51] and Sisler and Ragsdale.[53]

TABLE 28
Cases of Fungicide Antagonism

Chemical	Antagonist	Mechanism
Phthalimides	Thiols	Detoxification
Chlorothalonil	Thiols	Detoxification
Dithiocarbamates	Copper	Detoxification by complex formation
Polyene macrolids	Sterols	Detoxification
6-azauracil	Uracil	Compensation
2-aminopyrimidines	Adenine, quanine, folic acid, pyridoxal 5-phosphate, adenosine	Compensation?
2-aminopyrimidines and other fungicides	Riboflavin	Photo inactivation
Carboxamides	Glucose	Circumvention
Validamycin A	Meso-inositol	Compensation
Sterol biosynthesis inhibitors	Ergosterol	Compensation
Fenarimol	CaCl$_2$, tetracaine, carboxin, dialkyl-dithiocarbamates	?

No references see de Waard.[49]

only sensitize the host plants so that latter could respond quickly to invading pathogen by forming physical and/or chemical barriers at the point of contact with the pathogen. Compound as such should not induce the synthesis of antimicrobial substances in absence of the pathogen, because these are bound to be detrimental to the host, (6) sensitization effect should be prolonged and systemic in nature, and (7) induced resistance should be general in nature, i.e., effective against most of the pathogens.

Some of the above mentioned characteristics like water solubility, nonphytotoxicity, and ambimobility are linked to each other. None of the present day host immunizing chemicals possess all the characteristics outlined above for an ideal compound. However, antipathic agents are most close to them. They fulfil almost all the criteria except for not being broad spectrum in nature. At the same time there is need to critically assess the phytotoxicity of these chemicals which may occur whether or not there is any visible change in plant's appearance.

Chemical host immunization offers potentially powerful tool in crop protection. It extends our ability to control numerous diseases for which no alternate procedure currently exists, e.g., viruses, bacteria, soil borne fungi and those fungi affecting crown portion or deep inside the host tissue.[51] There is urgent need for extensive research on these compounds, to work out their exact mode of action and to establish structure activity relationship for them.

I. SYNERGISM AND ANTAGONISM IN FUNGICIDES

Synergism is defined as the simultaneous action of two or more compounds in which the total response of an organism to the pesticide combination is greater than sum of the individual components. Antagonism is the opposite. Different examples of antagonistic and synergistic interactions in fungicides are listed in Tables 28 and 29, respectively.

1. Mechanism

In insecticides action of synergists is often based on inhibition of detoxification reactions which cause resistance to insecticides. However, detoxification plays only minor role in fungicide toxicity and resistance. In the case of fungicides, synergism or antagonism may result from chemical interaction between the mixture components resulting in increased or decreased (detoxification) toxicity or it may relate to the interaction with physiological

TABLE 29
Cases of Fungicide Synergism

Components in mixture	Pathogen/Fungus
Anilazine/zinc or copper	*Botrytis cinerea*
	Colletotrichum coccodes
Zineb/polyram	*Plasmopara vizicola*
Carboxin/mancozeb	Various
Chloroneb/thiram	*Pythium ultimum*
Copper/zineb	*Plasmopara viticola*
Dimethyldithiocarbamates/complex-forming agents	*Botrytis cinerea*
Dodine/captan	*Venturia inaequalis*
	Xanthomonas prunei
Elemental sulfur/surfactants	*Botrytis cinerea*
Ethazole/pentachloronitro-benzene	*Pythium aphanidermatum*
Metalaxyl/mancozeb or cymoxanil or fosetyl-Al	*Phytophthora infestans*
	Plasmopara viticola
	Pseudoperonospora cubense
Fenarimol/captafol or captan or folpet or dodine or sodium lauryl sulfate	*Aspergillus nidulans*
Fenpropimorph/Chlorothalonil	*Pyrenophora teres*

For references see de Waard.[49]

processes of the target organisms. From a theoretical point of view interaction may involve: (1) nonmediated diffusion across the plasma membrane, (2) carrier-mediated transport across the plasma membrane, (3) energy-dependent efflux from the fungal cell, (4) transport to the target site, (5) activation, (6) detoxification, (7) affinity for the target site, (8) circumvention of the target site, or (9) compensation of the target site. The companion chemical may affect these processes in such a way that the efficacy or toxicity of the fungicide is enhanced (synergism) or reduced (antagonism).

2. Exploitation in Plant Disease Management

Optimization of the field performance of fungicide composition implies minimization of antagonistic and exploitation of synergistic interactions. Formulating agents also improve field performance but are not regarded as synergists. In fact, true fungicide synergists are not yet commercially available, in contrast to insecticide synergists. However, this situation may change in future since synergistic interactions with fungicides have recently been demonstrated in quite a few commercial and experimental fungicides. For an ideal combination both fungicide and synergist should be systemic—a goal probably difficult to realize. Other problems which may be involved in the development of fungicide synergists are[49] (1) difficulty in finding safe compounds, (2) the costs of potential synergists in relation to the price of the fungicide, (3) the ratio of fungicide and synergist necessary in the combination, (4) the simultaneous formulation of synergist and fungicide, and (5) variation in the spectrum of antifungal activity. Despite these difficulties it is believed that synergistic mixtures of fungicides may be developed in future which would be able to provide better disease control not only by enhancing the efficacy of individual components (if both are fungicidal) but also by prolonging the life of fungicide "at risk" (from point of the view of resistance development).

J. FUNGICIDE RESISTANCE
1. Introduction

Development of resistant strains of target fungi is now one of the major problems with selective fungicides. It causes unexpected crop losses for the growers, and may put him in

TABLE 30
**Prominent Cases of Field Resistance Resulting in Failure
of Disease Control**

Fungicide Group	Pathogen	Host
Benzimidazoles	*Cercospora beticola*	Sugarbeet
	Botrytis cinerea	Grapes
	Venturia inaequalis	Apple
	Pseudocercosporella herpotrichoides	Cereals
	Colletotrichum cofceanum	Coffee
Phenylamides	*Phytophthora infestans*	Potato
	Bremia lactucae	Lettuce
	Plasmopara viticola	Grapes
	Peronospora tabacina	Tobacco
	Pseudoperonospora cubensis	Cucurbits
Dicarboximides	*Botrytis cinerea*	Grape
Polyoxin	*Alternaria kikuchiana*	Pear

difficult position if no adequate substitutes are available. It may reduce the profits of manufacturer, which has developed the fungicide at high cost.[2] It may also have consequences for the extension officers, the regulatory authorities, and in some cases even for the consumer and national economy. Some important cases of field resistance are listed in Table 30.

2. Definitions

A guideline of fungicide resistance terminology was proposed in 1985.[60] According to this proposal "resistance should be used only to define a stable and heritable adjustment by a fungus to a fungicide resulting in less than normal sensitivity to that fungicide". It is different from adaptation of a fungal pathogen to a fungicide. Adaptations are neither heritable nor stable and are not expected to cause severe problems. Furthermore, insufficient field performance of a fungicide is not necessarily related to the presence of resistant strains in field. Thus, the term "field resistance" should be used only when decreased fungicide efficacy is correlated with increased frequency of resistant strains.[45]

3. Mechanism of Resistance
a. Origin of Resistance

Resistant strains of the fungus may arise either due to physiological adaptation or gene mutation. Since resistance due to physiological adaptation is unstable and disappears again when organism is no longer exposed to the fungicide, gene mutation is the main mechanism for the origin of stable, inheritable resistance. Gene mutation is a spontaneous phenomenon. The fungicide itself does not induce resistance, but acts solely as a selecting agent.

b. Nonselective vs. Selective Fungicides

Most of the nonselective conventional fungicides interfere with many metabolic processes in the fungal cell, hence, are called "multisite inhibitors". On the contrary, selective fungicides are site specific. Their primary site of action is restricted to only one, mono- or oligogene regulated metabolic activity. Even a single gene mutation may result in the development of a resistant strain against a specific-site fungicide. Whereas change in several genes may be required for the development of such strains against multisite inhibitors. Thus, the resistance problem is far more common and severe in selective fungicides as compared to nonselective one.

c. Biochemistry of Resistance

Selective fungitoxicity generally stems from site specific binding of fungicides to either an enzyme involved with respiration, with biosynthetic processes in fungi (e.g., synthesis of nucleic acids, proteins, ergosterol, chitin, etc.), or with structural proteins associated with membrane structure or nuclear function. Development of resistance to such specific-site inhibitors is most oftenly due to single gene mutation resulting in slightly altered target site, with reduced affinity to the fungicide (Table 31). Apart from alteration in target site, there are other ways also by which a fungus may develop the resistance not only against selective fungicides but also sometimes against multisite conventional fungicides (Table 31). These are: (1) nonaccumulation of a fungicide inside the fungal cell to a toxic level. This may happen by reduced uptake, increased efflux, binding at cell wall or at places other than target site, conversion of a fungicide into nonfungitoxic compound(s) or lack of conversion of an itself nonfungitoxic compound into a fungitoxic form (Table 31) and (2) in principle, compensation for the inhibiting effect, either by an increase in the production of a inhibited target protein (enzyme) or by changes in metabolism resulting in circumvention of the blocked site by an alternate pathway, may also impart resistance against a fungicide. However, no clear cut example of such a mechanism seems available in fungicides.

d. Genetics of Resistance

Due to lack of the methods to generate recombinant progeny and handling difficulties genetic studies on fungicide resistance in plant pathogenic fungi are limited. Resistance to fungicides is controlled by nuclear genes, which may be major or minor in nature. Major gene controlled resistance may develop in one step as a result of mutation of one gene which has major effect on the phenotype. However, upto 10 major genes may control resistance to fungicides and each gene may have several alleles.[70] Occassionally, action of each gene may be influenced by unrelated modifier genes or may itself exert effects on other genes. Usually resistance against benzimidazoles, phenylamides, carboximides, and aromatic hydrocarbons is controlled by major gene(s).[71] Minor genes (polygenes) controlled resistance are reported for dodine, ergosterol biosynthesis inhibitors and cycloheximides. Effect of minor genes are additive but too small to be identified individually. Depending upon the fungus-fungicide combination resistance gene(s) may be dominant, semidominant or recessive.[71]

4. Epidemiology of Resistance

Fungicide resistant individuals multiply faster in the presence of fungicides than those lacking resistance. Several mathematical models have been developed which describe the development of qualitative (mono- or oligogenic) resistance.[72,73] Despite varying assumptions, the conclusions drawn from all the models are broadly similar. These are[70] (1) resistance spreads rapidly where large, fast developing, pathogen populations are exposed to persistent and effective chemicals, and (2) fungicide mixtures are likely to be more effective than alternating sequences, especially where spray coverage is poor. However, only fragmentary experimental evidence exists to support these conclusions. At least for three diseases, a reasonable correlation exists between the time it actually took for fungicide resistance to emerge as a practical problem, and the time predicted by the models (Table 32). In certain instances, fungicide mixtures have been shown to delay the spread of resistance.[74,75] However, it is still a difficult problem to design and monitor field experiments to assess relative fitness through measurements of population size or apparent infection rate.[70]

5. Monitoring for Resistance

Since a reliable prediction system is yet to be developed for any fungicide, there is a need to keep watch in order to detect the resistant strains well advance in time to counter

TABLE 31
Some Mechanisms of Fungicide Resistance

Fungicide	Site of action	Mode of action	Mechanism of resistance	Comments	Ref.
Carbendazin	B-tubulin	Inhibition of assembly of microtubules to form spindle, and as a consequence mitosis and other cellular processes involving microtubules are inhibited	Altered site of action	Decreased affinity of tubulin to bind carbendazim, negatively correlated cross resistance with N-phenylcarbamates	61,62
Kasugamycin	Ribosomes	Inhibition of protein synthesis	Altered site of action	Reduced affinity of ribosome towards antibiotic	63
Metalaxyl	RNA polymerase	Inhibition of RNA synthesis	Altered site of action	Insensitivity of enzyme to metalaxyl	36
Carboxin	Succinate-ubiquinone reductase complex (Complex II)	Inhibition of electron transport and subsequently respiration	Altered site of action	Decreased affinity of complex II to carboxin	64
Polyoxin B	Chitin synthetase	Inhibition of chitin synthesis	Reduced uptake	Alteration in membrane resulting in reduced uptake	65
Imazalil Fenarimol Triazoles	Microsomal Cytochrome P-450	Inhibition of sterol biosynthesis	Reduced uptake	Increased energy dependent efflux of fungicide resulting in its low intracellular concentration	65
IBP Edifenphos		Inhibition of phospholipid (i.e. phosphatidyl choline) biosynthesis	Metabolism/detoxification	Conversion of fungicide by cleavage of the S-C bond into more polar non-fungi toxic derivatives	67
Pyrazophos		Inhibition of phospholipid biosynthesis and accumulation of free fatty acids	Non-conversion to fungi toxic form	Non-conversion of pyrazophos into the actual toxic principle: 3-hydroxy-5-methyl-6-ethoxycarbonylpyrazolo (1,5a) pyrimidine (PP-pyrazophos)	68
Fenpropimorph	Sterol 14 reductase	Inhibition of sterol biosynthesis	Utilization of intermediates	Unusual sterols accumudate and are used in membranes	69

TABLE 32

Predicted and Observed Time for Resistance Subpopulations of Three Pathogens to Cause Resistance Outbreaks in Field

Pathogen	Chemical	Standard selection time (days)	Selection-pressure duration (d: days, Y: years)	
			Predicted	Observed
Phytophthora infestans	Metalaxyl	3.7—3.8	51—70 d	1—2 seasons
Cercospora beticola	Benomyl	9.5—14.3	130—263 d	140—200 d
Ustilago nuda	Carboxin	158.5	5—7 Y	12—14 Y

*   Taken from Hollomon.[16]

their effects. However, when resistance development is very fast, the information may come too late to leave any room for action to minimize crop losses.

In vivo failure of a fungicide to control diseases incited by *in vitro* sensitive pathogen is not always associated with the development of resistant strains. In order to ascribe poor disease control to fungicide resistance one must:[76] (1) prove the existence of the pathogen which are less sensitive to the fungicide used, than what is considered normal for the species, (2) show that such strains represent a considerable fraction of the pathogen population in the problem area or field, and (3) demonstrate experimentally that the strains considered resistant are able to cause disease on treated plants of the same species and variety.

Monitoring for resistance in the field can be used in the following situations,[77] (1) Early monitoring during the introduction of a new fungicide to assess the resistance risk and to establish base-line sensitivity data. (2) Evaluating strategies to minimize resistance risk (mixtures, alternation programmes, limited number of applications); (3) determining persistence of resistance in absence of selection pressure (after withdrawal of fungicide), and (4) analyzing product failure.

Present monitoring system is entirely based on bioassay procedures to measure phenotypes. Considering the phenomenal advancements being made in the field of biotechnology and molecular biology, time is not far away when rapid and highly sensitive techniques would be available to monitor the resistance at protein level through immunological (using monoclonal antibodies) methods and at the DNA level by hybridization. A rapid immunological technique has already been developed to monitor insecticide resistance in aphids,[70] but fungicide resistance mechanisms are at present less well understood and comparable techniques are not so well advanced. However, knowledge of β-tubulin coupled with isolation and cloning of *ben A* gene[16] suggest that similar techniques for monitoring carbendazim resistance in a wide range of pathogens might soon be possible. These biochemical monitoring techniques would only be possible where resistance mechanics are understood at molecular level, but they are likely to be simple, rapid and more sensitive. Using these techniques changes in resistance could be detected at much lower gene frequencies, which would be invaluable when evaluating the effects of various fungicide resistance strategies on the critical early stages of the evolution of resistance.

6. Buildup of Resistance

Mere emergence of fungicide resistant strains in laboratory or field experiments does not guarantee the failure of disease control in field, because such strains initially constitute a very small, practically insignificant fraction of the total pathogen population. Failure of disease control will not arise before a considerable proportion of the pathogen population has become resistant and in order to achieve this a resistant mutant should be able to multiply at a faster rate than wild type by taking advantage of selection pressure exerted by the fungicide. Even in the absence of fungicide, mutant strains must be able to compete well

TABLE 3
Factors Affecting Build-Up of Fungicide Resistant Strains in Field

I.   Parasitic fitness of resistant strain
- spore germination
- chances of infection
- rate of colonization
- latent period
- degree of sporulation
- infectious period

II.   Type of disease and life cycle of pathogen
- fecundity and spread of the pathogen
- life cycle of the pathogen (complete on single or alternate hosts)
- soil borne or aerial pathogen
- Mono-, oligo-, or multi-nucleate pathogen cells
- mono- or multi-cycle disease

III.   Selection pressure by the fungicide
- efficiency of the fungicide
- dose
- coverage
- frequency of application
- persistence

IV.   Persistence of resistance particularly in absence of the fungicide

V.   Epidemiological factors
- environmental conditions favoring disease outbreak
- duration of high disease pressure
- influx of untreated pathogen propagules

VI.   Management practices (fungicide usage, crop management)
- duration of exposure
- presence of othe controlling factors (effective mixture partner, host resistance, biocontrol agent)
- size of target population, escape, over kill (protective vs. curative use)
- proportion of crop area treated

with wild-type strains. In some cases failure of the disease control occurred shortly after introduction, e.g., with phenylamides and benzimidazoles, but in other cases it took many years, e.g., Kitazin-P resistance in *Pyricularia oryzae* and dodine resistance in *Venturia inaequalis*.[78] There are specific-site fungicides (e.g., sterol biosynthesis inhibitors) yet to face the serious field resistance problem even after quite a few years of their introduction.[45] Depending upon the speed of development of resistance fungicides are classified as of "high-risk" (e.g., phenylamides and benzimidazoles), "moderate risk" (e.g., dicarboximides) or "low-risk" (e.g., sterol biosynthesis inhibitors).

The likelihood that resistance will develop seems to be determined by the chemical rather than the target organism. However, the buildup of resistant mutant population is influenced by several factors (Table 33).

7. Cross Resistance

The same mutation altering sensitivity of an organism against one fungicide may also influence its sensitivity toward others. This relationship between two or more fungicides, mediated by the same gene(s) of a fungus, is termed as cross resistance. This relationship may be positive or negative. The term (positive) cross resistance designates resistance to two or more toxicants, mediated by the same genetic factor. When such a factor mediates resistance to one fungicide at the same time increases sensitivity to other fungicide, the term negative cross resistance is used. The term "multiple resistance" is used for those situations where resistance to two or more toxicants is governed by different genes.

Usually a mutant selected on one fungicide is resistant to fungicides that are chemically related or share the common mode of action. This positively correlated cross resistance has been frequently observed among benzimidazoles, phenylamides and sterol biosynthesis inhibitors (SBIs).

The phenomenon of negatively correlated cross resistance is frequently observed for the benzimidazoles with N-phenyl carbamate fungicides. In several fungi benzimidazole-resistant isolates are reported to exhibit high degree of sensitivity toward N-phenylcarbamates.[79] Like benzimidazoles, N-phenylcarbamates inhibit mitosis by interfering with microtubule function. However, phenylcarbamates probably interact with α-tubulins whereas benzimidazoles bind with β-tubulin.[62] In order to explain negative cross resistance between these two groups of the fungicides, it is proposed that hyperstability of microtubules imparts resistance to benzimidazoles and that N-phenylcarbamates has a destabilizing effect on the mutant microtubules. Negative cross-resistance has further been reported for two sterol biosynthesis inhibitors: fenarimol-resistant mutants of *P. italicum*, but not of *A. nidulans*, appeared more sensitive to fenpropimorph than the wild-type strain.[80]

8. Management of Resistance

The fungicide resistance management tactics are primarily aimed at to avoid or delay the development of resistance. Once a major fraction of the pathogen population has become resistant, the most sensible approach is to change to other fungicides with a different mode of action, or to nonchemical control measures, if available.

Different strategies which could be adopted, preferably in integration, to manage the fungicide resistance problem are briefly described below.

a. Prediction and Monitoring

While working with new fungicide-pathogen combination, first it is desirable to obtain some information about the associated resistance risk like *in vitro* frequency of emergence of resistant strains, both in presence as well as in absence of mutagenic agents, parasite fitness of resistant strains etc. Based on these informations and those available on epidemiology and disease cycle, a model is simulated for the prediction of resistance risk under field condition. It would be a sort of tentative prediction just to gain some before hand idea about the resistance risk involved. It may or may not happen in practice but it would be useful in devising preventive measures. While fungicide is in use, monitoring is desirable to keep track of level of resistance, which may increase in the course of years, and the frequency of resistance.

b. Reduction of Selection Pressure

In cases where a fungicide is resistance prone, a reduction of selection pressure by the fungicide concerned should be considered, when necessary and possible in combination or alternation with other fungicide with different mode of action. On the other hand, when only a moderate level of resistance is likely to occur, particularly when resistance development is stepwise (additive) under polygenic control, an increase of the selection pressure, as far as technically and economically feasible, may be considered in order to kill also the moderately resistant strains.

Different tactics which could, in principle, reduce the selection pressure by the high-risk fungicides are (1) apply minimum possible required number of sprays and quantity of fungicide, (2) avoid using the same type of chemical for treatment of seed or plant material, spraying of the crop and post harvest treatment, since development of resistance in earlier treatments may jeopardize the effect of later applications, (3) avoid treatment of a crop in very large area, in a whole region or country, with the same type of chemical, and (4) avoid exclusively curative use of the fungicide.

A very thorough treatment of the crop will increase the selection pressure and, therefore, favour the buildup of resistance, however, avoiding it may not provide the satisfactory control unless a fungicide is ambimobile inside the plant and/or has high vapor phase activity. Unfortunately availability of such compounds in the arsenal of plant disease control chemicals is very limited.

As a part of strategies to reduce selection pressure 2 applications per season, 2 to 4 applications per season, and 4 to 8 applications per year are recommended for dicarboximides,[81] metalaxyl (or other phenylamides),[82] and DMI fungicides,[83] respectively. However, possibility to reduce selection pressure without use of other fungicides are very limited.

c. Combination and Alternation of Fungicides

The most widely advocated method to tackle the resistance problem is the use of resistance prone fungicide in a mixture with a low risk "no cross-resistant" multisite conventional fungicides, "synergistically acting" fungicides or "negatively correlated cross resistant" fungicides, depending upon the availability. A combination fungicide should not have positively correlated cross resistance with other fungicides. Sufficient theoretical[72,73] and experimental[74-76] data are now available to support the view that fungicide mixtures can delay or even reverse,[70] the spread of resistance.

Use of mancozeb and phthalimides has been advocated to prolong life of specific-site fungicides, phenylamides[74,75,77] and fenarimol,[65] respectively, due to synergistic interaction. Resistance to fenarimol depends on an energy-dependent efflux, and this efflux is inhibited by captan or its degradation products.[84]

Samoucha and Gisi[74] advocated the use of three-way mixtures (oxadixyl/mancozeb/ cymoxanil) to control phenylamide sensitive and resistant strains of *Phytophthora infestans* on tomato or potato and *Plasmopara viticola* on grape particularly under condition when proportion of resistant population is quite high. They demonstrated that as compared to two way mixtures, three way mixtures are more flexible in their use under difficult disease situation; they not only minimize the buildup of resistance, but also contribute to control resistant populations.[85]

In recent years several prepacked mixtures of specific-site and contact fungicides have been introduced by different manufacturers for the use by the farmers. Theoretically the use of prepacked mixtures offer several advantages: (1) the protectant component should control the resistant isolates, (2) the dose of the systemic component may be reduced to additive or synergistic levels, (3) reduced concentration of systemic component should reduce the selection pressure for resistance, (4) multiple disease control and most importantly, (5) the use of prepacked mixtures is an enforceable strategy.[77,86]

Alternate use of systemic-contact (S-C) or systemic-systemic (S-S) (with different mode of action) fungicides is another approach proposed to check or lower down the build-up of resistant strains. This approach may be specially useful for disease-fungicide combinations when the basic mixture is judged too risky to be used throughout the season.

In addition to combination or simple alternation more complex application sequences may be followed. A mixture has disadvantage that a compound at risk is continuously present, and with alternation, application of the nonrisky compound is interrupted for no good reason. Better results might therefore, be obtained by a sequential scheme in which mixtures and rotation are combined in such a way that the nonrisky fungicide is constantly present and that only the application of the risky compound is interrupted. One such assumed combination, $(S+C)-C-(S+C)-C$, was found to be most effective by calculations.[78]

d. Use of Antipathic Agents

Antipathic agents are the compounds which are not directly fungitoxic but provide protection against the disease by inducing general host defense system.[51,57] Since these compounds do not directly attack the pathogen, selection pressure should be low. At the same time induced general defense system of the host involve physical barriers like lignin and/or chemical barriers like phytoalexins, which are multisital in their mode of action. So theoretically such compounds should face low resistance risk. Unfortunately at present only few such compounds are known, they too are specific to only few diseases.

e. Integrated Control

Combination of chemical and nonchemical methods, e.g., host resistance, cultural practices, biological control, etc, may be considered for the avoidance of resistance.[86]

Combination of site-specific fungicides with resistant cultivars of the host may prolong the life of both the components. In lettuce downy mildew (*Bremia lactucae*) metalaxyl resistance is associated with the same (B2) sexual compatibility type (all resistant isolates have an identical virulence phenotype), and incompatibility on lettuce cultivars with the *Dm*11, *Dm*16, and *Dm*18 alleles. Since *Dm*11 allele is present in several widely grown lettuce cultivars (*Dm*16 and *Dm*18 are more recent in their introduction and are only now being introduced into the cultivars of commercial importance), a strategy for delaying resistance was adopted where metalaxyl was applied only on *Dm*11 cultivars.[87,88] This strategy, as anticipated, has proved to be very effective in providing control even against metalaxyl resistant isolates of *B. lactucae*.[88]

9. Conclusions

Fungicide resistance has become one of the major problems for which no easy solution is in sight. Tactics to prolong the life of the badly needed fungicides and long term strategies to ensure efficient disease control are urgent need of the hour. However, to do this will require more detailed information on biology, biochemistry, genetics, and epidemiology (population dynamics) of resistance. Of particular importance are the development of rapid resistance monitoring techniques and suitable model for the resistance prediction and effective evaluation of management strategies under field conditions.

To be able to avoid the development of fungicide resistance in future it is essential to make available a broad arsenal of fungicides, conventional as well as systemic, with different sites of action, low resistance risk and variable mode of resistance development so that enough flexibility exists in the design of counter measures. Although some useful novel fungicides may be found by current screening methods, biorational approach must be adopted for the designing and synthesis of such chemicals, but it will require considerably more information about the biochemical mechanism of fungicide resistance.

II. ANTIBIOTICS

Antibiotics are microbial products which are antimicrobial in nature. Antibiotics have done miracles in medical world. However, their use has been very restricted (in majority of the cases only up to the experimental level) in plant disease control. As per information contained in the 8th edition of *The Pesticide Manual*[23] currently only 6 antibiotics are in use, at limited scale, for the control of plant diseases (Figure 24, Table 34). Pimaricin, a polyene antibiotic, forms complexes with membrane sterols (ergosterol or cholesterol) resulting in rapid loss of cell permeability and cell function.[89] Blasticidin S, kasugamycin and cycloheximide are protein synthesis inhibitors; the first two are quite specific to *Pyricularia oryzae* while cycloheximide binds with 60S subunit of 80S ribosome of eukaryotic system and exhibits a wide spectrum of antifungal activity. Due to its structural similarity with UDP-*N*-acetylglucosamine (the chitin precursor) polyoxins act as competitive inhibitor for chitin synthetase.[90]

TABLE 34
Antibiotics in Plant Disease Control

Antibiotic (Produced by)	Chemical name	Year of introduction (by)	Main uses	Acute oral toxicity (LD_{50} at mg/kg)	Formulations
Blasticidin-S (*Streptomyces griseochromogenes*)	1-(4-amino-1,2-dihydro-2-oxo-pyrimidin-1-yl)-4-[(S)-3-amino-5-(1-methylguanidino) valeramido]-1,2,3,4-tetradeoxy-β-D-erythro-hex-2-enopyranuronic acid	1955 (Kaken/Kumi-ai/Nihon Nohyaku)	*Pyricularia oryzae*; due to phytotoxicity formulation used as benzylaminobenzene-sulfonate	53.3	'Bla-S' EC(20 g/l); WP(40 g/kg); DP (1.6 g/kg)
Cycloheximide (*S. griseus*)	4[(2R)-2-[(1S,3S,5S)-(3,5-dimethyl-2-oxocyclohexyl)]-2-hydroxyethyl] piperidine-2,6-dione	1958 (Upjohn)	Wide antifungal spectrum; mainly used against powdery mildews on ornamentals; rusts and leafspots on grasses; also used as growth regulator to promote fruit abcission in olives and oranges	2	'Acti-dione PM' (270 mg/kg); 'Acti-dione TGF' (21 g/kg); 'Acti-Aid', (42 g/l); 'Acti-dione', WP (1 g/kg) *Mixtures:* 'Actidione RZ' (13 g cyclo-heximide + 750 g quintozene/kg)
Kasugamycin (*S. kasugaensis*)	1L-1,3,4/2,5,6-1-deoxy-2,3,4,5,6-pentahydroxycyclohexyl-2-amino-2,3,4,6-tetradeoxy-4-(α-iminoglycino)-α-D-arabino- hexopyranoside	1965 (Institute of Microbial Chemistry; Hokko)	*Pyricularia oryzae* (rice), *Cercospora beticola* (sugarbeet), *Erwinia carotovora* (potato), *Fulvia fulva* (tomato), *Pseudomonas phaseolicola* (kidneybean), *P. lachrymans* (cucumber), *P glumae* (rice); *Xanthomonas citri* (citrus)	22	'Kasumin', DP(2 g/kg), WP (20 g/kg); liquid (20 g/l); GR (20 g/kg) *Mixture:* with copper oxychloride ('Kasuran', 'Kasumin'-Bordeaux, WP) or bis(2,4 dichlorophenyl) ethyl phosphate

Common name (organism)	Chemical name	Activity (mg/kg)	Disease/use	Year (manufacturer)	Trade name/formulation
Pimaricin/tennecetin/natamycin (*S. natalensis* and *S. chattanoogensis*)	(8E,14,E,16E,18E,20E)-(1S,3R,5S,7S,12R,24R,26R)-22-(3-amino-3,6-dideoxy-β-D-mannopyranosyloxy)-1,3,26-trihydroxy-12-methyl-10-oxo-6,11,28-trioxatricyclo[22.3.1·0⁵·⁷]octacosa-8,14,16,18,20-pentaene-25-carboxylic acid	2730—4670	Diseases of bulbs, especially basal rot of daffodils	1968 (Gist-Brocades NV)	'Delvolan', WP (68 g/kg)
Polyoxins (*S. cacaoi* var. *asoensis*)	*Polyoxin B* 5-(2-amino-5-O-carbamoyl-2-deoxy-L-xylonamido)-1,5-dideoxy-1-(1,2,3,4,-tetrahydro-5-hydroxymethyl-2,4-dioxopyrimidin-1-yl)-β-D-allofuranuronic acid	21000	*Alternaria* spp., *Botrytis cinerea*, Erysiphales	1968 (Hokko/Haken Pharmaceutical/Kumiai)	'Piomycin', 'Polyoxin AL', WP (100 g/kg)
	Polyoxin D 5-(2-amino-5-O-carbamoyl-2-deoxyl-L-xylonamido)-1-(5-carboxy-1,2,3,4-tetrahydro-2,4-dioxopyrimidin-1-yl)-1,5-dideoxy-β-D-allofuranuronic acid	>9600	*Rhizoctonia solani*	1968 (Kaken; Kumai; Nihon Nohyaku)	Polyoxin Z (20 g (as Zn salt)/kg)
Valdamycin	See section on antipenetrant compounds				

Information collected from Worthing.[23]

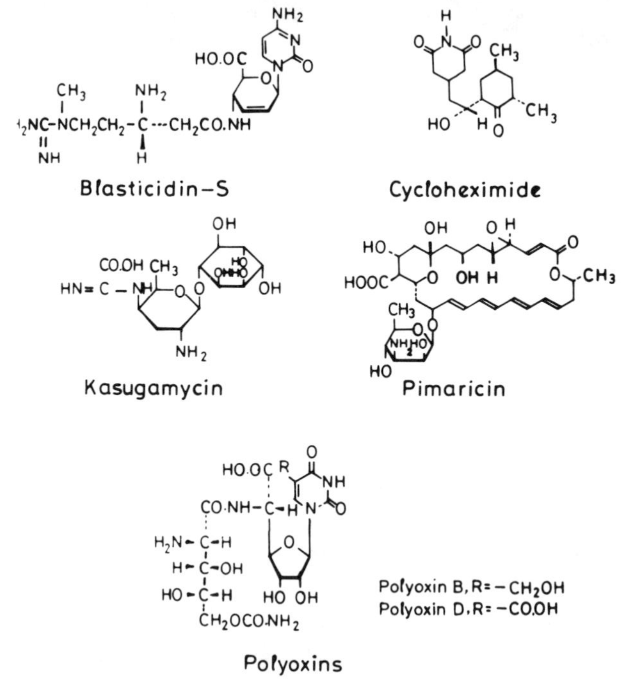

FIGURE 24.   Chemical structure of antibiotics in agricultural use for the control of plant diseases.

III. NEMATICIDES

The importance of the control of nematode damage is well accepted. Estimated losses due to nematodes in 1967—1968 in the U.S. were approximately 1.5 billion U.S. dollars.[91] Apart from causing direct losses nematodes may serve as vectors of viral, bacterial, and fungal pathogens. Nematode parasitism often reduces the natural resistance of the host to other soil borne plant pathogens particularly to wilt causing fungi. Depending upon the specific nematode, host crop and seasonal fluctuations, phytoparasitic nematode populations of from 0 to $1 \times 10^5/l$ of soil can be extracted from field situations. In deep-rooted perennials, as many as 1×10^{10} parasitic nematode/ha can be found in upper 150 cm of a sandy loam soil.[92]

Sometimes, it is important to attempt to kill every individual plant parasitic nematode (e.g., *Xiphinema index*, as it serves as vector for grape fan leaf virus).[93]

Since the discovery of 1,3 dichloropropene (1,3-D) and ethylene bromide in the 1940s, both the number and use of nematicides have grown widely. By 1970 the nematicide business in the U.S. had grown to a $51 million sales.[94] Nematicides are diverse in their chemical and biological activity and in their behaviour in soil. The toxicants considered as nematicides were developed primarily as general biocides or insecticides. Two general types, fumigants and nonfumigants can be distinguished (Table 35). Chemically they can be categorized as halogenated aliphatic hydrocarbons, methyl isothiocyanate liberators, organophosphates and organocarbamates (Table 35).

Nematicides are used primarily to protect root system since relatively few nematodes attack above ground plant parts and most foliar nematodes are controlled by sanitation, clean planting stock, or hot water treatments. The use and action of nematicides is diagrammatically illustrated in Figure 25. All fumigant nematicides are phytotoxic and must be used as a preplant treatment. Nonfumigant nematicides are rarely phytotoxic at recommended dosages, but they exhibit a high degree of mammalian toxicity.

TABLE 35
Nematicides

Common name	Chemical name	Year of introduction (by)	Uses	Acute oral toxicity (LD_{50} rat mg/kg)	Formulations
I. Fumigants					
A. Haloginated aliphatic hydrocarbons					
Methylbromide	Bromomethane	1932 ? (Dow)	Potent insecticide, some acaricidal properties; soil fumigant against nematode, fungi, weeds, etc.	Highly toxic	Packed in glass ampoules (up to 50 ml), metal cans or cylinders; chloropicrin may be added as a warning agent
Chloropicin	Trichloronitromethane	1908	As an insecticide for fumigation of stored grain; soil insecticide, nematicide, and fungicide; highly phytotoxic	Highly toxic	No specific formulation *Mixture*: with 1,3-D, methyl isothiocyanate, etc.
1,3-dichloropropene	(EZ)- 1,3-dichloro-1-propene	1956 (Dow)	Soil fumigant and nematicide, particularly root knot nematode	127—250	Without formulation: 'D-D92', 'Dedisol C', AL (1107 g/l); 'Nemotox II', 'Telone II', *Mixtures* 'Nematox' ((EZ)-1,3 dichloropropene + 1,2-dichloropropane)
EDB	1,2-Dibromoethane	1925 ? (Dow)	Basically insecticidal fumigant; also used against certain nematodes	146	Solutions in inert solvent
B. Methyisothio-cyanate liberators					
Methylisothio-cyanate	Isothiocyanato-methane	1959 ? (Schering)	Soil fumigant for control of soil fungi, insects, nematodes and weed seeds	175	'Trapex', EC (175 g/l) *Mixtures* with 1,3-D, 1,2-dichloropropane, chloropicrin, etc.

TABLE 35 (continued)
Nematicides

Common name	Chemical name	Year of introduction (by)	Uses	Acute oral toxicity (LD$_{50}$ rat mg/kg)	Formulations
II. Nonfumigants					
A. Organophosphates					
Ehoprophos	O-ethyl S.S-dipropyl phosphorodithioate	1966 ? (Mobile)	Non-systemic nematicide; also effective against soil dwelling insects	62	'Mocap', GR (100 g/kg); EC (700 g/l)
Fensulfothion	O,O-diethyl O-4 methylsulfinylphenyl phosphorothioate	1967 (Bayer)	Against free living, cyst forming and root knot nematodes; insecticide; some systemic activity	4.7—10.5	'Dasanit', 'Terracur P' SL (600 g/l); WP (250 g/kg); DP (100 g/kg); GR (25, 50 or 100 g/kg) for foliar and soil application
Fenamiphos	Ethyl 4-methylthio-m-tolyl isopropylphosphoramidate	1971 (Baychem Corp.)	Systemic nematicide, active against ecto- and endoparasitic, free living, cyst forming and root knot nematodes; used in banana, cacao, cereals, citrus, coffee, pome fruit, peanuts, potatoes, stone fruit, tobacco and vegetables	15.3—19.4	'Nemacur', EC (400 g/l), GR (50 or 100 g/kg)
B. Organocarbamates					

299

Name	Chemical name	Introduced (Company)	Activity	Toxicity	Formulations / trade names
Carbofuran	2,3-Dihydro-2,2-dimethyl benzofuran-7-yl methyl carbamate	1965 ? (FMC/Bayer)	Systemic acaricide, insecticide and nematicide	8-14 (in corn oil)	'Furadan', 'Curaterr', 'Yaltox', WP (750 g/kg); flowable paste (480 g/l); GR (20, 30, 50 or 100 g/kg)
Aldicarb	2-Methyl-2-(methylthio) propionaldehyde O-methylcarbamoyloxime	1965 ? (Union Carbid)	Systemic pestide used against mites, nematodes and insects; soil application (seed furrow)	0.93	'Temic', GR (50, 100 or 150 g/kg) Mixtures: 'Sentry', 'Temic LD' (aldicarb + gamma-HCH)
Oxamyl	N,N-dimethyl-2-mehyl-carbamoyloxyimino-2-(methylthio) acetamide	(du Pont)	Systemic insecticide, acaricide, and nematicide; used on many field crops, fruits, ornamentals and vegetables; foliar and soil applications		'Vydate', L, SL (240 g/l) 'Vydate' G, GR (100 g/kg)

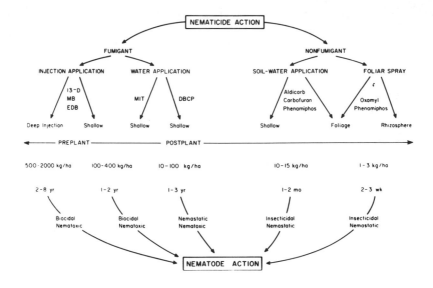

FIGURE 25.   A diagrammatic representation of the action of the commonly used nematicides illustrating application method, time of application, quantity of toxicant used, benefit time, and biological toxicity (from Gundy and McKenry, 1977).

Systemic activity of nematicides in plants is confined to the organophosphates and organocarbamates. Most of these compounds with the exception of ethoprop (nonsystemic) are translocated apoplastically. Oxamyl and phenamiphos show some downward movement in plants.[95]

Movement of fumigant nematicides in soil is by mass flow through the air passages and diffusion through the water films. It is influenced by soil texture, porosity, organic matter, temperature and moisture. As the molecules move, many disappear through sorption on soil particles and by chemical and biological degradation. The usefulness of soil fumigants varies also with nematode species. The halogenated aliphatic hydrocarbons are best used against noncyst-forming nematodes, whereas methyl iso-thiocyanate liberators are more useful for killing cyst-forming nematodes.

REFERENCES

1. **Nene, Y. L. and Thapliyal, P. N.** *Fungicides in Plant Disease Control,* 2nd ed., Oxford and IBH Publishing Co., New Delhi, 1979, 507.
2. **Woodburn, A.,** Agrochemical products and agrochemical-overview, in *Agrochemical Service,* Woodburn, A., Ed., Wood Mackenjie, Edinburgh, 1986, 27.
3. **Schwinn, F.,** unpublished data, 1984.
4. **Shephard, M. C.,** Screening for fungicides, *Ann. Rev. Phytopathol.,* 25, 189, 1987.
5. **Ordish, G. and Mitchell, J. F.,** World fungicide usage, in *Fungicides—An Advanced Treatise,* Vol. 1, Torgeson, D. C., Ed., Academic Press, New York, 1967, 39.
6. **Somers, E.,** Formulations, in *Fungicides An Advanced Treatise,* Vol. I., Torgeson, D. C., Ed., Academic Press, New York, 1967, 153.
7. **Jacob, F. and Neumann, St.,** Principle of uptake and systemic transport of fungicides within the plant, in *Modern Selective Fungicides—Properties, Applications, Mechanisms of Action,* Longman Group UK Ltd., London, and VEB Gustav Fischer Verlag, Jena, 1987, Chap. 1.
8. **Singh, U. S. and Tripathi, R. K.,** Physico-chemical and biological properties of metalaxyl. I. Octanol number, absorption spectrum and effect of different physico-chemical factors on stability of metalaxyl *Indian J. Mycol. Plant Pathol.,* 12, 287, 1982.

9. **Singh, U. S., Tripathi, R. K., Kumar, J., and Dwivedi, T. S.** Uptake, translocation, distribution and persistence of ^{14}C-metalaxyl in pearl millet (*Pennisetum americanum* [L.] Leeke), *J. Phytopathol.*, 117, 122, 1986.

10. **Singh, U. S. and Tripathi, R. K.,** Physico- chemical and biologic properties of metalaxyl. II. Absorption by excised maize roots, *Indian J. Mycol. Plant Pathol.*, 12, 295, 1982.

11. **Edgington, L. V. and Peterson, C. A.,** Systemic fungicides: Theory, uptake, and translocation, in *Antifungal Compounds*, Vol. 2, Siegel, M. R. and Sisler, H. D., Eds., Dekker, New York, 1977, 51.

12. **Edgington, L. V.,** Structural requirements of systemic fungicides, *Ann. Rev. Phytopathol,* 19, 107, 1981.

13. **Kumar, J., Singh, U. S., Beniwal, S. P. S., and Srivastava, P. C.,** Binding of carbendazim by lignin, *Indian J. Mycol. Plant Pathol.,* 17, 24, 1987.

14. **Peterson, C. A. and Edgington, L. V.,** Factors influencing apoplastic transport in plants, in *Systemic Fungicides,* Akademie-Verlag, Berlin, 1975, 287.

15. **Singh, U. S.,** Uptake, translocation, distribution and persistence of ^{14}C-metalaxyl in maize *(Zea mays), Z. Pflanzenkr. Pflanzenschutz.,* 94, 478, 1987.

16. **Singh, U. S., Kumar, J., and Tripathi, R. K.,** Uptake, translocation, distribution and persistence of ^{14}C-metalaxyl in pea *(Pisum sativum), Z. Pflanzenkr. Pflanzenschutz.,* 92, 164, 1985.

17. **Chaube, H. S., Singh, U. S., and Razdan, V. K.,** Studies on uptake, translocation and distribution of ^{14}C-metalaxyl in pigeonpea, *Indian Phytopathol.,* 40, 507, 1987.

18. **Singh, U. S.,** Studies on the systemicity of [^{14}C] metalaxyl in cowpea (*Vigna unguiculata* [L] Walp), *Pestic. Sci.,* 25, 145, 1989.

19. **Singh, U. S. and Kumar, J.,** unpublished data, 1989.

20. **Singh, U. S. and Saxena, A. K.,** unpublished data, 1989.

21. **Tripathi, R. K. and Singh, U. S.,** Metalaxyl: physico-chemical and biological properties, in *Recent Advances in Plant Pathology,* Hussain, A., Singh, K., Singh, B. P., and Agnihotri, V. P., Eds., Print House, Lucknow, 1983, 227.

22. **Lyr, H.,** Selectivity in modern systemic fungicides, in *Modern Selective Fungicides—Properties, Applications, Mechanisms of Action,* Lyr, H., Ed., Longman Group UK Ltd., London, VEB Gustav Fischer Verlag, Jena, 1987, chap. 2.

23. **Worthing, C. R.,** *The Pesticide Manual—A World Compendium,* 8th ed., British Crop Protection Council, Thornton, Heath, U.K.

24. **McCallan, S. E. A.,** History of fungicides, in. *Fungicides An Advanced Treatise,* Vol. I, Torgeson, D. C., Ed., Academic Press, New York, 1967, 1.

25. **Sisler, H. D. and Ragsdale, N. N.,** Mode of action and selectivity of fungicides, in *Agricultural Chemicals of the Future,* Hilton, J. L., Ed., Rowman & Allanheld, New Jersey, 1985, 175.

26. **Kaars Sijpesteijn, A.,** Mechanism of action of fungicides, in *Fungicide Resistance in Crop Protection,* Dekker, J. and Georpoulos, S. G., Eds., Pudoc, Wageningen, the Netherlands, 1982, 32.

27. **Vander Kirk, G. J. M. and Luijten, J. G. A,** Investigations on organotin compounds. III. The biocidal properties of organotin compounds, *J. Appl. Chem. London,* 4, 314, 1954.

28. **Lyr, H.,** Aromatic hydrocarbon fungicides, in *Modern Selective Fungicides—Properties, Applications, Mechanisms of Action,* Lyr, H., Ed., Longman Group UK Ltd., London, and VEB Gustav Fischer Verlag, Jena, 1987, chap. 5.

29. **Gasztonyi, M. and Lyr, H.,** Other fungicides, in *Modern Selective Fungicides—Properties, Applications, Mechanisms of Action,* Lyr, H., Ed., Longman Group UK Ltd., London, and VEB Gustav Fischer Verlag, Jena, 1987, chap. 21.

30. **Pommer, E. H. and Lorenz, G.,** Dicarboximide fungicides, in *Modern Selective Fungicides—Properties, Applications, Mechanisms of Action,* Lyr, H., Ed., Longman Group UK Ltd., London, and VEB Gustav Fischer Verlag, Jena, 1987, chap. 7.

31. **Holloman, D. W. and Schmidt, H. H.,** 2-Aminopyrimidine fungicides, in *Modern Selective Fungicides— Properties, Applications, Mechanisms of Action,* Lyr, H., Ed., Longman Group UK Ltd., London, and VEB Gustav Fischer Verlag, Jena, 1987, chap. 18.

32. **Schreiber, B.,** Organophosphorus fungicides, in *Modern Selective Fungicides—Properties, Applications, Mechanisms of Action,* Lyr, H., Ed., Longman Group UK Ltd., London, and VEB Gustav Fischer Verlag, Jena, 1987, chap. 20.

33. **Kurogochi, S., Katagiri, M., Tagase, I., and Uesugi, Y.,** Metabolism of edifenphos by strains of *Pyricularia oryzae* with varied sensitivity to phosphorothiolate fungicides, *J. Pestic. Sci.,* 10, 41, 1985.

34. **Kodama, O., Yamashita, K., and Akatsuka, T.,** Edifenphos, inhibitor of phosphatidylcholine biosynthesis, *Biol. Chem.,* 44, 1015, 1980.

35. **Delp, C. J.,** Benzimidazole and related fungicides, in *Modern Selective Fungicides—Properties, Applications, Mechanisms of Action,* Lyr, H., Ed., Longman Group UK Ltd., London, and VEB Gustav Fischer Verlag, Jena, 1987, chap. 15.

36. **Davidse, L. C.**, Biochemical aspects of phenylamide fungicides - action and resistance, in *Modern Selective Fungicides—Properties, Applications, Mechanisms of Action*, Lyr, H., Ed., Longman Group UK Ltd., London, and VEB Gustav Fischer Verlag, Jena, 1987, chap. 18.

37. **Schmeling, B. Von, and Kulka, M.**, Systemic fungicidal activity of 1,4-Oxathiin derivatives, *Science*, 152, 659, 1966.

38. **Kulka, M., and Von Schmeling, B.**, Carboxin fungicides and related compounds, in *Modern Selective Fungicides—Properties, Applications, Mechanisms of Action*, Lyr, H., Ed., Longman Group UK Ltd., London, and VEB Gustav Fischer Verlag, Jena, 1987, chap. 9.

39. **Pommer, E.-H.**, Morpholine fungicides, in *Modern Selective Fungicides—Properties, Applications, Mechanisms of Action*, Lyr, H., Ed., Longman Group UK Ltd., London, and VEB Gustav Fischer Verlag, Jena, 1987, chap. 11.

40. **Bohnen, K., Pfiffner, A., Siegle, H., and Zobrist, P.**, Fenpropidin a new systemic cereal mildew fungicide, in *Proc. 1986 Brit. Crop Prot. Conf. Pests Diseases*, 1, 27, 1986.

41. **Scheinpflug, H. and Kuck, K. H.**, Sterol biosynthesis inhibiting piperazine, pyridine, pyrimidine and azole fungicides, in *Modern Selective Fungicides - - Properties, Applications, Mechanisms of Action*, Lyr, H., Ed., Longman Group UK Ltd., London, and VEB Gustav Fischer Verlag, Jena, 1987, chap. 13.

42. **Gisi, U., Schaub, F., Wiedmer, H., and Ummel, E.**, SAN 619F, a new triazole fungicide, in *Proc. 1986 Brit. Crop Prot. Conf. Pests Diseases*, 1, 33, 1986.

43. **Reinecke, P., Kaspers, H., Scheinpflug, H., and Holmwood, G.**, BAY HWG 1608, a new fungicide for foliar spray and seed treatment use against a wide spectrum of fungal pathogens, in *Proc. 1986 Brit. Crop Prot. Conf. Pests Diseases*, 1, 41, 1986.

44. **Shephard, M. C., Noon, R. A., Worthington, P. A., McClellan, W. D., and Lever, B. G.**, Hexaconazole: a novel triazole fungicide, in *Proc. 1986 Brit. Crop Prot. Conf. Pests Diseases* 1, 19, 1986.

45. **Koller, W., and Scheinpflug, H.**, Fungal resistance to sterol biosynthesis inhibitors, a new challenge, *Plant Dis.*, 71, 1066, 1987.

46. **Gadher, P., Mercer, E. I., Baldwin, B. C., and Wiggins, T. E.**, A comparison of the potency of some fungicides as inhibitors of sterol C-14 demethylation, *Pestic. Biochem. Physiol.*, 19, 1, 1983.

47. **Schwinn, E. J. and Staub, T.**, Phenylamides and other fungicides against oomycetes, in *Modern Selective Fungicides—Properties, Applications, Mechanisms of Action*, Lyr, H., Ed., Longman Group UK Ltd., London, and VEB Gustav Fischer Verlag, Jena, 1987, chap. 17.

48. **Fenn, M. E. and Coffey, M. C.**, Studies on the *in vitro* and *in vivo* antifungal activity of fosetyl-Al and phosphorous acid, *Phytopathology*, 74, 606, 1984.

49. **de Waard, M. A.**, Synergism and antagonism in fungicides, in *Modern Selective Fungicides—Properties, Applications, Mechanisms of Action*, Lyr, H., Ed., Longman Group UK Ltd., London, and VEB Gustav Fischer Verlag, Jena, 1987, chap. 24.

50. **Sawant, S. D.**, personal communication, 1988.

51. **Lazarovits, G.**, Induced resistance - xenobiotics, in *Experimental and Conceptual Plant Pathology*, Vol. III, Singh, R. S., Singh, U. S., Hess, W. M. and Weber, D. J., Eds., Gordon and Breach Science Publishers, New York and Oxford and IBH Publishing Co., New Delhi, 1983, 575.

52. **Koller, W., Allam, C. R., and Kolattukudy, P. E.**, Inhibition of cutinase and prevention of fungal penetration into plants by benomyl - a possible protective mode of action, *Pestic. Biochem. Physiol.*, 18, 15, 1982.

53. **Sisler, H. D. and Ragsdale, N. N.**, Disease control by non-fungitoxic compounds, in *Modern Selective Fungicides—Properties, Applications, Mechanisms of Action*, Lyr, H., Ed., Longman Group UK Ltd., London, and VEB Gustav Fischer Verlag, Jena, 1987, chap. 23.

54. **Sisler, H. D.**, Control of fungal diseases by compounds acting as antipenetrants, *Crop Prot.* 5, 306, 1986.

55. **Singh, U. S. and Sunderlal, P. E.**, unpublished, 1989.

56. **Trinci, A. P. J.**, Effect of validamycin A and L-sorbose on the growth and morphology of *Rhizoctonia cerealis* and *Rhizoctonia solani*, *Exp. Mycol.*, 9, 20, 1985.

57. **Mukhopadhyay, A. N. and Singh, U. S.** Recent thoughts on plant disease control. I. Crop protection through host immunization by the chemicals, *Pesticides*, 14, 14, 1985.

58. **Nash, R. G.**, Phytotoxic interaction studies—Techniques for evaluation and presentation of results, *Weed Sci.*, 29, 147, 1981.

59. **Wilkinson, C. F.**, Insecticide Synergism, in *The Future for Insecticides*, Metcalf, R. L. and McKelvey, J. J., Eds., John Wiley & Sons, New York, 1976.

60. **Delp, C. J. and Dekker, J.**, Fungicide resistance: definitions and use of terms. *EPPO Bull.*, 15, 333, 1985.

61. **Tripathi, R. K., and Schlösser, E.**, The mechanism of resistance of *Botrytis cinerea* to methylbenzimidazol-2-yl-carbamate (MBC), *Z. Pflanzenkr. Pflanzenschutz.*, 89, 151, 1982.

62. **Davidse, L. C.**, Benzimidazole fungicides: mechanism of action and biological impact, *Ann. Rev. Phytopathol.*, 24, 43, 1986.

63. **Taga, M., Nakagava, H., Tsuda, M. and Ueyama, A.,** Ascospore analysis of Kasugamycin resistance in the perfect stage of *Pyricularia oryzae, Phytopathology,* 68, 815, 1978.

64. **Georgopoulos, S. G.,** Genetical and biochemical background of fungicide resistance, in *Fungicide Resistance in Crop Protection,* Dekker, J. and Georgopoulos, S. G., Eds., Pudoc, Wageningen, 1982, 46.

65. **Misato, T., Kakiki, K., and Hori, M.,** Mechanism of polyoxin resistance, *Neth. J. Plant Pathol.,* 83, Suppl. 1, 253, 1977.

66. **de Waard, M. A. and Van Nistelrooy, J. G. M.,** An energy dependent efflux mechanism for fenarimol in a wild-type strain and fenarimol resistant mutants of *Aspergillus nidulans, Pestic. Biochem. Physiol.,* 13, 255, 1981.

67. **Uesugi, Y. and Katagiri, M.,** Metabolism of phosphorothiolate fungicide IBP, by strains of *Pyricularia oryzae* with varied sensitivity, in *Pesticide, Human Welfare and Environment,* Vol. 3, Miyamoto, J., and Kearney, P. C., Eds., Pergamon Press, Oxford, 1983, 165.

68. **de Waard, M. A. and Van Nistelrooy, J. G. M.,** Mechanism of resistance to pyrazophos in *Pyriularia oryzae, Neth. J. Plant Pathol.,* 86, 251, 1980.

69. **Leroux, P., and Gredt, M.,** Characterization de sourches d' *Ustilago maydis* (DC) CDA resistantes aux fongicides inhibiteurs des sterols., *C. R. Acad. Sci. Paris, Series III,* 296, 191, 1983.

70. **Hollomon, D. W.,** Contribution of fundamental research to combating resistance, in *Proc. 1986 Brit. Crop Prot. Conf.—Pests Diseases* 2, 801, 1986.

71. **Georgopoulos, S. G.,** The genetics of fungicide resistance, in *Modern Selective Fungicides—Properties, Applications, Mechanisms of Action,* Lyr, H., Ed., Longman Group UK Ltd., London, and VEB Gustav Fischer Verlag, Jena, 1987, chap. 4.

72. **Skylakakis, G.,** The development and use of models describing outbreak of resistance to fungicides, *Crop Prot.,* 1, 249, 1982.

73. **Levy, Y., Levi, R., and Cohen, Y.,** Buildup of a pathogen subpopulation resistant to systemic fungicide under various control strategies: A flexible simulation model, *Phytopathology,* 73, 1475, 1983.

74. **Samoucha, Y. and Gisi, U.,** Use of two-and three- way mixtures to prevent buildup of resistance to phenylamide fungicides in *Phytophthora* and *Plasmopara, Phytopatholgy,* 77, 1405, 1987.

75. **Sanders, P. L., Houser, W. J., Parish, P. J., and Cole, H., Jr.,** Reduced-rate fungicide mixtures to delay fungicide resistance and to control selected turf-grass diseases, *Plant Dis.,* 69, 939, 1985.

76. **Georgopoulos, S. G.,** Fungicide resistance in crop protection: Detection and measurement principles, in *Fungicide Resistance in Crop Protection,* Dekker, J., Chiarappa, L., Delp, C. J., Georgopoulos, S. G., Schwinn, F. J., Verghese, G., and de Waard, M. A., Eds., University Pertanian Malaysia, Selangor, Malaysia, 1984, 31.

77. **Staub, T. and Sozzi, D.,** Fungicide resistance: a continuing challenge, *Plant Dis.,* 68, 1026, 1984.

78. **Dekker, J.,** Development of resistance to modern fungicides and strategies for its avoidance, in *Modern Selective Fungicides—Properties, Applications, Mechanisms of Action,* Lyr, H., Ed., Longman Group UK Ltd., London, and VEB Gustav Fischer Verlag, Jena, 1987, chap. 3.

79. **Gasztonyi, M., Josepovits, G., and Vegh, A.,** Cross-resistance relationships between benzimidazole fungicides, *N*-phenylcarbamates and other related compounds, in *Proc. 1986 Brit. Crop Prot. Conf. Pests Diseases,* 2, 547, 1986.

80. **de Waard, M. A. and Van Nistelrooy, J. G. M.,** Laboratory resistance to fungicides which inhibit eryosterol biosynthesis in *Penicillium italicum, Neth. J. Plant Pathol.,* 88, 99, 1982.

81. **Locher, F. J., Lorenz, G., and Beetz, K. J.,** Resistance management strategies for dicarboximide fungicides in grapes: results of six years' trial work, *Crop Prot.,* 6, 139, 1987.

82. **Urech, P. A. and Staub, T.,** The resistance strategy for acylalane fungicides, *OEPP/EPPO Bull.,* 15, 539, 1985.

83. **Anon.** Use of DMI- fungicides in bananas, special Report of Working Group Meeting, *ISSP Chem. Control Newslett.,* 11, 11, 1988.

84. **de Waard, M. A.,** Negatively correlated cross-resistance and synergism as strategies in coping with fungicide resistance, in *Proc. 1984 Br. Crop Prot. Conf.—Pests Diseases,* 1984, 573.

85. **Grabski, C., and Gisi, U.,** Mixtures of fungicides with synergistic interactions for protection against phenylamide resistance in *Phytophthora,* in *Fungicides for Crop Protection: BCPC Monogr. No. 31,* 1985, 315.

86. **Mukhopadhyay, A. N., and Singh, U. S.,** Recent thoughts in plant diseases control. II. Fungal resistance to fungicide, *Pesticides* 14, 23, 1985.

87. **Crute, I. R.,** The integrated use of genetic and chemical methods for control of lettuce downy mildew (*Bremia lactucae* Regel), *Crop Prot.,* 3, 223, chap. 13.

88. **Crute, I. R.,** unpublished data, 1989.

89. **Lyr, H.,** Mechanism of action of fungicides, in *Plant Disease* Vol. I, Horsfall, J. G. and Cowling, E. B., Eds., Academic Press, New York, 1977, 239.

90. **Ohta, N., Kakiki, K., and Misato, T.,** Studies on the mode of action of polyoxin D, *Agric. Biol. Chem.,* 34, 1224, 1970.

91. **Feldmesser, J., Chr.,** Estimated crop losses from plant-parasite nematodes in United States, *J. Nematol.,* Spec. Publ. No. 1, 1970.
92. **Gundy, S. D., Van, and McKenry, M. V.,** Action of nematicides, in *Plant Disease,* Vol. I, Horsfall, J. G. and Cowling, E. B., Eds., Academic Press, New York, 1977, chap. 14.
93. **Raski, D. J., Hewitt, W. B., and Schmitt, R. V.,** Controlling fan leaf virus-dagger nematode disease complex in vineyards by soil fumigation, *Calif. Agric.,* 25, 11, 1971.
94. **Hodges, L. R.,** *Nematodes and Their Control,* Union Carbide Corporation, Salinas, California, 1973.
95. **Tyree, M. T., Peterson, C. A., and Edgington, L. V.,** A simple theory regarding ambimobility of xenobiotics with special reference to the nematicide, oxamyl, *Plant Physiol.,* 63, 367, 1979.

Chapter 18

INTEGRATED PEST (DISEASE) MANAGEMENT (IPM)

I. INTRODUCTION

Since time immemorial biological species such as plant pathogens, nematodes, and weeds have exploited energy resources for their continued survival. Food production systems for human kind are no different: they are energy-limited and subject to all the laws of natural systems. To maintain stability toward specific production goals, human society must expand energy to exert control over other natural systems.

This is no trivial endeavor, despite our technological sophistication pest populations are not inert masses to be passively decimated by our arsenal of control technology. Often, if not always, the consequences of our control actions have been counterproductive. Heavy crop losses, pesticide resistance, adverse environmental effects, and low success rates with biological control strongly signal that we know very little about the biological interactions involved in our food production system. Methods for controlling pests affect not only agriculture, forestry, and natural ecosystems, but ultimately the consumers of these products. The widespread use of pesticides since World War II has created public concern over environment, human health, and human safety. Moreover, the agricultural sector is concerned about the increasing resistance of pests to pesticides and shifts in pest complex in food production systems. A renewed emphasis on developing or improving alternative pest control tactics (biological, genetic, and cultural) has in recent years fostered a new philosophy concerning the management of pests—integrated pest management (IPM)—which is based on ecological, sociological, and economic factors.[1]

II. CONCEPTS OF IPM

Since IPM philosophy is in continuous transition, it is definable only at specific times. "The 1977 USDA Policy on Management of Pest Problems", defined IPM as "a desirable approach to the selection, integration, and use of methods on the basis of their anticipated economic, ecological, and sociological consequences". An operational concept of IPM was developed for the Science and Education Administration (SEA) of the USDA in 1979. The concept includes a classification of pest management programs, including the major elements of (1) basic research, (2) control components research, (3) extension IPM systems level I, (4) extension IPM systems level II, and (5) IPM higher education.[1]

There are four reasons for using the term "integrated" in IPM. First, it calls for a multidisciplinary approach, which jointly considers all classes of pests and their interrelationships. Second, it requires that all available management tactics be coordinated into a unified program seeking an optional management strategy. Third, crop protection is treated as but one aspect, of the total management program of the agroecosystem. Finally, IPM recognizes the necessity of addressing economic, ecological, and social concerns.

IPM research, extension, and instructional programs, underway in every state in the U.S., probably represent the most widely recognized new agricultural program thrust in the world. Yet the definition of IPM is extremely pluralistic. It is interdisciplinary in nature, but each participating discipline functions under its own definition. Therefore, IPM has various meanings. Some disciplines see pesticides as the dominating control component in IPM, others focus on natural enemies, cultural practices, and host plant resistance. In these cases, the definitions could legitimately be paraphrased as integrated pesticide management, integrated biological control management, etc.

The diversified definition of IPM should not however, be used as a criterion for denigrating the concept. Instead, they should give new insight into the nature of the problem. IPM is a concept that is evolving. It was not the single creation of an individual mind. Instead it is philosophy that, if followed, leads to certain activities or conclusions. From the outset, the word integrated has generally referred to the use of two or more control tactics in a crop protection program. Virtually all definitions of IPM preclude crop protection systems that involve only a single control tactic. IPM forces multiplicity in control strategies.

III. THRESHOLD OF IPM

Literally, threshold indicates the point at which some effect begins. In IPM, it is used to describe[2,3] (1) the point at which damage is first seen (also called the damage threshold or tolerance limit); (2) the point of first economic loss for a given control cost (the point at which the value of the crop loss is equal to the cost of control, also referred to as the discrete choice threshold); and (3) the point of maximization of profit (the point at which the difference between the crop value and pest control cost is greatest). Thus, threshold is a measure of biological stress imposed by the pest or pathogen. It may be quantified in terms of population densities of individuals or as some measure of disease or damage severity. Normally threshold determination is straightforward for a single disease system in a crop; however, development of multiple-disease thresholds for a single crop is extremely complex because of the interactions that occur among the disease systems and with yield. Determination of thresholds is limited by constraints of the data available for any given disease system. Spore concentrations, meteorological parameters, disease severity estimates, and biometerological parameters have been used to develop thresholds for models of foliar disease development. Ferris[2] and Eversmeyer and Kramer[3] have proposed components and techniques of IPM threshold determinations for soil-borne and aerial pathogens, respectively.

IV. PHASES

In the development of IPM several phases have been conceived of. They may be stated as follows.[4]

1. Single tactic phase—In the initial stage, the best method for tackling or managing a pest/disease is found out.
2. Multiple tactic phase—In this phase, all the available methods—cultural, biological, chemical, and use of resistant cultivars are worked out in relation to ecology, inoculum density and damages. This phase is the most critical one and needs to be worked out carefully.
3. Biological monitoring phase—After the multiple tactics have been worked out, they are to be carefully monitored particularly biological ones in relation to beneficial organisms and host plants so that control measures may be suitably timed and phased out.
4. Modeling phase—Data obtained from phases 1, 2, and 3 need to be very carefully worked out so that a model can be built pointing out the major steps that need to be taken regarding the application of control measures and the critical stages in the host plant with reference to diverse measures. The entire approach has to be systematized through critical analysis of data.
5. Management phase—After the different models of control have been worked out in an integrated manner, the entire process has to be incorporated into the overall production techniques of the crop in question.
6. Acceptance phase—The most critical stage is the development of system approach

which will be acceptable to growers. The approach has to be technologically sound, economically viable and not too cumbersome for adoption by farmers.

Integrated pest/disease management is a new system approach which has been necessitated primarily out of the growing concern about the undesirable side effects of large-scale use of fungicides and often failure of the same to provide the suppression of the pathogen at economic level. The Food and Agriculture Organization of the United Nations has been interested in IPM since 1963. It has developed and implemented several field projects in different parts of the world. Global program has been proposed to be undertaken. Criteria of program priority have been developed which point out the necessity of IPM. These are[4] (1) crop must be of vital national and regional importance, (2) serious losses are caused by pests/pathogens, (3) inadequate control by use of organic pesticides, (4) use of pesticide is generating more problems, but the same cannot be given up otherwise food production will not be stepped up, and (5) an integrated approach can be developed which will yield desirable results and be acceptable to growers.

Emphasis has been laid on major crops namely, rice, maize, sorghum, cotton, etc. Potato, sugarcane, grain legumes, tapioca, coconut, etc., have also been considered as second order of priority.

V. IPM AND PLANT PATHOLOGY

The conventional approach to disease control has not stimulated an active holistic approach to disease management, i.e., disease control has historically depended upon diverse methods ranging from plant resistance and sanitation to intense fungicide application. IPM emphasizes a holistic approach. Interactions among pests and pest management activities are critical, and the value of pest management activities is determined increasingly on the basis of economic, ecological, and social considerations. There is now greater research emphasis on disease forecasting, and integrations of management methods are increasing so that disease management technology (whatever its form) will be efficient. Innovative biological, cultural, host resistance, and chemical means of management are currently being investigated, and the application of systems science to pest and disease management should provide new capabilities for refining the implementation of disease management technology.[5]

VI. INTEGRATED CROP MANAGEMENT SYSTEM

El-Zik and Frishie[6] defined an ICMS as ''a system whereby all interacting crop production and pest control tactics aimed at maintaining and protecting plant health are harmonized in the appropriate sequence to achieve optimum crop yield and quality and maximum net profit, in addition to stability in the agroecosystem, benefiting society and mankind.'' The main objective of an ICMS is to maintain the plant's health throughout the growing season. ''Plant health'' means relative freedom from biotic and abiotic stresses that limit the plant's productivity in both quantity and quality from its maximum genetic potential.

The examples where ICMS has been employed for suppression of plant diseases are very few. Some work has been done on the management of Verticillium wilt of cotton. Conceptual model of an integrated system for controlling this disease is given in the chart outlined below.[7]

VII. INTEGRATED DISEASE MANAGEMENT-SOME EXAMPLES

Since integrated disease management system has to be a part of the agroecosystem, it has to manage a system rather than a single crop or a single disease. It has to provide bases for ensuring crop health of the entire programme in the agroecosystem. It will be an ideal

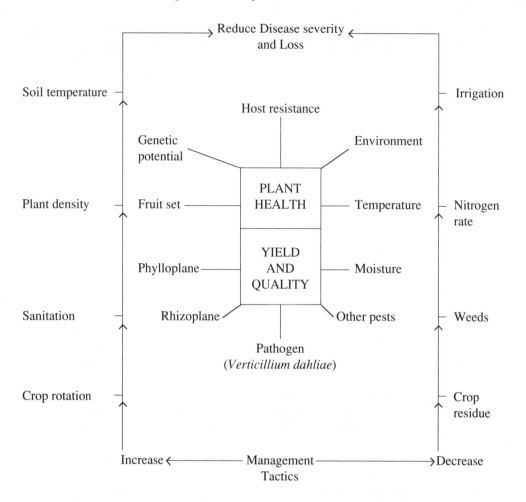

situation, but has not been practically demonstrated so far. In integrated disease management (IDM) there are two options. The first is to integrate the available methods and resources to suppress a highly destructive disease of a crop or secondly to integrate the methods for managing all the economically important diseases of a crop. Depending upon the importance, both the options require serious considerations. Several plant diseases, especially soil-borne diseases are mostly localized and highly destructive in certain pockets. Unfortunately no single method has so far given adequate protection against such diseases. In such situations, integration of available methods for controlling the disease will be both desirable and successful. Attempts to control all the major diseases of a crop too will be a positive step. It is simply because that all recorded diseases of a crop become serious only occassionally as ever varying environments influence the development of a disease. For example, the potato crop is reported to be attacked by at least 18 viruses, 46 fungal pathogens, 5 plant parasitic bacteria, and many species of nematodes. All these diseases do not occur in every potato growing country and intensity of their attack also varies in different regions. In such conditions only a few of them assume serious proportions and IDM could and should be tried.

A. INTEGRATED CONTROL OF BEAN ROOT-ROT

The root rot disease of bean (*Phaseolus vulgaris* in Wisconsin U.S., is often caused by a combination of several pathogens, including *Pythium* spp., *Fusarium solani* f.sp. *phaseoli*, and *Rhizoctonia solani*. Control of this disease has been investigated[8] using several approaches including studies on effects of cover crops, crop rotations, chemicals and disease resistance. Some of the results have already been used to reduce losses by this disease complex, i.e., oats are now widely used instead of rye as winter cover crop by many bean growers in the Central Sands of Wisconsin. The use of crop rotation as a control measure is also more common than before. In addition, timely supplementary fertilization with N sometimes has a beneficial effect. Control through the use of disease resistance has repeatedly been very effective.

B. INTEGRATED CONTROL OF *RHIZOCTONIA SOLANI*

The reduction of damping off and hypocotyl rot of snapbean caused by *R. solani* by an integrated ecosystem approach was studied for 3 years in the field.[9] The field infested with *R. solani* was divided into two subplots, one for discing the bean vines and weeds of the previous planting 5 to 7 cm deep and the other for ploughing with a moldboard plough to incorporate the vines and weeds 20 to 25 cm deep. Within each cultural practice, microorganisms antagonistic to *R. solani* (e.g., *Trichoderma harzianum*) were used as second control components. Seed treatments with single fungicide or combinations were used as the third component. Snapbean plant stand was significantly higher and hypocotyl rot was less severe with ploughing than with discing. Seed treatment with chloroneb significantly increased stand in the ploughed but not in the disced plot. The various antagonists used did not affect plant stand or disease severity.[9]

C. INTEGRATED MANAGEMENT OF DISEASES IN A SINGLE CROP

As mentioned earlier, all the recorded diseases of a single crop do not occur simultaneously or concurrently in a locality. However, the possibility of occurrence of any one or more of them in the field does exist. Taking example of potato crop and its diseases, IDM in a single crop is discussed here. In Indian subcontinent and elsewhere, late and early blight diseases, black scurf/stem canker, bacterial wilt and brown rot, root knot and mosaic and leaf curl are quite common diseases of this crop. Due to nature of propagation of the crop all the diseases are mainly seed (tuber) borne. Some of them are soil-borne also. Spread by soil, insects, water and wind is common. Therefore, in this crop the improvement of quality and quantity of the yield through disease control can be brought about by adopting a schedule prepared on the following lines.

Proper selection of the field for potato is the first step. Fields which had shown serious incidence of soil-borne diseases of potato in the near past should be avoided. The land must have proper drainage and high fertility level. The next step is the selection of seed. In areas where late blight regularly occurs every year it is always advisable to grow only resistant varieties even if they do not yield as much as a susceptible but high yielding variety. Since tubers are the main source of perennation and introduction of the pathogens in the field it is essential that tubers for seed should be obtained from sources which guarantee certification against presence of important diseases in the tuber. In absence of such an arrangement seed tubers can be obtained from a field where necessary steps for control of diseases (roguing, spraying, etc.) have been taken and tubers for seed have not been collected from diseased plants. The seed tubers thus obtained may be completely free from or will carry less inoculum of viruses, bacterial pathogens, black scurf fungus, and root knot nematodes. Additional precautions should be taken by treating the seed tubers with fungicides such as Aretan, Agallol, Benlate, etc. The chemical treatment helps in eradication of residual inoculum from seed and protects it during germination and emergence. The date of planting should be decided according to prevailing soil temperature and moisture. High soil temperature and moisture should be avoided. When the crop is established in the field protective steps to prevent secondary infection are essential. Plants showing viral infection are promptly removed and the crop sprayed with Metasystox or other insecticides when there is build up of populations of insect vectors of potato viruses. In addition, regular spraying with fungicides such as zineb, mancozeb, copper oxychloride, etc., is continued to prevent late and early blight. Normally spraying with Dithane Z-78, Dithane M-45, Lonacol, Difolatan, Blitox-50, etc., should be started when the crop is about 6 inch high and repeated at 10- to 21-d intervals depending upon occurrence of the foliar diseases in the neighborhood and upon weather.[10] If all these precautions are followed a disease-free crop can be successfully raised.

D. IDM IN A PERENNIAL CROP

In an integrated disease control program of an orchard crop like apple, peach, citrus, etc., one must first consider the nursery stock to be used and location where it will be planted. In case the trees of the concerned orchard are known to be susceptible to viruses, MLO, crown-gall bacterium or nematodes, the nursery stock (both the root-stock and scion) must be free from these agents. Even after obtaining them from a nursery where the saplings are scientifically inspected and certified, the stock must be appropriately fumigated, especially to eliminate nematodes. The next step is selection of planting sites. Several pathogens like *Armillaria, Fusarium, Phytophthora,* etc., and several nematodes are known to survive in soil. If the experience and history suggest their presence in soil, the soil should be treated with desired pesticide, particularly with fumigants before planting. As far as possible roots-tock resistant to these pathogens should be preferred. The drainage of the location should be checked and improved, if necessary. Finally, the young trees should not be planted between or next to old trees that are heavily infected with canker fungi and bacteria, insect-transmitted viruses and MLO, pollen transmitted viruses or with other pathogens.[11]

Once the trees have established and until they begin to bear fruits, they should be fertilized, irrigated, pruned, and sprayed for the most common insects and diseases so that they will grow vigorously and free of infections. Any tree that develops symptoms of a disease caused by a systemic pathogen should be removed at the earliest.

Later on when trees bear fruits, the care should be taken to control diseases that affect them. The following step wise treatments ensure healthy crop of fruits if followed with care.[11]

1.   Removal through pruning and destruction of dead twigs, branches and fruits. This will reduce amount of initial inoculum of the pathogens that will start infection.
2.   Pruning shears and saws should be disinfected before moving to new trees to avoid the spread of the inoculum.
3.   A dormant spray containing a fungicide-bactericide (such as Bordeaux mixture) or a plain fungicide plus a acaricide-insecticide (such as Superior oil), is applied before bud break.
4.   After the buds open, blossoms and leaves must be protected with sprays containing a fungicide and/or a bactericide and, possibly, an insecticide and/or acaricide that does not harm bees.
5.   Once blossoming is over, young fruits appear. If they are infected, appropriate chemical sprays should be undertaken.
6.   Usually the fruits become susceptible to several fruit-rotting fungi at early maturity through harvest and storage. Therefore, fruits must be sprayed every fortnight with chemicals that will control these pathogens till harvest. Most fruit rots start at wounds made by insects, and, therefore, insect control must continue.
7.   Also, wounding of fruits during harvesting and handling must be avoided to prevent fungus infection.
8.   Harvested fruit should be washed in a water solution containing a fungicide to further protect the fruits during storage and transit. During packing infected fruits must be discarded.

REFERENCES

1. **Allen, G. E. and Bath, J. E.,** The conceptual and institutional aspects of integrated pest management, *BioScience,* 30, 658, 1980.
2. **Ferris, H.,** Components and techniques of integrated pest management threshold determinations for soil-borne pathogens, *Plant Dis.,* 71, 452, 1987.
3. **Eversmeyer, M. G. and Kramer, C. L.,** Components and techniques of integrated pest management threshold determinations for aerial pathogens, *Plant Dis.,* 71, 456, 1987.
4. **Chattopadhyay, S. B.,** *Principles and Procedures of Plant Protection,* Oxford & IBH Publishing Co., New Delhi, 1985, 547.
5. **Fry, W. E. and Thurston, H. D.,** The relationship of plant pathology to integrated pest management, *BioScience,* 665, 1980.
6. **El-zik, K. M. and Frisbie, R. E.,** Integrated crop management systems for pest control and plant protection, in *Handbook of Natural Pesticides: Methods,* Vol. I., Mandava, N. B., Ed., CRC Press, Boca Raton, FL, 1985, 534.
7. **El-zik, K. M.,** Integrated control of Verticillium wilt of cotton, *Plant Dis.,* 69, 1025, 1985.
8. **Hagedorn, D. J. and Rand, R. E.,** Research for integrated control of bean root-rot, in *Soil-Borne Plant Pathogens,* Schippers, B. and Gams, W., Eds., Academic Press, New York, 1979, 686.
9. **Papavizas, G. C. and Lewis, J. A.,** Integrated control of *Rhizoctonia solani,* in *Soil-Borne Plant Pathogens,* Schippers, B. and Gams, W., Eds., Academic Press, London, 1979, 686.
10. **Singh, R. S.,** *Introduction to Principles of Plant Pathology,* Oxford & IBH Publishing Co., New Delhi, Bombay, Calcutta, 1984, 534.
11. **Agrios, G. N.,** *Plant Pathology,* 3rd ed., Academic Press, New York, 1988, 803.

INDEX